Signaling and Communication in Plants

Series Editors

František Baluška
Department of Plant Cell Biology, IZMB, University of Bonn, Kirschallee 1,
D-53115 Bonn, Germany

Jorge Vivanco
Center for Rhizosphere Biology, Colorado State University, 217 Shepardson Building,
Fort Collins, CO 80523-1173, USA

For further volumes:
http://www.springer.com/series/8094

Jorge M. Vivanco • František Baluška
Editors

Secretions and Exudates in Biological Systems

Springer

Editors
Jorge M. Vivanco
Colorado State University
Center for Rhizosphere Biology
Colorado
USA

František Baluška
Universität Bonn
Inst. Zelluläre und Molekulare
Botanik (IZMB)
Bonn
Germany

ISSN 1867-9048 e-ISSN 1867-9056
ISBN 978-3-642-23046-2 e-ISBN 978-3-642-23047-9
DOI 10.1007/978-3-642-23047-9
Springer Heidelberg Dordrecht London New York

Library of Congress Control Number: 2011945428

Printed on acid-free paper

Springer is part of Springer Science+Business Media (www.springer.com)

Preface

Secretions and emissions in biological systems play important signaling roles within the organism but also in its communication with the surrounding environment. This relatively recent knowledge is in stark contrast with the view of secretions that is available in most text books in different biological science disciplines. Not too long ago, secretions and emissions were considered biological waste products that were simply discharged out of plants through a chemical gradient with no function to the environment whatsoever. The realization of this void of information was the driving force behind the compilation of this volume. This volume brings together state-of-the-art information about the role of secretions and emissions in different organs and organisms ranging from flowers and roots of plants to human organs.

The plant chapters will relate information regarding the biochemistry of flower volatiles and root exudates, and their role in attracting pollinators and interacting with soil microbial communities, respectively. Furthermore, these chapters will illustrate information about the fine molecular and biochemical machinery that is involved in the biosynthesis and secretion of these compounds; which suggests that the organism actively coordinates the release of these chemical signals. The release of compounds by roots is further highlighted by the most economically important root–microbe association in agriculture. The rhizobium–legume root association forms an organ called the nodule that can fix the nitrogen from the air and entirely eliminates the need of nitrogen fertilization in legume plants such as soybean. Proteoid roots release proteases for improving nitrogen and phosphorus availability for their mineral nutrition. Roots of plants not only release compounds, but also entire cells and the detailed mechanisms and functions of this phenomenon will be highlighted. Plants emit fine bouquets of smell not only through their flowers but through several organs; the biosynthesis and function of volatile organic compounds (VOCs) in plants are also covered in this volume. Moreover, in order to manipulate their animal pollinators, plants provide them with nutritive exudates.

Microbial chapters will explain the biochemistry and ecology of quorum sensing and how microbial communities aggregate in different environments through the continuous release and sensing of compounds that regulate the "quorum" in the community. A related chapter will touch upon highly coadapted association between plants and soil microbes that can aid in bioenergy applications by degrading lignocellulosic materials.

Other chapters will explain the biology of secretions by algae and humans, among other organisms. All in all, this volume will be a welcome addition to the literature as no other book covers aspects related to biological secretion in such a holistic and integrative manner.

Fort Collins, CO, USA Jorge Vivanco
Bonn, Germany František Baluška

Contents

Plant Root Secretions and Their Interactions with Neighbors

Clelia De-la-Peña, Dayakar V. Badri, and Víctor M. Loyola-Vargas

Abstract The rhizosphere biology at the molecular level has advanced dramatically since last decade. The continuous supply of carbon compounds from plant roots engages complex interactions among rhizosphere organisms including interactions between microbes and plants and between plants with other plants being these of the same or different species. Root exudation is part of the rhizodeposition process, which is a major source of soil organic carbon released by plant roots which clearly represents a significant carbon cost to the plant. Root exudates also play a role in soil nutrient availability by altering soil chemistry and soil biological processes. Different studies have highlighted that the rhizosphere soil surrounded by plant roots is more abundant in microbes than the nonrhizosphere soils. Therefore, the major responses in the interaction between plants and microbes must happen in that limited zone. Plants respond to the presence of microbes by releasing a mixture of phytochemicals, volatiles, and high-molecular-weight compounds. Soil microbes, on the other hand, modulate the secretion of root exudates to positively regulate plant growth and disease resistance. Several negative interactions are mediated by root exudates including antimicrobial, biofilm inhibitors, and quorum-sensing mimics to prevent soil-borne pathogens. There is a need to understand these rhizospheric multitrophic interactions in the realistic field conditions to improve the plant growth at species and community level. In addition, studies should be conducted in the field

---

C. De-la-Peña
Unidad de Biotecnología, Centro de Investigación Científica de Yucatán, Calle 43 No. 130, Col. Chuburná de Hidalgo, CP 97200, Mérida, Yucatán, México

D.V. Badri
Department of Horticulture and Landscape Architecture and Center for Rhizosphere Biology, Colorado State University, Fort Collins, CO 80523, USA

V.M. Loyola-Vargas (✉)
Unidad de Bioquímica y Biología Molecular de Plantas, Centro de Investigación Científica de Yucatán, Calle 43 No. 130, Col. Chuburná de Hidalgo, CP 97200 Mérida, Yucatán, México
e-mail: vmloyola@cicy.mx

J.M. Vivanco and F. Baluška (eds.), *Secretions and Exudates in Biological Systems*, Signaling and Communication in Plants 12, DOI 10.1007/978-3-642-23047-9_1,

conditions to understand the rhizospheric complex interactions in monocultures and polycultures. This will help to understand the dynamics of interactions and their outcome in influencing the plant's success when they are in monocultures and in polycultures. The combination of techniques and the continuous development of new techniques in the field of rhizosphere biology coupled with systems approach will allow us partly to elucidate these complex interactions under field conditions.

1 Introduction

Until recently, the difficulty of working underground has kept the rhizosphere in a scientific state of "out of sight, out of mind." However, our understanding of the rhizosphere biology at the molecular level has advanced dramatically since last decade, thanks in large part to the completion of the Arabidopsis genome and the experimental tools and resources that have resulted from this key event. Rhizosphere processes are driven mainly by photosynthetically fixed carbon which is either directly transferred to symbionts or released as root exudates and is considered as a major factor in regulating soil microbiota. Soil microbiota regulates carbon storage via mineralization and immobilization of soil organic carbon (Paterson et al. 1997) and in terrestrial ecosystems the mineralization is not only due to the activity of microorganisms. The continuous supply of carbon compounds from plant roots engages complex interactions among rhizosphere organisms including interactions between microbes and plants, among microbes, between microfauna and microbes, between animals and plants, among animals, and among plants. Rhizosphere ecological interactions are broadly classified into two types: indirect ecological interactions and direct ecological interactions. Interactions between organisms that involve physical contact are considered direct ecological interactions and indirect ecological interactions include any mechanism of interaction between organisms mediated by a number of steps, where one organism affects another one without direct contact (Strauss 1991). These types of indirect interactions that occur in the rhizosphere can be grouped by considering the nature of the interacting organisms as plant–plant, plant–microbe, microbe– microbe, microbe–fauna, plant–fauna, etc., which are mediated by their secretions or mediator species. A large body of literature exists about rhizosphere interactions (Badri and Vivanco 2009; Badri et al. 2009; Bais et al. 2004, 2006, 2008; Bertin et al. 2003; Lambers et al. 2009; Prithiviraj et al. 2007). In this chapter, we focus on the current knowledge of the indirect ecological interactions mediated by their secretions.

2 Root-Secreted Components

For the last decade the field of rhizosphere biology has discovered the importance of root exudates in mediating complex rhizospheric interactions (Bais et al. 2004; Broeckling et al. 2008; Walker et al. 2003; Weir et al. 2004). Root exudation is part of the rhizodeposition process, which is a major source of soil organic carbon released by plant roots (Hütsch et al. 2002; Nguyen 2003). The quantity and quality of the compounds secreted by the roots depends on the plant species, the physiological stage of the plant, presence or absence of neighbors, plant nutritional status, mechanical impedance, sorption characteristics of the soil, and the microbial activity in the rhizosphere. Typically root exudation clearly represents a significant carbon cost to the plant (Marschner 1995) and with young seedlings typically exuding about 30–40% of their fixed carbon as root exudates (Lynch and Whipps 1990). Root exudates encompass ions (i.e., H^+), inorganic acids, oxygen, and water but mainly consist of carbon-based containing compounds (Bais et al. 2006; Uren 2000). These carbon-containing compounds can often be separated into two classes of compounds: low-molecular-weight compounds, which include amino acids, organic acids, sugars, phenolics, secondary metabolites, and volatile compounds such as terpenoids, and high-molecular-weight compounds, which include mucilage and proteins. The classes of compounds secreted by roots are listed in Table 1.

Table 1 Classes of compounds released in plant root exudates

Class of compounds	Single components[a]
Carbohydrates	Arabinose, glucose, galactose, fructose, sucrose, pentose, rhamnose, raffinose, ribose, xylose and mannitol, alanine, and threonine
Amino acids	All 20 proteinogenic amino acids, L-hydroxyproline, homoserine, mugineic acid, and aminobutyric acid
Organic acids	Acetic acid, succinic acid, L-aspartic acid, malic acid, L-glutamic acid, salicylic acid, shikimic acid, isocitric acid, chorismic acid, sinapic acid, caffeic acid, *p*-hydroxybenzoic acid, gallic acid, tartaric acid, ferulic acid, protocatacheuic acid, and *p*-coumaric acid
Flavonols	Naringenin, kaempferol, quercitin, myricetin, naringin, rutin, genistein, strigolactone, and their substitutes with sugars
Lignins	Catechol, benzoic acid, nicotinic acid, phloroglucinol, cinnamic acid, gallic acid, ferulic acid, syringic acid, sinapoyl aldehyde, chlorogenic acid, coumaric acid, vanillin, sinapyl alcohol, quinic acid, and pyroglutamic acid
Coumarins	Umbelliferone
Aurones	Benzyl aurones synapates and sinapoyl choline
Glucosinolates	Cyclobrassinone, desuphoguconapin, deslphoprogoitrin, desulphonapoleiferin, and desulphoglucoalyssin
Anthocyanins	Cyanidin, delphinidin, pelargonidin, and their substitutes with sugar molecules
Indole compounds	Indole-3-acetic acid, brassitin, sinalexin, brassilexin, methyl indole carboxylate, and camalexin glucoside
Fatty acids	Linoleic acid, oleic acid, palmitic acid, and stearic acid
Sterols	Campestrol, sitosterol, and stigmasterol
Allomones	Jugulone, sorgoleone, 5,7,4′-trihydroxy-3′, 5′-dimethoxyflavone, DIMBOA, and DIBOA
Proteins and enzymes	PR proteins, lectins, proteases, acid phosphatases, peroxidases, hydrolases, and lipase
Volatile organic compounds (VOCs)	Carbon dioxide, ethanol, methanol, acetone, acetaldehyde, β-phellanderene, 1,8-cineol, and longifolene

[a]List of compounds presented in this table are mostly reported from model plant Arabidopsis (see Narasimhan et al. 2003) and this list is not complete. This table was adopted and modified from Badri and Vivanco (2009)

3 Plant–Plant Interactions by Root-Secreted Phytochemicals

Plants are sessile and therefore cannot move in response to biotic or abiotic attack. However, they respond to these attacks by releasing a mixture of chemical compounds. Communication between plants has not been studied in detail. The best known example is the communication mediated by volatile compound methyl salicylate (Shulaev et al. 1997). Plant hormones, such as ethylene and jasmonic acid, play an indispensable role in mediating plant–plant communication, and plant communication with other organisms (Lou et al. 2005; Ruther and Kleier 2005). Similarly, below-ground plant communication is orchestrated by roots through secreting phytochemicals and emitting volatiles. The most widely studied below-ground chemical mediated plant–plant interference is called allelopathy by which plants gain an advantage over their neighbors by using interfering chemicals called as allamones. Plant produced allamones vary considerably in structure, mode of action, and their effect on plants. Different compounds in root exudates affect metabolite production, respiration, photosynthesis, membrane transport, and inhibition of root and shoot growth in susceptible plants (Einhelling 1995; Weir et al. 2004). For example, a potent allamone juglone produced by black walnut (*Juglans nigra*) plants act as an electron donor and acceptor in photosynthesis and respiration reactions, affecting these processes in susceptible plants (Jose and Gillespie 1998). Recently, a flavonoid called catechin was identified in the root exudates of *Centaurea maculosa*, an invasive spotted knapweed exhibits a strong inhibitory effect on a number of plant species (Bais et al. 2003; Weir et al. 2003), and considered as a potent factor for its successful invasion in a nonnative range. Root exudates are also playing a big role in establishing associations between parasitic plants and their hosts. There are several examples that demonstrate the chemical cross talk to establish the parasitic association, including *Striga* spp. and *Orobanche* spp. (Palmer et al. 2004). Very recently, a root-secreted allelochemical identified as gallic acid from the roots of the noxious weed *Phragmites australis* which is considered a potent factor for its successful invasion in marsh and wetland communities by displacing the native species was identified (Rudrappa et al. 2007).

Besides these negative interactions, root exudates can also have positive effects in plant–plant interactions. However, these positive interactions are less frequently reported. The best studied interaction is the root exudates that induce herbivore resistance in neighboring plants. For example, when *Hordeum vulgare* (barley) plants were treated with *Elytrigia repens* (couch-grass) root exudates or the phytotoxic compound identified from *E. repens* exudates called carboline, *H. vulgare* were chosen less by aphids than the control (Glinwood et al. 2003). Besides, having direct effect on herbivore behavior, root exudates have an indirect effect by inducing defense responses in neighboring plants resulting in reduced herbivore populations indirectly by attracting predators and parasites of the offending herbivore (Du et al. 1998; Guerrieri et al. 2002).

Root exudates also play a role in soil nutrient availability by altering soil chemistry and soil biological processes (Hopkins et al. 1998). Certain compounds such as phytosiderophores, mugineic acid, and malate improve iron availability (Dakora and Phillips 2002; Fan et al. 2001). Roots secrete a range of chemicals including the secretion of organic acids and acid phosphatases and the production of proteoid roots to survive in P-deficient soils (Ascencio 1997; Raghothama 1999). For example, several plants including *Lupinus alba, Brassica napus*, and *Medicago sativa* increase the release of organic acids in P-deficient soils (Hoffland et al. 1992; Johnson et al. 1994; Lipton et al. 1987).

3.1 Plant–Plant Interactions Mediated by Root-Emitting Volatiles

Many interactions between organisms are based on the emission and perception of volatiles. These volatiles act as communication signals for chemoattractant or repellent for species-specific interactions or mediators for cell-to-cell recognition. These volatiles do not only function as signals in the above-ground interactions, but below-ground volatile interactions are similarly complex. The majority of volatile organic compounds (VOCs) tend to be lipophilic, small in molecular mass (less than 300 Da), and have a high vapor pressure (0.01 kPa or higher at 20°C). Most of the volatile compounds belong to the following three chemical groups: terpenoids, phenylpropanoids, or fatty acid derivatives. Unlike, the root-secreted phytochemicals, volatiles can travel long distances in the atmosphere and also in the soil by permeating through air-filled pores. The efficiency of volatile penetration in the soil depends on the type of mineral, texture, and particle architecture (Aochi and Farmer 2005). Also, different VOCs exchange rates indicate that soils have the potential to act as VOC sinks rather than VOC sources (Asensio et al. 2007). Volatiles emitted in the underground enable plants to influence directly or indirectly the community of soil-dwelling organisms and combat competitive plant species (Nardi et al. 2000). Several studies demonstrated that the emission of terpenoids by plant roots and particularly obvious in forest soils (Hayward et al. 2001; Lin et al. 2007; Rohloff 2002). Furthermore, a blend of unidentified root volatiles of *Echinacea angustifolia* showed allelopathic effect on different plant species such as *Lactuca sativa, Panicum virgatum*, and *Sporobolus heterolepis* (Viles and Reese 1996).

4 Plant–Microbe Interactions Mediated by Root-Secreting Phytochemicals

The rhizosphere soil surrounded by plant roots is more abundant in microbes than the nonrhizosphere soils (Bending 2003; Lynch 1987; Rouatt and Katznelson 1960; Rouatt et al. 1960). However, more recently the term "rhizosphere" has broadened to include both the volume of soil influenced by the root and the root tissues colonized by microorganisms (Pinton et al. 2001). Microorganisms in the rhizosphere react to the many metabolites secreted by plant roots. The microorganisms and their products also interact with plant roots or root-secreting compounds in a variety of positive, negative, and neutral ways. The positive interactions include classic symbioses, association with biocontrol agents, epiphytes, and mycorrhizal fungi. The negative interactions include association with parasitic plants, pathogenic bacteria, fungi, and invertebrate herbivores. Colonization and dominance of specific microbe species in the rhizosphere is very critical for pathogenic soil microbes and also important in the application of beneficial microorganisms for plant protective purposes. Although a general increase in microbes in the rhizosphere is always noted, the community structure and functional consequences of this increase are poorly understood.

The well-known classical example for positive plant–microbe interaction is the interaction between legume roots and *Rhizobia* bacteria, which are capable of forming dinitrogen-fixing nodules in the roots of legumes. However, in this chapter we intended to focus only on the nonlegumes–microbes interactions because legumes–microbe interactions are discussed as a separate chapter in this book. Similarly to *Rhizobia*, arbuscular mycorrhizal fungi (AMF) and plant roots form associations in more than 80% of terrestrial plants. Mycorrhizal fungus and bacterial rhizobial associations are thought to derive from a common-ancestral plant–microbe interaction likely of fungal origin and it was demonstrated that the activity of some host proteins regulates both mycorrhizal and rhizobial associations (Lévy et al. 2004). Root exudates play a role in the recognition of mycorrhizal fungi with the compatible host plant (Nagahashi and Douds 1999; Tamasloukht et al. 2003). Although root exudates have long been suspected to play a communicative role in mycorrhizal associations, the identification of specific molecule interactions from AMF and host still remains elusive. Recently, a sesquiterpene called strigolactone 5-deoxystrigol was identified in the root exudates of *Lotus japonicus* which is responsible for inducing AMF hyphal branching in germinating spores (Akiyama et al. 2005). In the presence of AMF symbiosis, plants trade carbon with phosphate from their fungal partners (Harrison 2005; Karandashov and Bucher 2005; Paszkowski 2006). Molecular data and fossil studies suggest that AMF have facilitated the adaptation and evolution of primitive plant species to life on land demonstrating more than 400 million years of coevolution which shows that plants and AMF are highly interdependent (Remy et al. 1994; Simon et al. 1993). Although less understood, similar processes are thought to control symbioses between nitrogen-fixing *Frankia* bacteria and their actinorhizal plant hosts (Wall 2000). Recently, a leucine-rich-repeat receptor kinase (SymRK) was identified as a common genetic basis for plant root endosymbioses with AMF, *Rhizobia*, and *Frankia* bacteria (Gherbi et al. 2008).

Soil microbes can also modulate the secretion of root exudates to positively regulate plant growth and disease resistance by indirect mechanisms. Plant growth promoting rhizobacteria (PGPR) have been found to positively influence plants through a wide variety of direct and indirect mechanisms (Gray and Smith 2005). Bacteria are likely to locate plant roots through cues extended from the root and the carbohydrates and amino acids stimulate PGPR chemotaxis on root surfaces (Somers et al. 2004). A very recent report demonstrated that the rhizobacterial elicitor acetoin from *Bacillus subtilis* induces systemic resistance in Arabidopsis to reduce plant's susceptibility by pathogen attack (Rudrappa et al. 2010).

Several negative interactions are mediated by root exudates including antimicrobial, biofilm inhibitors, and quorum-sensing mimics to prevent soil-borne pathogens (Bais et al. 2006, 2008). Plants are known to use diverse chemical molecules for defense, although some groups of compounds (phenylpropanoids) are used for defensive function across taxa (Bouarab et al. 2002). Recent years of research have clarified the antimicrobial properties of root exudates. For example, rosamarinic acid from hairy root cultures of sweet basil (*Ocimum basilicum*), pigmented napthoquinones from *Lithospermum erythrorhizon* hairy root cultures, and aromatic phenolic compounds from the exudates of *Gladiolus* spp. have shown potent antimicrobial activity against an array of soil-borne pathogens (Bais et al. 2002; Brigham et al. 1999; Taddei et al. 2002).

Plant–microbe interactions in the rhizosphere are responsible for a number of intrinsic processes such as carbon sequestration, ecosystem functioning, and nutrient cycling (Singh et al. 2004). A great variety of biotic and abiotic factors shape soil and plants associated habitats, as well as modify the composition and activities of their microbial communities (Bever et al. 1997). Bacterial communities in root-associated habitats respond specifically with respect to density and composition of root exudates, eventually yielding plant species-specific microfloras which may also vary depending upon the plant developmental stage (Mahaffee and Kloepper 1997; Wieland et al. 2001; Yang and Crowley 2000). Recent evidence suggests that specific plant species are responsible for driving their own soil fungal community composition and diversity mediated by root-secreting compounds (Broeckling et al. 2008). In addition, a recent report demonstrated that a mutation in the ABC transporter (*At*PDR2) dramatically changes the composition of root-secreted phytochemicals which influenced the qualitative and quantitative changes in the Arabidopsis native soil microbiota by culturing more beneficial microbes compared to the wild type (Badri et al. 2009). Plant root exudates also affect the level of contamination found in soil and groundwater from various environmental pollutants by a process called "rhizoremediation." For example, *Pseudomonas putida* from the rhizosphere of corn and wheat helps to effectively decontaminate 3-methylbenzoate and 2,4-D, respectively (Kingsley et al. 1994; Ronchel and Ramos 2001). Several lines of evidences demonstrated the role of root exudates in metal remediation either directly by solubilizing the metals by root-secreting organic acids or indirectly with the help of soil microbes (Do Nascimento and Xing 2006).

4.1 *Plant–Microbe Interactions Influenced by Root Volatiles*

The microbial community in the rhizosphere is limited by carbon availability but carbon-containing root volatiles especially monoterpenes contribute significantly to the below-ground carbon cycle (Owen et al. 2007). Becard and Piche (1989) first demonstrated that the carbon dioxide is a crucial root volatile that stimulates hyphal growth of vesicular–arbuscular mycorrhizal (VAM) fungus. They also showed that elevated carbon dioxide promotes hyphal length of VAM and depresses the growth of non-VAM at low nitrogen availability. In addition, Scher et al. (1985) reported that *Pseudomonas fluorescens* was attracted by carbon dioxide. Besides carbon dioxide, there are other simple compounds such as acetaldehyde, acetone, acetic acid, ethanol, and methanol emitted by Arabidopsis roots (Steeghs et al. 2004). Root volatiles are important for defense response against microbial populations. For example, the root volatile β-phellanderene was effective against the root fungal pathogen *Fomes annosus* (Cobb et al. 1968), which was emitted from the roots of *Smyrnium olusatrum* (Bertoli et al. 2004) and *Rhodiola rosea* (Rohloff 2002).

Similarly, the monoterpene 1,8-cineol emitted from Arabidopsis roots in response to pathogen infection (Steeghs et al. 2004) had antimicrobial property against several microbes (Kalemba et al. 2002; Vilela et al. 2009). Besides the role of volatiles in plant defense, they also influence in symbiotic association either positively or negatively. For instance, the sesquiterpene longifolene from the roots of *Pinus sylvestris* inhibits the vegetative growth of mycorrhizal fungus *Boletus varigatus* and *Rhizopogon roseolus* (Melin and Krupa 1971).

5 Plant–Faunal Interactions

Protozoa and microbial feeding nematodes are known to be the most important grazers of the microflora in the terrestrial ecosystems (Ingham et al. 1985). Thus, grazing of the microflora by microbivores is considered as a critical mechanism to maintain the balance in the competition between microbes and plants. Despite the critical importance of interactions between roots (root exudates), microbes, and their predators for plant growth, knowledge of these interactions is still fragmentary and the mechanisms are poorly understood (Zwart et al. 1994). The outcome of the rhizosphere plant–faunal, plant–microbial, and faunal–microflora interactions may be either positive (e.g. mutualistic and associative) or negative (predatory and competitive) (Bonkowski et al. 2000). Much attention has been drawn only to the negative plant–faunal interactions (parasitic nematodes) (Curtis 2008). It should also be noted that root-feeding by nematodes may increase allocation of carbon below-ground and increase significantly the leaking of carbon from roots that stimulate rhizosphere microbial processes (Bardgett et al. 1998). Although most nematodes are free-living organisms that consume bacteria, there are some nematodes that are pathogenic for plants causing important economic losses each year (Barker and Koenning 1998). Some of the most harmful plant–parasitic nematodes include root-knot and cyst nematodes such as *Meloidogyne* spp., *Heterodera* spp., and *Globodera* spp. (Chitwood 2003).

During coevolution with the host plant, parasitic nematodes have developed the capacity to recognize and respond to the chemical signals of particular host species.

Understanding the complexity of the chemical signal exchange and response during the early stage of host–parasite interactions is important to identify the critical steps in the parasite life cycle to disrupt the host–nematode recognition. Plant signals are indispensible for nematodes to locate hosts and feeding sites (Robinson et al. 1987). However, the identities of the plant signals involved in the key stages of the plant–nematode interactions are not yet clearly dissected. Chemical components of root exudates may deter one organism while attracting another and these compounds alter the nematode behavior by attracting the nematodes to the roots or result in repellence, motility inhibition, or even death (Robinson 2002; Wuyts et al. 2006). For instance, root exudates of cucumber and their fractions having both repellent and attractant activity to *M. incognita* were reported (Castro et al. 1989). Similarly, the root cap exudates that include enzymes, antibiotics, and other soluble chemicals and mucilage repelled both plant parasitic nematodes and free-living nematode *Caenorhabditis elegans* and resulted in reversible state of immobility in these nematodes (Hubbard et al. 2005; Wuyts et al. 2006; Zhao et al. 2000). This study indicates that the root tip delivered products has the potential to temporarily immobilize nematodes.

The best examples describing the role of plant signals in synchronizing host–parasite life cycle are the two species of potato cyst nematodes (*Globodera* spp.), as these nematodes are completely dependent on root exudates for hatching. Several hatching factors have been identified in crop plants to explore the potential of using these compounds for agrochemical use (Devine and Jones 2001; Timmermans et al. 2007). For example, *Solanum sisymbriifolium* is being used successfully as a trap crop for potato cyst nematodes in Europe, because the plant root exudates stimulate hatching of the second-stage juveniles but does not support their development to complete their life cycle (Timmermans et al. 2007). Mostly, the root exudates act as attractants for nematodes to move closer to individual host roots; these are considered "short distance attractants." However there are "long-distance attractants" that enable the nematodes to locate roots. So far, only carbon dioxide has been identified as a prime long-distance attractant to nematodes including *M. incognita* (Robinson 2002). Other additional short-distance attractants such as amino acids, sugars, and metabolites are also reported (Bird 1959; Perry 2001; Robinson 2002). In addition, plant roots also produce allelochemicals to defend other plant species or soil-borne pathogens, which have been shown to function as nematodes antagonists (Guerena 2006). For example, cucurbitacin A from cucumber plants repels nematodes and β-terthienyl from *Tagetes erecta* acts as repellent as well as nematotoxic (Castro et al. 1989). Other compounds such as cyclic hydroxamic acid from maize root exudates affect the behavior of *M. incognita, Pratylenchus zea*, and *Xiphinema americanum* (Friebe et al. 1998; Zasada et al. 2005). Root exudate compounds not only induce nematode hatching, attraction, and repellence, but also induce the exploratory behavior of nematodes including stylet thrusting and aggregation and increase in nematode mobility (Curtis 2007; Robinson 2002). In addition, root exudate compounds and phytohormones such as IAA induce the changes in the surface cuticle of nematodes (Curtis 2008). The surface changes in the cuticle might allow the nematodes (*M. incognita*) to adapt and survive plant defense responses (Curtis 2007). Identifying the signaling and perception process executed by root exudate compounds occurring in plant–nematode interactions will reveal targets for chemical or genetic intervention.

Besides the role of root exudates compounds in attracting and repelling nematodes, nematodes also respond to the host by secreting specific proteins to complete their life cycle within the host. The way nematodes secrete proteins is mainly through their stylet, a hollow, protrusible spear at the anterior of the worm (Davis et al. 2000). The stylet secretions are studied widely for their role in host penetration, feeding site induction, and disease induction (Hussey 1989). The first analysis of the stylet-secreted proteins was realized in one-dimensional electrophoresis (Robertson et al. 1999; Veech et al. 1987). However, with the modern techniques of proteomics, knowledge of the nematode secretome has been extended (Bellafiore et al. 2008).

In the secretome of *M. incognita* were identified 486 proteins mainly required for invasion, immune suppression, and host cell reprogramming (Bellafiore et al. 2008). In another proteomic study on *M. incognita*, calreticulin a calcium-binding protein was identified as the most abundant stylet-secreted protein (Jaubert et al. 2002). Calreticulin has already been studied for its role in many host–parasite interactions (Nakhasi et al. 1998; Pritchard et al. 1999). In plants, calreticulin has been involved in cell-to-cell trafficking and pressure support (Baluska et al. 2001; Sivaguru et al. 2000). The fact that nematode protein homologues were found in plants indicates that nematodes could manipulate plant cell functions during the compatible interaction (Caillaud et al. 2008) to elude the defense plant response. Most of the nematode proteins secreted in the root–nematode interaction have been found in the first 18 h of infection being some of these glycoproteins (Veech et al. 1987) proteases and superoxide dismutases (Robertson et al. 1999).

Some of the secreted proteins have been determined to have important roles in parasitism (Davis et al. 2000; De Meutter et al. 1984; Popeijus et al. 2000). Enzymes such as β-1,4-endoglucanases, cellulases, pectate lyase, and polygalacturonase are likely to be used by nematodes in softening the cell wall in order to facilitate their movement through the root (Davis et al. 2000; Doyle and Lambert 2002; Goellner et al. 2000; Popeijus et al. 2000; Smant et al. 1998). Mawuenyega et al. (2003) found, by two-dimensional liquid chromatography (2DLC) coupled with electrospray ionization (ESI) tandem mass spectrometry (MS/MS), 110 secreted/targeted proteins and 242 transmembrane proteins. Also, it was found that many peptides of these proteins have *N*-terminal glycosylation and phosphorylation, which suggest the importance of posttranslational modification for recognition and infection. Depending on the modification that the protein has, it would have critical effects on cell regulatory and signaling processes (Mann and Jensen 2003). For instance, it was found in a human–filanial nematode parasite *Acanthocheilonema viteae* that the covalent attachment of phosphorylcholine to a major secreted protein named ES-62 is likely involved in the interference of the host immune system (Houston et al. 1997). This effect may also happen in plant nematodes that secrete numerous posttranslational modified proteins (Bellafiore et al. 2008; Caillaud et al. 2008; Mawuenyega et al. 2003).

Reports are also available in the model plant *A. thaliana* response to the nematodes. Huang et al. (2006) found that the root-knot nematode *M. incognita* secretes a peptide named 16D10 that interacts with the plant SCARECROW-like transcription factor; this peptide–protein interaction probably represents an early signaling event in the plant–nematode interaction. Some nematodes induce the well-known salicylic acid-related defense response during incompatible interactions. The fact that salicylic acid inhibits the parasitism of *H. schachtii* by inducing the expression of PRP genes in Arabidopsis roots (Wubben et al. 2008) suggests that pathogenesis-related proteins play a role in signaling and perception process in the host–nematode interactions.

6 Plant-Root-Secreting Proteins Involve in Neighbors Interaction

Studies on proteins involved in root–microorganism interaction have provided strong evidence about the importance of root-secreted proteins during the recognition between pathogenic and nonpathogenic interactions. One of the most studied common proteins found in root exudates are lectins (De-la-Peña et al. 2008; Wen et al. 2007). Lectins are a diverse group of carbohydrate binding proteins that are found in dual systems, functioning in defense with some pathogens, and in recognition of a compatible symbiosis (De Hoff et al. 2009; Sharon and Lis 2004). De Hoff et al. (2009) illustrated a hypothetical model of the perception of pathogenic and symbiotic bacteria where the lectin gradient is secreted from the root to permit the growth of symbiotic bacteria and agglutination of the pathogenic bacteria. However, those bacteria that elude the first line of the plant defense can be recognized by specific receptors in the root triggering a cascade of MAP kinase signaling leading to antipathogen response or prosymbiotic response depending on the microbe that is in contact with the root. Another set of proteins found highly secreted in the rhizosphere are the PRP. PRP, such as chitinases, osmotin, and thaumatin-like proteins, have been found in root exudates under pathogen contact as well as secreted constitutively (Basu et al. 2006; De-la-Peña et al. 2008; Nóbrega et al. 2005). Root-secreted proteins are not only important for defense, but also for attracting microbes to the roots, a process known as chemotaxis (Currier and Strobel 1977). Chemotaxis is one of the earliest essential events in the interaction between plants and bacteria (Hawes and Smith 1989; Manson 1990). Proteins in the rhizosphere are so important for chemotaxis that even a glycoprotein, named trefoil chemotactin, from *Lotus corniculatus* has been identified (Currier and Strobel 1977, 1981). After that, other secreted proteins, able to recognize bacterial surface carbohydrate structures, which help to adhere to root hairs of many plants, have been identified in Rhizobium (Ausmees et al. 2001).

In order to colonize the roots, bacteria usually congregate together by using of quorum-sensing (QS) signals. The most common type of QS signal in proteobacteria are the *N*-acyl-homoserine lactones (AHL) (von Bodman et al. 2003), which affect the expression of more than 600 genes in bacteria (Schuster et al. 2003). Biofilm-forming bacteria is a dense population that perform many biological responses as community, including production of extracellular polysaccharides, degradative enzymes, antibiotics, Hrp protein secretion, Ti plasmid transfer, and other functions (von Bodman et al. 2003). QS stimulates the production of extracellular enzymes which has been related to pathogenesis in *P. aeruginosa* PAO1 (Passador et al. 1993), *P. fluorescens* (Worm et al. 2000), and *Aeromonas hydrophila* (Swift et al. 1999). More recent evidence suggests that QS-related enzymes such as chitinases and proteases could be involved in nitrogen mineralization process instead of pathogenesis (DeAngelis et al. 2008). On the other hand, *M. truncatula* roots are able to detect low concentrations of bacterial QS signals from the pathogenic bacterium *P. aeruginosa* by change and accumulation of 154 proteins, from which 21 are related to defense and stress responses (Mathesius et al. 2003). Based on these information one can easily predict that considerable percentage of root-secreted proteins function in the rhizosphere still remains elusive.

Studies on the secretome of Gram-positive bacteria *B. subtilis* started 10 years ago (Hirose et al. 2000; Tjalsma et al. 2000) revealed nearly 300 possible proteins secreted into the soil, among them more than a half have not been yet identified (Antelmann et al. 2006). The secretome analysis of pathogenic bacteria is very important in revealing new virulence proteins (Desvaux and Hébraud 1978; Kaffarnik et al. 2009; Kazemi-Pour et al. 2004; Saarilahti et al. 1992; Watt and Wilke 2005). For instance, Kazemi-Pour et al. (2004) analyzing the secretome of *Erwinia chrysanthemi*, a well-known plant pathogenic bacterium, found proteins related to virulence, disease symptoms, and pathogenicity: Avr-like protein, elongation factor EF-Tu, flagellin, pectate lyases, and metalloproteases. Furthermore, *E. chrysanthemi* and *E. carotovora* secrete proteases, polygalacturonase, and proteins that degrade plant cell walls, such as pectin lyase and cellulose (Collmer and Keen 1986; Perombelon and Kelman 1980). Polyglacturonases are considered to be key enzymes involved in pathogenesis (Palomski and Saarilahti 1997). Another secretome study on *Xanthomonas campestris* revealed 97 proteins; some of these are involved in element acquisition, protein maintenance and folding, compound degradation, and proteins with unknown functions (Watt and Wilke 2005). It is worth noting that in both bacteria, *E. chrysanthemi* and *X. campestris*, there are some shared secreted proteins that could be linked with pathogenicity. *E. chrysanthemi* (Kazemi-Pour et al. 2004) secreted an elongation factor, a chaperonin GroEL, flagellin, and celluloses that also were found in the secretome of *X. campestris* (Watt and Wilke 2005). Furthermore, the secretion of these proteins has found to increase in the presence of plants and plant compounds (De-la-Peña et al. 2008; Kazemi-Pour et al. 2004).

As far as root–fungus interactions are concerned, several secretomes have been annotated (Choi et al. 2010), being these from symbiotic to pathogenic, proteomics have been applied to complement the genomics analysis of the fungal secretome (Phalip et al. 2005). Pathogenic fungus in comparison to symbiotic mycorrhizal fungi represents serious damage and loss to agriculture. Furthermore, some fungi such as *Trichoderma* spp. represent a group of fungi that have been used as plant disease control against a wide diversity of phytopathogenic fungi and even they have positive effect on plants as a plant growth enhancer (Harman and Björkman 1998; Harman et al. 2004; Yedidia et al. 2003). This mycoparasitic activity has been attributed to the secretion of complex mixture of hydrolytic enzymes such as chitinases, glucanases, and proteases able to degrade different cell wall systems (Benítez et al. 1978; Suárez et al. 2005; Szekeres et al. 2004). The secretome of *T. harzianum* revealed to vary both qualitatively and quantitatively on different ascomycetes, oomycetes, and basidiomycetes cell walls (Suárez et al. 2005).

In pathogenic fungi the most important molecules that promote the infection process are the extracellular effectors that produce the elicitation of plant defense responses (Birch et al. 2006; Colditz et al. 2004; Dean et al. 2005; Hahn 1996; Rose et al. 2002). These effectors have been found in the secretome of different oomycetes (Kamoun 2006). Among the secretomes studied on plant pathogenic fungi are those from *Fusarium graminearum*, and *Sclerotinia sclerotiorum* (Phalip et al. 2005; Yajima and Kav 2006). In *F. graminearum* secretome, in the presence of *Humulus lupulus* L. cell wall, were identified 84 proteins (Phalip et al. 2005) which 45% of them are actually involved in cell wall degradation and the most abundant proteins were cellulases, endoglucanases, proteases, and chitinases. In *S. sclerotiorum* were identified 18 secreted proteins from the liquid culture of this fungus (Yajima and Kav 2006) where L-arabinofuranosidase was one of the most abundant in the secretome but not in the mycelia. This protein is much known for its function in the virulence process of this fungus. The study of secreted proteins not only in pathogens alone, but also in the presence of the host, should be persuaded in order to know the principal signals involved between fungal and roots interaction. Some of these proteins could lead to the identification of new effector proteins produced by fungi and defense-related proteins produced by roots specific-secreted to a given fungus.

The way plants and pathogenic fungi cross talk is very dynamic and complex, and it is not known which one of the organisms emits the first signal. Fungus as much as plants turn on their genetic and biochemical systems to generate a series of signals to invade or defend (Nürnberger and Brunner 2002). In the case of fungus, once contact with the plant host root has been established, elicitors start to be produced and secreted by the fungus (Nürnberger and Brunner 2002). These elicitors are perceived by the plant and plant-specific proteins rapidly phosphorylated in response to fungus signals (Peck et al. 2001). Dietrich et al. (1990) and after Felix et al. (1991) found that protein kinase-mediated phosphorylation might be the first trigger immediately after the fungal is perceived by the plant cells. Posttranslational modifications by phosphorylation/dephosphorylation in the signal transduction cascade produced by pathogens have been studied in other plants as an early plant defense response (Grant and Mansfield 1999; Stone and Walker 1995). The possible existence of an extracellular phosphorylation network (Ndimba et al.

2003) has opened the possibility to investigate the plant–microbe interaction signaling in the rhizosphere mediated by kinases and phosphatases.

The principal way that roots combat the fungal invasion is through enzymes such as PRP (Colditz et al. 2004; Fagoaga et al. 2001), which some have been seen to be effective in repressing the growth of root pathogenic fungi (Nóbrega et al. 2005). For instance, the root protein profile of *M. truncatula* infected with the pathogen *Aphanomyces euteiches* (Colditz et al. 2004) showed an induction of proteins belonged to the family of class 10 of PRP (PR10). Because the fungal cell walls are build with chitin and β-glucans, proteins belonging to the family PR2 such as β-1,3-endoglucanases and family PR-3, -4, -5, and -11 such as several types of endo-chitinases are effective to inhibit the growth of fungi while depolymerizing poly-saccharides of mycelia walls and disturbing intracellular targets (Abad et al. 1996; Ferreira et al. 1984; Joosten and De Wit 1989; Li et al. 2000). PRPO have been studied extensively and they are found in both pathogenic and nonpathogenic interactions. This is the case of osmotin, a PR5 protein found in the root exudates of alfalfa inoculated with *Sinorhizobium meliloti* (De-la-Peña et al. 2008), which also participates in plasma membrane permeabilization, which is associated with patho-genic fungal spore lysis (Abad et al. 1996). This specificity is very important for the plant in order to avoid the killing of beneficial microbes or the free invasion of pathogenic organisms. How the plants provoke such specificity is a research that needs to be persuaded. Although mycorrhiza are beneficial fungi for plants, the induction of plant-defense-related genes still takes place at early stages of the interaction (Gianinazzi-Pearson et al. 1996; Harrison 1999, 2005). During the earlier stages of the development of the VAM symbiosis association between *Allium porrum* L. and *Glomus versiforme*, the root chitinase activity was almost twice as high as in uninfected roots (Spanu et al. 1989). However, once the symbiosis was fully established, the chitinase activity in mycorrhizal roots was even lower than in the control roots. The possible explanation for this observation is that at the earliest stages of the interaction with the fungus, roots respond with a defense response. However, once the symbiotic interaction is established, the fungus is able to suppress the plant stress reaction and grow inside the roots.

7 Methods for Studying Rhizosphere Interactions

The rhizosphere is complex; quantitatively a single gram of soil has over 10,000 distinct microbes (Kent and Triplett 2002). The traditional culture-based techniques are inadequate to study the actual goings-on of the microbes of the interest because most of the rhizosphere organisms are unculturable (Kent and Triplett 2002). Novel approaches are needed to probe this complex environment and recently a broad range of techniques and strategies have been used to study the rhizosphere interactions. The increasing applications of molecular techniques will provide a basis for studying rhizosphere interactions at broad-scale (community level) to fine-scale (species level) investigations. Previously, Biolog has been used to characterize the differences in microbial communities between contrasting habitat and soil types (Zak et al. 1994). The Biolog assay uses microtiter plates consisting of 96 wells containing separate sole carbon sources and a redox indicator dye, which produce patterns of potential carbon utilization for microbial communities. However, this method is completely dependent on the growth of microbial population in artificial media and also biased toward faster growing microbes (Paterson et al. 1997). To overcome this problem, phospholipid fatty acid analysis (PLFA) has been used to analyze the microbial population at community level and this method is totally culture-independent analyses and provides the broad number of bacterial taxes present in the samples (Zelles 1997). The combination of Biolog and PLFA techniques has shown differences in the microbial community composition of bulk and rhizosphere soils (Söderberg et al. 2004), but these two methods cannot identify certain microbial species at community level. Later, polymerase chain reaction (PCR) amplification of rDNA genes combined with fingerprinting techniques such as denaturing gradient gel electrophoresis (DGGE), terminal restriction fragment analyses (T-RFLP), and amplified rDNA restriction analysis (ARDRA) is proven to study the species composition of whole communities in detail (Nicol et al. 2003; Torsvik and Ovreás 2002). In addition, development of novel methods such as fluorescence in situ hybridization (FISH) and microautoradiography allows to determine the phylogenetic identification of uncultured bacteria in natural environments by using fluorescence-specific phylogenetic probes (targeting rRNA) and fluorescence microscopy to detect and quantify the active population utilizing a specific substrate (Gray et al. 2000). The recent development of metagenomics coupled with bioinformatics will allow studying the genomics analyses of uncultured microbes in the rhizosphere (Rondon et al. 2000), but it does not make any sense to gather data on every microbe present in the rhizosphere instead of the organisms of interest which are actively interacting in the rhizosphere. The discovery of an elegant technique called stable isotope probing (SIP) allows studying the organisms of interest which are actively interacting in the rhizosphere with the root exudates. In this technique, plants are exposed to $^{13}CO_2$, which has a heavier carbon atom than regular CO_2, metabolized by the plant, and deposited in the rhizosphere through rhizodeposition and utilized by the microbes present in the rhizosphere. The nucleic acids of the microbes utilizing the $^{13}CO_2$ will be heavier than the noninteracting microbes and analyzed using density gradient centrifugation (Kiely et al. 2006) and also yield the entire genome of all the participating microbes in the rhizosphere (Singh et al. 2004). In addition, the recent development of "omics" technologies coupled with bioinformatics studies is appropriate to study the rhizospheric soil microbe's interactions at community levels to species level. The "omics" techniques such as transcriptomics, proteomics, and metabolomics allow studying the microbial interactions in a given environment as a part of functional genomics. The recent development of next-generation sequencing methods will complement these "omics" techniques to study the rhizospheric microbial interactions in detail to detect and quantify the unculturable microbes that actively participate in the rhizosphere. Finally, the rhizosphere is a complex system and no single method is sufficient to describe the complex nature of the rhizosphere. Therefore, there exists a need to develop systems approach to describe the complex nature of the rhizosphere. In addition, the multitude of interactions in the rhizosphere requires high-throughput techniques in order that they can be elucidated in a reasonable time frame.

8 Concluding Remarks and Future Perspectives

Competition between plants is high because plant growth in the natural environment frequently takes place in dense stands of established vegetation. The complex interactions that take place in the rhizosphere between plants and microbes and their regulation by soil faunal activity may be of fundamental importance for individual plant success but also at community level.

Although there is a large body of literature available to prove the significance of plant–microbes, plant–microfaunal, and microbial–microfaunal interactions to enhance plant growth, but still there is a lacuna on the knowledge of multitropic interactions occurring in the rhizosphere. In addition, most information about important processes in the rhizosphere comes from studies in controlled environments where roots are grown in simple uniform media and organisms of interest are applied. There is a need to understand these rhizospheric multitrophic interactions in the realistic field conditions to improve the plant growth at species and community level. In addition, studies should conduct in the field conditions to understand the rhizospheric complex interactions in monocultures and polycultures. This will help to understand the dynamics of interactions and their outcome in influencing the plant's success when they are in monocultures and in polycultures.

Obviously, studying these complex rhizospheric interactions by employing single method is impossible. However, the combination of techniques and the continuous development of new techniques in the field of rhizosphere biology coupled with systems approach will allow us partly to elucidate these complex interactions under field conditions. For many years, ecologists have viewed soil organisms and plants as relatively independent from each other. But to unravel these complex interactions, further research requires multidisciplinary system approach, which includes involving and exchange of knowledge between plant physiologists, soil scientists, microbiologists, and zoologists with the help of bioinformatics specialists. In addition, the differences in plant growth and plant community composition can only be understood in relation to indirect microbial–faunal, plant–microbial, faunal–plant, and microbial–microbial interactions in the rhizosphere.

Acknowledgments The work in our laboratories was supported by CONACYT grants (VMLV, 61415 and CDS, 121768).

References

Abad LR, D'Urzo MP, Liu D, Narasimhan ML, Reuveni M, Zhu JK, Niu X, Singh NK, Hasegawa PM, Bressan RA (1996) Antifungal activity of tobacco osmotin has specificity and involves plasma membrane permeabilization. Plant Sci 118:11–23

Akiyama K, Ki M, Hayashi H (2005) Plant sesquiterpenes induce hyphal branching in arbuscular mycorrhizal fungi. Nature 435:824–827

Antelmann H, van Dijl JM, Bron S, Hecker M (2006) Proteomic survey through secretome of *Bacillus subtilis*. In: Humphery-Smith I, Hecker M (eds) Microbial proteomics: functional biology of whole organisms. Wiley, New Jersey, pp 179–208

Aochi YO, Farmer WJ (2005) Impact of soil microstructure on the molecular transport dynamics of 1,2-dichloroethane. Geoder 127:137–153

Ascencio J (1997) Root secreted acid phosphatase kinetics as a physiological marker for phosphorus deficiency. J Plant Nutr 20:9–26

Asensio D, Peñuelas J, Filella I, Llusiá J (2007) On-line screening of soil VOCs exchange responses to moisture, temperature and root presence. Plant Soil 291:249–261

Ausmees N, Jacobsson K, Lindberg M (2001) A unipolarly located, cell-surface-associated agglutinin, RapA, belongs to a family of *Rhizobium*-adhering proteins (Rap) in *Rhizobium leguminosarum* bv. trifolii. Microbiology 147:549–559

Badri DV, Vivanco JM (2009) Regulation and function of root exudates. Plant Cell Environ 32: 666–681

Badri DV, Quintana N, El Kassis EG, Kim HK, Choi YH, Sugiyama A, Verpoorte R, Martinoia E, Manter DK, Vivanco JM (2009) An ABC transporter mutation alters root exudation of phytochemicals that provokes an overhaul of natural soil microbiota. Plant Physiol 151: 2006–2017

Bais HP, Walker TS, Schweizer HP, Vivanco JM (2002) Root specific elicitation and antimicrobial activity of rosmarinic acid in hairy root cultures of *Ocimum basilicum*. Plant Physiol Biochem 40:983–995

Bais HP, Vepachedu R, Gilroy S, Callaway RM, Vivanco JM (2003) Allelopathy and exotic plant invasion: From molecules and genes to species interactions. Science 301:1377–1380

Bais HP, Park S-W, Weir T, Callaway RM, Vivanco JM (2004) How plants communicate using the underground information superhighway. Trends Plant Sci 9:26–32

Bais HP, Weir TL, Perry LG, Gilroy S, Vivanco JM (2006) The role of root exudates in rhizosphere interactions with plants and other organisms. Annu Rev Plant Biol 57:233–266

Bais HP, Broeckling CD, Vivanco JM (2008) Root exudates modulate plant-microbe interactions in the Rhizosphere. In: Karlovsky P (ed) Secondary metabolites in soil ecology. Springer, Berlin, pp 241–252

Baluska F, Cvrckova F, Kendrick-Jones J, Volkmann D (2001) Sink plasmodesmata as gateways for phloem unloading. Myosin VIII and calreticulin as molecular determinants of sink strength? Plant Physiol 126:39–46

Bardgett RD, Wardle DA, Yeates GW (1998) Linking above-ground and below-ground interactions: how plant responses to foliar herbivory influence soil organisms. Soil Biol Biochem 30:1867–1878

Barker KR, Koenning SR (1998) Developing sustainable systems for nematode management. Annu Rev Phytopathol 36:165–205

Basu U, Jennafer L, Whittal RM, Stephens JL, Wang Y, Zaiane O, Taylor G (2006) Extracellular proteomes of *Arabidopsis thaliana* and *Brassica napus* roots: analysis and comparison by MudPIT and LC-MS/MS. Plant Soil 286:357–376

Becard G, Piche Y (1989) Fungal growth stimulation by CO_2 and root exudates in vesicular-arbuscular mycorrhizal symbiosis. Appl Environ Microbiol 55:2320–2325

Bellafiore S, Shen Z, Rosso MN, Abad P, Shih P, Briggs SP (2008) Direct identification of the *Meloidogyne incognita* secretome reveals proteins with host cell reprogramming potential. PLoS Pathog 4:e1000192

Bending GD (2003) The rhizosphere and its microorganisms. In: Thomas B, Murphy DJ, Murray BG (eds) Encyclopaedia of applied plant sciences. Academic, London, pp 1123–1129
Benítez T, Rincón AM, Limón MC, Codón AC (1978) Biocontrol mechanisms of Trichoderma strains. Int Microbiol 7:249–260
Bertin C, Yang X, Weston LA (2003) The role of exudates and allelochemicals in the rhizosphere. Plant Soil 256:67–83
Bertoli A, Pistelli L, Morelli I, Fraternale D, Giamperi L, Ricci D (2004) Volatile constituents of different parts (roots, stems and leaves) of *Smyrnium olusatrum* L. Flavour Fragr J 19:522–525
Bever JD, Westover KM, Antonovics J (1997) Incorporating the soil community into plant population dynamics: the utility of the feedback approach. J Ecol 85:561–573
Birch PRJ, Rehmany AP, Pritchard L, Kamoun S, Beynon JL (2006) Trafficking arms: oomycete effectors enter host plant cells. Trends Microbiol 14:8–11
Bird F (1959) The attractiveness of roots to the plant parasitic nematodes *Meloidogyne javanica* and *M. hapla*. Nematologica 4:322–335
Bonkowski M, Cheng W, Griffiths BS, Alphei J, Scheu S (2000) Microbial-faunal interactions in the rhizosphere and effects on plant growth. Eur J Soil Biol 36:135–147
Bouarab K, Melton R, Peart J, Baulcombe D, Osbourn A (2002) A saponin-detoxifying enzyme mediates suppression of plant defences. Nature 418:889–892
Brigham LA, Michaels PJ, Flores HE (1999) Cell-specific production and antimicrobial activity of naphthoquinones in roots of *Lithospermum erythrorhizon*. Plant Physiol 119:417–428
Broeckling CD, Broz AK, Bergelson J, Manter DK, Vivanco JM (2008) Root exudates regulate soil fungal community composition and diversity. Appl Environ Microbiol 74:738–744
Caillaud MC, Dubreuil G, Quentin M, Perfus-Barbeoch L, Lecomte P, de Almeida EJ, Abad P, Rosso MN, Favery B (2008) Root-knot nematodes manipulate plant cell functions during a compatible interaction. J Plant Physiol 165:104–113
Castro CE, Belser NO, McKinney HE, Thomason IJ (1989) Quantitative bioassay for chemotaxis with plant parasitic nematodes. J Chem Ecol 15:1297–1309
Chitwood DJ (2003) Research on plant-parasitic nematode biology conducted by the United States Department of Agriculture-Agricultural Research Service. Pest Manag Sci 59:748–753
Choi J, Park J, Kim D, Jung K, Kang S, Lee YH (2010) Fungal secretome database: integrated platform for annotation of fungal secretomes. BMC Genomics 11:105
Cobb FW, Krstic M, Zavarin E (1968) Inhibitory effects of volatile oleoresin components on *Fomes annosus* and four Ceratocystis species. Phytopathology 58:1327–1335
Colditz F, Nyamsuren O, Niehaus K, Eubel H, Braun HP, Krajinski F (2004) Proteomic approach: Identification of *Medicago truncatula* proteins induced in roots after infection with the pathogenic oomycete *Aphanomyces euteiches*. Plant Mol Biol 55:109–120
Collmer A, Keen NT (1986) The role of pectic enzymes in plant pathogenesis. Annu Rev Phytopathol 24:383–409
Currier WW, Strobel GA (1977) Chemotaxis of rhizobium spp. to a glycoprotein produced by birdsfoot trefoil roots. Science 196:434–436
Currier AW, Strobel GA (1981) Characterization and biological activity of trefoil chemotactin. Plant Sci Lett 21:159–165
Curtis RHC (2007) Do phytohormones influence nematode invasion and feeding site establishment? Nematology 9:155–160
Curtis RHC (2008) Plant-nematode interactions: environmental signals detected by the nematode's chemosensory organs control changes in the surface cuticle and behaviour. Parasite 15:310–316
Dakora FD, Phillips DA (2002) Root exudates as mediators of mineral acquisition in low-nutrient environments. Plant Soil 245:35–47
Davis EL, Hussey RS, Baum TJ, Bakker J, Schots A, Rosso MNl, Abad P (2000) Nematode parasitism genes. Annu Rev Phytopathol 38:365–396
De Hoff P, Brill L, Hirsch A (2009) Plant lectins: the ties that bind in root symbiosis and plant defense. Mol Genet Genomics 282:1–15

De Meutter J, Vanholme B, Bauw G, Tytgat T, Gheysen G, Gheysen G (1984) Preparation and sequencing of secreted proteins from the pharyngeal glands of the plant parasitic nematode Heterodera schachtii. Mol Plant Pathol 2:297–301

Dean RA, Talbot NJ, Ebbole DJ, Farman ML, Mitchell TK, Orbach MJ, Thon M, Kulkarni R, Xu JR, Pan H, Read ND, Lee YH, Carbone I, Brown D, Oh YY, Donofrio N, Jeong JS, Soanes DM, Djonovic S, Kolomiets E, Rehmeyer C, Li W, Harding M, Kim S, Lebrun MH, Bohnert H, Coughlan S, Butler J, Calvo S, Ma LJ, Nicol R, Purcell S, Nusbaum C, Galagan JE, Birren BW (2005) The genome sequence of the rice blast fungus *Magnaporthe grisea*. Nature 434: 980–986

DeAngelis KM, Lindow SE, Firestone MK (2008) Bacterial quorum sensing and nitrogen cycling in rhizosphere soil. FEMS Microbiol Ecol 66:197–207

De-la-Peña C, Lei Z, Watson BS, Sumner LW, Vivanco JM (2008) Root-microbe communication through protein secretion. J Biol Chem 283:25247–25255

Desvaux M, Hébraud M (1978) The protein secretion systems in *Listeria*: inside out bacterial virulence. FEMS Microbiol Rev 30:774–805

Devine KJ, Jones PW (2001) Potato cyst nematode hatching activity and hatching factors in interspecific Solanum hybrids. Nematology 3:141–149

Dietrich A, Mayer JE, Hahlbrock K (1990) Fungal elicitor triggers rapid, transient, and specific protein phosphorylation in parsley cell suspension cultures. J Biol Chem 265:6360–6368

Do Nascimento CWA, Xing B (2006) Phytoremediation: a review on enhanced metal availability and plant accumulation. Sci Agric (Piracicaba, Braz) 63:299–311

Doyle EA, Lambert KN (2002) Cloning and characterization of an esophageal-gland-specific pectate lyase from the root-knot nematode *Meloidogyne javanica*. Mol Plant Microbe Interact 15:549–556

Du Y, Poppy GM, Powell W, Pickett JA, Wadhams LJ, Woodcock CM (1998) Identification of semiochemicals released during aphid feeding that attract parasitoid *Aphidius ervi*. J Chem Ecol 24:1355–1368

Einhelling FA (1995) Mechanisms of actions of allelochemicals in allelopathy. In: Inderjit, Dakshini KMM, Einhelling FA (eds) Allelopathy: organisms, processes and applications. American Chemical Society, Washington DC, pp 96–116

Fagoaga C, Rodrigo I, Conejero V, Hinarejos C, Tuset JJ, Arnau J, Pina JA, Navarro L, Peña L (2001) Increased tolerance to *Phytophthora citrophthora* in transgenic orange plants constitutively expressing a tomato pathogenesis related protein PR-5. Mol Breed 7:175–185

Fan TW, Lane AN, Shenker M, Bartley JP, Crowley D, Higashi RM (2001) Comprehensive chemical profiling of gramineous plant root exudates using high-resolution NMR and MS. Phytochemistry 57:209–221

Felix G, Grosskopf DG, Regenass M, Boller T (1991) Rapid changes of protein phosphorylation are involved in transduction of the elicitor signal in plant cells. Proc Natl Acad Sci USA 88:8831–8834

Ferreira RB, Monteiro S, Freitas R, Santos CN, Chen Z, Batista LM, Duarte J, Borges A, Teixeira AR (1984) The role of plant defence proteins in fungal pathogenesis. Mol Plant Pathol 8:677–700

Friebe A, Lever W, Sikora R, Schnabl H (1998) Allelochemical in root exudates of maize. Effects on root lesion nematode Pratylenchus zea. In: Romeo JT, Downum KR, Verpoorte R (eds) Phytochemical signals and plant microbe interactions. Springer, Heidelberg, pp 71–93

Gherbi H, Markmann K, Svistoonoff S, Estevan J, Autran D, Giczey G, Auguy F, Péret B, Laplaze L, Franche C, Parniske M, Bogusz D (2008) SymRK defines a common genetic basis for plant root endosymbioses with arbuscular mycorrhiza fungi, rhizobia, and *Frankia* bacteria. Proc Natl Acad Sci USA 105:4928–4932

Gianinazzi-Pearson V, Dumas-Gaudot E, Gollotte A, Tahiri-Alaoui A, Gianinazzi S (1996) Cellular and molecular defence-related root responses to invasion by arbuscular mycorrhizal fungi. New Phytol 133:45–57

Glinwood R, Pettersson J, Ahmed E, Ninkovic V, Birkett M, Pickett J (2003) Change in acceptability of barley plants to aphids after exposure to allelochemicals from couch-grass (*Elytrigia repens*). J Chem Ecol 29:261–274

Goellner M, Smant G, De Boer JM, Baum TJ, Davis EL (2000) Isolation of Beta-1,4-endoglucanase genes from *Globodera tabacum* and their expression during parasitism. J Nematol 32:154–165

Grant M, Mansfield J (1999) Early events in host-pathogen interactions. Curr Opin Plant Biol 2: 312–319

Gray EJ, Smith DL (2005) Intracellular and extracellular PGPR: commonalities and distinctions in the plant-bacterium signaling processes. Soil Biol Biochem 37:395–412

Gray ND, Howarth R, Pickup RW, Jones JG, Head IM (2000) Use of combined microautoradiography and fluorescence *in situ* hybridization to determine carbon metabolism in mixed natural communities of uncultured bacteria from the genus *Achromatium*. Appl Environ Microbiol 66:4518–4522

Guerena M (2006) Nematodes: alternative control. National sustainable agriculture information service. ATTRA Publication 1–20

Guerrieri E, Poppy GM, Powell W, Rao R, Pennacchio F (2002) Plant-to-plant communication mediating in-flight orientation of *Aphidius ervi*. J Chem Ecol 28:1703–1715

Hahn MG (1996) Microbial elicitors and their receptors in plants. Annu Rev Phytopathol 34: 387–412

Harman GE, Björkman T (1998) Potential and existing uses of Trichoderma and Gliocladium por plant disease control and plant growth enhancement. In: Kubicek CP, Harman GE (eds) Trichoderma and Gliocladium: enzymes, biological control and commercial applications. Taylor & Francis, London, pp 229–265

Harman GE, Howell CR, Viterbo A, Chet I, Lorito M (2004) Trichoderma species – opportunistic, avirulent plant symbionts. Nat Rev Microbiol 2:43–56

Harrison MJ (1999) Molecular and cellular aspects of the arbuscular mycorrhizal symbiosis. Annu Rev Plant Physiol Plant Mol Biol 50:361–389

Harrison MJ (2005) Signaling in the arbuscular mycorrhizal symbiosis. Annu Rev Microbiol 59: 19–42

Hawes MC, Smith LY (1989) Requirement for chemotaxis in pathogenicity of *Agrobacterium tumefaciens* on roots of soil-grown pea plants. J Bacteriol 171:5668–5671

Hayward S, Muncey RJ, James AE, Halsall CJ, Hewitt CN (2001) Monoterpene emissions from soil in a Sitka spruce forest. Atmosph Environ 35:4081–4087

Hirose I, Sano K, Shioda I, Kumano M, Nakamura K, Yamane K (2000) Proteome analysis of *Bacillus subtilis* extracellular proteins: a two-dimensional protein electrophoretic study. Microbiology 146:65–75

Hoffland ELLI, van den Boogaard RIKI, Nelemans JAAP, Findenegg G (1992) Biosynthesis and root exudation of citric and malic acids in phosphate-starved rape plants. New Phytol 122: 675–680

Hopkins BG, Whitney DA, Lamond RE, Jolley VD (1998) Phytosiderophore release by sorghum, wheat, and corn under zinc deficiency. J Plant Nutr 21:2623–2637

Houston KM, Cushley W, Harnett W (1997) Studies on the site and mechanism of attachment of phosphorylcholine to a filarial nematode secreted glycoprotein. J Biol Chem 272:1527–1533

Huang G, Dong R, Allen R, Davis EL, Baum TJ, Hussey RS (2006) A root-knot nematode secretory peptide functions as a ligand for a plant transcription factor. Mol Plant Microbe Interact 19:463–470

Hubbard JE, Flores-Lara Y, Schmitt M, McClure MA, Stock SP, Hawes MC (2005) Increased penetration of host roots by nematodes after recovery from quiescence induced by root cap exudate. Nematology 7:321–331

Hussey RS (1989) Disease-inducing secretions of plant-parasitic nematodes. Annu Rev Phytopathol 27:123–141

Hütsch BW, Augustin J, Merbach W (2002) Plant rhizodeposition – an important source for carbon turnover in soils. J Plant Nutr Soil Sci 165:397–407

Ingham RE, Trofymow JA, Ingham ER, Coleman DC (1985) Interactions of bacteria, fungi, and their nematode grazers: effects on nutrient cycling and plant growth. Ecol Monogr 55:119–140

Jaubert S, Ledger TN, Laffaire JB, Piotte C, Abad P, Rosso M-N (2002) Direct identification of stylet secreted proteins from root-knot nematodes by a proteomic approach. Mol Biochem Parasitol 121:205–211

Johnson JF, Allan DL, Vance CP (1994) Phosphorus stress-induced proteoid roots show altered metabolism in *Lupinus albus*. Plant Physiol 104:657–665

Joosten MHAJ, De Wit PJGM (1989) Identification of several pathogenesis-related proteins in tomato leaves inoculated with *Cladosporium fulvum* (syn. Fulvia fulva) as 1,3-b-glucanases and chitinases. Plant Physiol 89:945–951

Jose S, Gillespie AR (1998) Allelopathy in black walnut (*Juglans nigra* L.) alley cropping. I. Spatio-temporal variation in soil juglone in a black walnut-corn (*Zea mays* L.) alley cropping system in the midwestern USA. Plant Soil 203:191–197

Kaffarnik FAR, Jones AME, Rathjen JP, Peck SC (2009) Effector proteins of the bacterial pathogen *Pseudomonas syringae* alter the extracellular proteome of the host plant, *Arabidopsis thaliana*. Mol Cell Proteomics 8:145–156

Kalemba D, Kusewicz D, Swiader K (2002) Antimicrobial properties of the essential oil of *Artemisia asiatica* Nakai. Phytother Res 16:288–291

Kamoun S (2006) A catalogue of the effector secretome of plant pathogenic oomycetes. Annu Rev Phytopathol 44:41–60

Karandashov V, Bucher M (2005) Symbiotic phosphate transport in arbuscular mycorrhizas. Trends Plant Sci 10:22–29

Kazemi-Pour N, Condemine G, Hugouvieux-Cotte-Pattat N (2004) The secretome of the plant pathogenic bacterium *Erwinia chrysanthemi*. Proteomics 4:3177–3186

Kent AD, Triplett EW (2002) Microbial communities and their interactions in soil and rhizosphere ecosystems. Annu Rev Microbiol 56:211–236

Kiely P, Haynes J, Higgins C, Franks A, Mark G, Morrissey J, O'Gara F (2006) Exploiting new systems-based strategies to elucidate plant-bacterial interactions in the rhizosphere. Microb Ecol 51:257–266

Kingsley MT, Fredrickson JK, Metting FB, Seidler RJ (1994) Environmental restoration using plant-microbe bioaugmentation. In: Hinchee RE, Leeson A, Semprini L, Ong SK (eds) Bioremediation of chlorinated and polycyclic aromatic hydrocarbon. Lewis, Boca Raton, FL, pp 287–292

Lambers H, Mougel C, Jaillard B, Hinsinger P (2009) Plant-microbe-soil interactions in the rhizosphere: an evolutionary perspective. Plant Soil 321:83–115

Lévy J, Bres C, Geurts R, Chalhoub B, Kulikova O, Duc G, Journet EP, Ané JM, Lauber E, Bisseling T, Dénarié J, Rosenberg C, Debellé F (2004) A putative Ca^{2+} and calmodulin-dependent protein kinase required for bacterial and fungal symbioses. Science 303:1361–1364

Li S, Hartman GL, Bs L, Widholm JW (2000) Identification of a stress-induced protein in stem exudates of soybean seedlings root-infected with *Fusarium solani* f. sp. glycines. Plant Physiol Biochem 38:803–809

Lin C, Owen SM, Peñuelas J (2007) Volatile organic compounds in the roots and rhizosphere of *Pinus* spp. Soil Biol Biochem 39:951–960

Lipton DS, Blanchar RW, Blevins DG (1987) Citrate, malate, and succinate concentration in exudates from P-sufficient and P-stressed *Medicago sativa* L. seedlings. Plant Physiol 85: 315–317

Lou Y-G, Du MH, Turlings TCJ, Cheng J-A, Shan W-F (2005) Exogenous application of jasmonic acid induces volatile emissions in rice and enhances parasitism of *Nilaparvata lugens* eggs by the parasitoid *Anagrus nilaparvatae*. J Chem Ecol 31:1985–2001

Lynch JM (1987) The rhizosphere. Wiley Interscience, Chichester

Lynch JM, Whipps JM (1990) Substrate flow in the rhizosphere. Plant Soil 129.1–10

Mahaffee WF, Kloepper JW (1997) Temporal changes in the bacterial communities of soil, rhizosphere, and endorhiza associated with field-grown cucumber (*Cucumis sativus* L.). Microb Ecol 34:210–223

Mann M, Jensen ON (2003) Proteomic analysis of post-translational modifications. Nat Biotechnol 21:255–261

Manson MD (1990) Introduction to bacterial motility and chemotaxis. J Chem Ecol 16:107–118
Marschner H (1995) Mineral nutrition of higher plants. Academic, London
Mathesius U, Mulders S, Gao M, Teplitski M, Caetano-Anolles G, Rolfe BG, Bauer WD (2003) Extensive and specific responses of a eukaryote to bacterial quorum-sensing signals. Proc Natl Acad Sci USA 100:1444–1449
Mawuenyega KG, Kaji H, Yamauchi Y, Shinkawa T, Saito H, Taoka M, Takahashi N, Isobe T (2003) Large-scale identification of *Caenorhabditis elegans* proteins by multidimensional liquid chromatography-tandem mass spectrometry. J Proteome Res 2:23–35
Melin E, Krupa S (1971) Studies on ectomycorrhizae of pine II. Growth inhibition of mycorrhizal fungi by volatile organic constituents of *Pinus silvestris* (Scots Pine) roots. Physiol Plant 25: 337–340
Nagahashi G, Douds DD (1999) Rapid and sensitive bioassay to study signals between root exudates and arbuscular mycorrhizal fungi. Biotechnol Tech 13:893–897
Nakhasi HL, Pogue GP, Duncan RC, Joshi M, Atreya CD, Lee NS, Dwyer DM (1998) Implications of calreticulin function in parasite biology. Parasitol Today 14:157–160
Narasimhan K, Basheer C, Bajic VB, Swarup S (2003) Enhancement of plant-microbe interactions using a rhizosphere metabolomics-driven approach and its application in the removal of polychlorinated biphenyls. Plant Physiol 132:146–153
Nardi S, Concheri G, Pizzeghello D, Sturaro A, Rella R, Parvoli G (2000) Soil organic matter mobilization by root exudates. Chemosphere 5:653–658
Ndimba BK, Chivasa S, Hamilton JM, Simon WJ, Slabas AR (2003) Proteomic analysis of changes in the extracellular matrix of Arabidopsis cell suspension cultures induced by fungal elicitors. Proteomics 3:1047–1059
Nguyen C (2003) Rhizodeposition of organic C by plants: mechanisms and controls. Agronomy 23:375–396
Nicol GW, Glover LA, Prosser JI (2003) Spatial analysis of archaeal community structure in grassland soil. Appl Environ Microbiol 69:7420–7429
Nóbrega FM, Santos IS, Cunha MD, Carvalho AO, Gomes VM (2005) Antimicrobial proteins from cowpea root exudates: inhibitory activity against *Fusarium oxysporum* and purification of a chitinase-like protein. Plant Soil 272:223–232
Nürnberger T, Brunner F (2002) Innate immunity in plants and animals: emerging parallels between the recognition of general elicitors and pathogen-associated molecular patterns. Curr Opin Plant Biol 5:318–324
Owen SM, Clark S, Pompe M, Semple KT (2007) Biogenic volatile organic compounds as potential carbon sources for microbial communities in soil from the rhizosphere of *Populus tremula*. FEMS Microbiol Lett 268:34–39
Palmer AG, Gao R, Maresh J, Erbil WK, Lynn DG (2004) Chemical biology of multi-host/ pathogen interactions: chemical perception and metabolic complementation. Annu Rev Phytopathol 42:439–464
Palomski T, Saarilahti HT (1997) Isolation and characterization of new C-terminal substitution mutations affecting secretion of polygalacturonase in *Erwinia carotovora* ssp. carotovora. FEBS Lett 400:122–126
Passador L, Cook JM, Gambello MJ, Rust L, Iglewski BH (1993) Expression of *Pseudomonas aeruginosa* virulence genes requires cell-to-cell communication. Science 260:1127–1130
Paszkowski U (2006) Mutualism and parasitism: the yin and yang of plant symbioses. Curr Opin Plant Biol 9:364–370
Paterson E, Hall JM, Rattray EAS, Griffiths BS, Ritz K, Killham K (1997) Effect of elevated CO_2 on rhizosphere carbon flow and soil microbial processes. Glob Chang Biol 3:363–377
Peck SC, Nühse TS, Hess D, Iglesias A, Meins F, Boller T (2001) Directed proteomics identifies a plant-specific protein rapidly phosphorylated in response to bacterial and fungal elicitors. Plant Cell 13:1467–1475
Perombelon MCM, Kelman A (1980) Ecology of the soft rot Erwinias. Annu Rev Phytopathol 18: 361–387

Perry RN (2001) An evaluation of types of attractants enabling plant-parasitic nematodes to locate plant roots. Russ J Nematol 13:83–88
Phalip V, Delalande F, Carapito C, Goubet F, Hatsch D, Leize-Wagner E, Dupree P, Dorsselaer A, Jeltsch JM (2005) Diversity of the exoproteome of *Fusarium graminearum*; grown on plant cell wall. Curr Genet 48:366–379
Pinton R, Varanini Z, Nannipieri P (2001) The rhizosphere: biochemistry and organic substances at the soil-plant interface. Marcel Dekker, New York
Popeijus H, Overmars H, Jones J, Blok V, Goverse A, Helder J, Schots A, Bakker J, Smant G (2000) Enzymology: degradation of plant cell walls by a nematode. Nature 406:36–37
Pritchard DI, Brwon A, Kasper G, Mcelroy P, Loukas A, Hewitt C, Berry C, Füllkrug R, Beck E (1999) Par Immunol 2:439–450
Prithiviraj B, Perry LG, Dayakar BV, Vivanco JM (2007) Chemical facilitation and induced pathogen resistance mediated by a root-secreted phytotoxin. New Phytol 173:852–860
Raghothama KG (1999) Phosphate acquisition. Annu Rev Plant Physiol Plant Mol Biol 50: 665–693
Remy W, Taylor TN, Hass H, Kerp H (1994) Four hundred-million-year-old vesicular arbuscular mycorrhizae. Proc Natl Acad Sci USA 91:11841–11843
Robertson L, Robertson WM, Jones JT (1999) Direct analysis of the secretions of the potato cyst nematode *Globodera rostochiensis*. Parasitology 119:167–176
Robinson F (2002) Nematodes behaviour and migration through soil and host tissue. In: Chen C, Chen S, Dickson DW (eds) Basis of behaviour. CABI, Wallingford, pp 331–401
Robinson MP, Atkinson HJ, Perry RN (1987) The influence of soil moisture and storage time on the motility, infectivity and lipid utilization of second stage juveniles of the potato cyst nematodes *Globodera rostochiensis* and *G. pallida*. Rev Nematol 10:343–348
Rohloff J (2002) Volatiles from rhizomes of *Rhodiola rosea* L. Phytochemistry 59:655–661
Ronchel MC, Ramos JL (2001) Dual system to reinforce biological containment of recombinant bacteria designed for rhizoremediation. Appl Environ Microbiol 67:2649–2656
Rondon MR, August PR, Bettermann AD, Brady SF, Grossman TH, Liles MR, Loiacono KA, Lynch BA, MacNeil IA, Minor C, Tiong CL, Gilman M, Osburne MS, Clardy J, Handelsman J, Goodman RM (2000) Cloning the soil metagenome: a strategy for accessing the genetic and functional diversity of uncultured microorganisms. Appl Environ Microbiol 66:2541–2547
Rose JKC, Ham KS, Darvill AG, Albersheim P (2002) Molecular cloning and characterization of glucanase inhibitor proteins: coevolution of a counterdefense mechanism by plant pathogens. Plant Cell 14:1329–1345
Rouatt JW, Katznelson H (1960) Influence of light on bacterial flora of roots. Nature 186:659–660
Rouatt JW, Katznelson H, Payne TMB (1960) Statistical evaluation of the rhizosphere effect. Soil Sci Soc Am J 24:271–273
Rudrappa T, Bonsall J, Gallagher J, Seliskar D, Bais H (2007) Root-secreted allelochemical in the noxious weed *Phragmites australis* deploys a reactive oxygen species response and microtubule assembly disruption to execute rhizotoxicity. J Chem Ecol 33:1898–1918
Rudrappa T, Biedrzycki ML, Kunjeti SG, Donofrio N, Czymmek KJ, Paré PW, Bais HP (2010) The rhizobacterial elicitor acetoin induces systemic resistance in *Arabidopsis thaliana*. Commun Integr Biol 3:130–138
Ruther J, Kleier S (2005) Plant-plant signaling: ethylene synergizes volatile emission In *Zea mays* induced by exposure to (Z)-3-hexen-1-ol. J Chem Ecol 31:2217–2222
Saarilahti HT, Pirhonen M, Karlsson MB, Flego D, Palva ET (1992) Expression of *pehA-bla* gene fusions in *Erwinia carotovora* subsp. *carotovora* and isolation of regulatory mutants affecting polygalacturonase production. Mol Gen Genet 234:81–88
Scher FM, Kloepper JW, Singleton CA (1985) Chemotaxis of fluorescent *Pseudomonas* spp. to soybean seed exudates *in vitro* and in soil. Can J Microbiol 31:570–574
Schuster M, Lostroh CP, Ogi T, Greenberg EP (2003) Identification, timing, and signal specificity of *Pseudomonas aeruginosa* quorum-controlled genes: a transcriptome analysis. J Bacteriol 185:2066–2079

Sharon N, Lis H (2004) History of lectins: from hemagglutinins to biological recognition molecules. Glycobiology 14:53R–62R
Shulaev V, Silverman P, Raskin I (1997) Airborne signaling by methyl salicylate in plant pathogen resistance. Nature 386:718–721
Simon L, Bousquet J, Levesque RC, Lalonde M (1993) Origin and diversification of endomycorrhizal fungi and coincidence with vascular land plants. Nature 363:67–69
Singh BK, Millard P, Whiteley AS, Murrell JC (2004) Unravelling rhizosphere-microbial interactions: opportunities and limitations. Trends Microbiol 12:386–393
Sivaguru M, Fujiwara T, Samaj J, Baluska F, Yang ZM, Osawa H, Maeda T, Mori T, Volkmann D, Matsumoto H (2000) Aluminum-induced 1– > 3-b-D-glucan inhibits cell-to-cell trafficking of molecules through plasmodesmata. A new mechanism of aluminum toxicity in plants. Plant Physiol 124:991–1005
Smant G, Stokkermans JPWG, Yan Y, De Boer JM, Baum TJ, Wang X, Hussey RS, Gommers FJ, Henrissat B, Davis EL, Helder J, Schots A, Bakker J (1998) Endogenous cellulases in animals: isolation of b-1,4-endoglucanase genes from two species of plant-parasitic cyst nematodes. Proc Natl Acad Sci USA 95:4906–4911
Söderberg KH, Probanza A, Jumpponen A, Baath E (2004) The microbial community in the rhizosphere determined by community-level physiological profiles (CLPP) and direct soil- and cfu-PLFA techniques. Appl Soil Ecol 25:135–145
Somers E, Vanderleyden J, Srinivasan M (2004) Rhizosphere bacterial signalling: a love parade beneath our feet. CRC Crit Rev Microbiol 30:205–240
Spanu P, Boller T, Alexander L, Wien S-V, Faccio A, Bonfante-Fasolo P (1989) Chitinase in roots of mycorrhizal *Allium porrum*: regulation and localization. Planta 177:447–455
Steeghs M, Bais HP, de Gouw J, Goldan P, Kuster W, Northway M, Fall R, Vivanco JM (2004) Proton-transfer-reaction mass spectrometry as a new tool for real time analysis of root-secreted volatile organic compounds in Arabidopsis. Plant Physiol 135:47–58
Stone JM, Walker JC (1995) Plant protein kinase families and signal transduction. Plant Physiol 108:451–457
Strauss SY (1991) Indirect effects in community ecology: their definition, study and importance. Trends Ecol Evol 6:206–210
Suárez MB, Sanz L, Chamorro MI, Rey M, González FJ, Llobell A, Monte E (2005) Proteomic analysis of secreted proteins from *Trichoderma harzianum*: Identification of a fungal cell wall-induced aspartic protease. Fungal Genet Biol 42:924–934
Swift S, Lynch MJ, Fish L, Kirke DF, Tomas JM, Stewart GSAB, Williams P (1999) Quorum sensing-dependent regulation and blockade of exoprotease production in *Aeromonas hydrophila*. Infect Immun 67:5192–5199
Szekeres A, Kredics L, Antal Z, Kevei F, Manczinger L (2004) Isolation and characterization of protease overproducing mutants of *Trichoderma harzianum*. FEMS Microbiol Lett 33: 215–222
Taddei P, Tugnoli V, Bottura G, Dallavalle E, Zechini D'Aulerio A (2002) Vibrational, ^{1}H-NMR spectroscopic, and thermal characterization of gladiolus root exudates in relation to *Fusarium oxysporum* f. sp. *gladioli* resistance. Biopolymers 67:428–439
Tamasloukht M'B, Sejalon-Delmas N, Kluever A, Jauneau A, Roux C, Becard G, Franken P (2003) Root factors induce mitochondrial-related gene expression and fungal respiration during the developmental switch from asymbiosis to presymbiosis in the arbuscular mycorrhizal fungus *Gigaspora rosea*. Plant Physiol 131:1468–1478
Timmermans BGH, Vos J, Stomph TJ, Van Nieuwburg J, Van der Putten PEL (2007) Field performance of *Solanum sisymbriifolium*, a trap crop for potato cyst nematodes. II. Root characteristics. Ann Appl Biol 150:99–106
Tjalsma H, Bolhuis A, Jongbloed JDH, Bron S, van Dijl JM (2000) Signal peptide-dependent protein transport in *Bacillus subtilis*: a genome-based survey of the secretome. Microbiol Mol Biol Rev 64:515–547

Torsvik V, Ovreás L (2002) Microbial diversity and function in soil: from genes to ecosystems. Curr Opin Microbiol 5:240–245

Uren NC (2000) Types, amounts and possible functions of compounds released into the rhizosphere by soil-grown plants. In: Pinton R, Varanini Z, Nammipieri P (eds) The rhizosphere, biochemistry and organic substances at the soil-plant interface. Marcel Dekker, New York, pp 19–40

Veech JA, Starr JL, Nordgren RM (1987) Production and partial characterization of stylet exudate from adult females of *Meloidogyne incognita*. J Nematol 19:463–468

Vilela GR, de Almeida GS, D'Arce MABR, Moraes MHD, Brito JO, da Silva MFd, Silva SC, de Stefano Piedade SM, Calori-Domingues MA, da Gloria EM (2009) Activity of essential oil and its major compound, 1,8-cineole, from Eucalyptus globulus Labill., against the storage fungi Aspergillus flavus Link and Aspergillus parasiticus Speare. J Stor Prod Res 45:108–111

Viles AL, Reese RN (1996) Allelopathic potential of *Echinacea angustifolia* D.C. Environ Exp Bot 36:39–43

von Bodman SB, Bauer WD, Coplin DL (2003) Quorum sensing in plant-pathogenic bacteria. Annu Rev Phytopathol 41:455–482

Walker TS, Bais HP, Grotewold E, Vivanco JM (2003) Root exudation and rhizosphere biology. Plant Physiol 132:44–51

Wall LG (2000) The actinorhizal symbiosis. J Plant Growth Regul 19:167–182

Watt SA, Wilke A (2005) Comprehensive analysis of the extracellular proteins from *Xanthomonas campestris* pv. campestris B100. Proteomics 5:153–167

Weir T, Bais H, Vivanco J (2003) Intraspecific and interspecific interactions mediated by a phytotoxin, (−)-catechin, secreted by the roots of *Centaurea maculosa* (Spotted Knapweed). J Chem Ecol 29:2397–2412

Weir TL, Park S-W, Vivanco JM (2004) Biochemical and physiological mechanisms mediated by allelochemicals. Curr Opin Plant Biol 7:472–479

Wen F, VanEtten HD, Tsaprailis G, Hawes MC (2007) Extracellular proteins in pea root tip and border cell exudates. Plant Physiol 143:773–783

Wieland G, Neumann R, Backhaus H (2001) Variation of microbial communities in soil, rhizosphere, and rhizoplane in response to crop species, soil type, and crop development. Appl Environ Microbiol 67:5849–5854

Worm J, Jensen LE, Søndergaard M, Nybroe O (2000) Interactions between proteolytic and non-proteolytic *Pseudomonas fluorescens* affect protein degradation in a model community. Plant Sci Lett 32:103–109

Wubben MJE, Jin J, Baum TJ (2008) Cyst nematode parasitism of *Arabidopsis thaliana* is inhibited by salicylic acid (SA) and elicits uncoupled SA-independent pathogenesis-related gene expression in roots. Mol Plant Microbe Interact 21:424–432

Wuyts N, Swennen R, Waele D (2006) Effects of plant phenylpropanoid pathway products and selected terpenoids and alkaloids on the behaviour of the plant parasitic nematodes adopholis similis, *Pratylenchus penetrans* and *Meloidogyne incognita*. Nematology 8:89–101

Yajima W, Kav NNV (2006) The proteome of the phytopathogenic fungus *Sclerotinia sclerotiorum*. Proteomics 6:5995–6007

Yang CH, Crowley DE (2000) Rhizosphere microbial community structure in relation to root location and plant iron nutritional status. Appl Environ Microbiol 66:345–351

Yedidia I, Shoresh M, Kerem Z, Benhamou N, Kapulnik Y, Chet I (2003) Concomitant induction of systemic resistance to *Pseudomonas syringae* pv. lachrymans in cucumber by *Trichoderma asperellum* (T-203) and accumulation of phytoalexins. Appl Environ Microbiol 69:7343–7353

Zak JC, Willig MR, Moorhead DL, Wildman HG (1994) Functional diversity of microbial communities: A quantitative approach. Soil Biol Biochem 26:1101–1108

Zasada IA, Meyer SLF, Halbrendt JM, Rice C (2005) Activity of hydroxamic acids from *Secale cereale* against the plant-parasitic nematodes *Meloidogyne incognita* and *Xiphinema americanum*. Phytopathology 95:1116–1121

Zelles L (1997) Phospholipid fatty acid profiles in selected members of soil microbial communities. Chemosphere 35:275–294

Zhao X, Schmitt M, Hawes M (2000) Species-dependent effects of border cell and root tip exudates on nematode behaviour. Phytopathology 90:1239–1245

Zwart KB, Kuikman PJ, van Veen JA (1994) Rhizosphere protozoa: their significance in nutrient dynamics. In: Darbyshire JF (ed) Soil protozoa. CAB International, Wallingford, pp 93–122

Root Exudates of Legume Plants and Their Involvement in Interactions with Soil Microbes

Akifumi Sugiyama and Kazufumi Yazaki

Abstract Plants secrete both high- and low-molecular weight compounds from their roots, and these root exudates function not only as nutrients for soil microbes but as signal molecules in plant–microbe interactions. Legume plants establish symbiotic interactions with rhizobia and arbuscular mycorrhizal fungi to obtain several nutrients such as nitrogen and phosphate. In these interactions, flavonoids and strigolactones in root exudates serve as signal molecules to establish the symbiotic interactions. Root exudates from some legume plants also function to acidify surrounding soils to acquire phosphate. Here, we provide an overview of the functions of legume root exudates with emphasis on the interaction between legume plants and soil microbes and also on the acquisition of nutrients from surrounding soil.

1 Introduction

The legume family (Fabaceae) is composed of more than 700 genera containing approximately 20,000 species (Doyle and Luckow 2003). It represents the third largest plant family next to Orchidaceae and Asteraceae. Legume plants have

A. Sugiyama (✉)
Center for Rhizosphere Biology, Colorado State University, 1173 Campus Delivery, Fort Collins, CO 80523, USA

Research Institute for Sustainable Humanosphere, Kyoto University, Gokasho, Uji 611-0011, Japan
e-mail: akifumi_sugiyama@rish.kyoto-u.ac.jp

K. Yazaki
Research Institute for Sustainable Humanosphere, Kyoto University, Gokasho, Uji 611-0011, Japan
e-mail: yazaki@rish.kyoto-u.ac.jp

J.M. Vivanco and F. Baluška (eds.), *Secretions and Exudates in Biological Systems*, Signaling and Communication in Plants 12, DOI 10.1007/978-3-642-23047-9_2,

a significant agricultural importance because many crop plants such as soybean (*Glycine max*), pea (*Pisum sativam*), and bean (*Phaseolus vulgaris*) belong to this family, and legume crops are cultivated on 12–15% of the arable land throughout the world. Legume species, including soybean and peanut (*Arachis hypogeae*), provide more than one third of the processed vegetable oil throughout the world's market (Graham and Vance 2003). In addition to these crop legumes, nongrain legumes, such as licorice (*Glycyrrhiza galbra*) provide glycyrrhizin, which is a useful phytotherapeutic and sweating agent, while legume plants such as indigo (*Indigofera tinctoria* and *I. suffruticosa*) and logwood (*Haematoxylon campechianum*) provide natural dyes. Forage legumes such as alfalfa (*Medicago sativa*) and birdsfoot trefoil (*Lotus corniculatus*) are widely used for livestock feed, and tree legume such as various Acacia and Robinia species supply large-scale biomaterials like tannin and pulp (Dixon and Sumner 2003; Downs et al. 2003; Christie 2007).

A hallmark feature of legume plants is their ability to establish a mutualistic symbiosis with bacteria belonging to Rhizobiaceae family for the utilization of atmospheric nitrogen as a nitrogen source. It is estimated that approximately 40–60 million metric tons of atmospheric nitrogen is fixed by cultivated legume plants annually (Smil 1999), which is important not only for agriculture, but also for the environment because nitrogen fixation can supplement the use of synthetic nitrogen fertilizers which require a large amount of energy input during production that can contribute to environmental pollution. Most of the legume plants also establish a symbiotic interaction with arbuscular mycorrhizal fungi which help to obtain mineral nutrients, as well as water.

In addition to these mutualistic interactions, legume plants interact with diverse soil microbes both positively and negatively. Both culture-dependent and culture-independent methods were employed to analyze the microbial communities and unique communities were found depending on the plant species, cultivars, and ecotypes (Priha et al. 1999; Yang and Crowley 2000; Narasimhan et al. 2003; Innes et al. 2004; Mazzola et al. 2004; Batten et al. 2006; Kowalchuka et al. 2006; Mougel et al. 2006; Micallef et al. 2009). It has also been reported that symbiotic mutants (i.e., hypernodulating mutants, mutants defective in *Rhizobium* symbiosis, and mutants defective in both rhizobial and arbuscular mycorrhizal symbiosis) have different microbial communities contained in the rhizosphere soil as well as the roots than do wild type plants (Offre et al. 2007; Ikeda et al. 2008). Among the various methods plants employ to interact with soil microbes, root exudates are a key factor that influences the microbial communities (Uren 2007; Badri and Vivanco 2009; Badri et al. 2009).

This chapter provides an overview of the literature on the root exudates of legume plants with emphasis on the interaction between legume plants and soil microbes. We first describe the characteristics of legume root exudates and then discuss the functions of these root exudates in the interaction with soil microbes such as rhizobia and arbuscular mycorrhizal fungi. Finally, we describe some features of root exudates in the acquisition of nutrients from soil.

2 Characteristics of Root Exudates from Legume Plants

Metabolites in legume plants have been intensively studied due to biological interests such as symbiotic nitrogen fixation, as well as their importance to human health. Recently, metabolomic analyses have been performed on various legume species such as *M. truncatula*, *L. japonicus*, soybean, and the common bean (Desbrosses et al. 2005; Farag et al. 2008; Suzuki et al. 2008; Farag et al. 2009; Hernandez et al. 2009; Brechenmacher et al. 2010; Rispail et al. 2010). Many of these studies are focused, at least in part, on flavonoids and other phenolic compounds. Profiling of flavonoids in *M. truncatula* roots and cell suspension culture showed *M. truncatula* contains at least 40 flavonoids, most of which are glycosides (glucoside or glucoside malonate), and that methyl jasmonate induced the accumulation of the phytoalexin medicarpin and decreased the isoflavone glucosides, while a yeast elicitor coordinately increased isoflavonoid precursors and medicarpin (Farag et al. 2008). In *L. japonicus*, 61 flavonoid compounds have been identified from leaves, stems, flowers, and seeds. Most of these are glycosides and about half of them are flower specific (Suzuki et al. 2008). Another group recently reported that *M. loti* inoculation induced quantitative rather than qualitative changes of the phenolic compounds in roots of *L. japonicus* and that phytoalexin vestitol and sativan were not found in the inoculated roots (Rispail et al. 2010).

In contrast to the metabolite profiling in plants, metabolite profiling of root exudates are limited to the targeted analysis of particular compounds such as organic acids, flavonoids, and fatty acids (Smit et al. 1992; Lucas Garcia et al. 2001). There are several reports on the metabolome and proteome analyses of Arabidopsis root exudates identifying various metabolites such as sugars, amino acids, fatty acids, as well as various classes of proteins (De-la-Pena et al. 2008, 2010; Badri and Vivanco 2009; Badri et al. 2009). Proteome and metabolome analyses of root exudates in legume species will be of particular interest to target compounds that function in plant–microbe interactions, such as the recruitment of rhizobia, arbuscular mycorrhizal fungi, and PGPR (plant growth promoting rhizobacteria) species that potentially improve crop yields. It was shown that *L. japonicus* secreted strigolactone for the establishment of arbuscular symbiosis and more recently was reported that *L. japonicus* secretes a phytoalexin vestitol as a chemical barrier against parasitic weeds (Ueda and Sugimoto 2010). We also performed metabolite profiling of the root exudates of *L. japonicus* and found that methyl jasmonate induced several unknown compounds in the exudates (unpublished results). It would be interesting to characterize the function of these induced compounds in response to pathogens. The following chapters provide the functions of legume root exudates in plant–microbe interactions in soils.

3 Functions of Root Exudates in Symbiosis with Rhizobia

3.1 Signaling Molecules from Legume Plants to Rhizobia

Interaction between legume plants and rhizobia in the soil is of particular importance in agriculture and ecology and many studies have been performed to characterize the molecular mechanisms on how this species-specific interaction is established (Werner 2007). In soils, *Rhizobium* spp. can find its host legume plant from a distance because of the chemotactic nature of rhizobia. Rhizobia have been shown to be attracted to the root exudates of legume plants (Currier and Strobel 1976). It was reported that flavonoids present in root exudates are responsible for the attraction of rhizobia in alfalfa (Parke et al. 1985; Caetano-Anolles et al. 1988; Dharmatilake and Bauer 1992), and a *Rhizobium* mutant defective in chemotaxis was shown to be less competitive in forming the functional nodules (Yost et al. 1998). Because phytochemicals such as flavonoids may not diffuse for long distances in soils, it is presumed that root volatiles are also involved in the attraction of rhizobia to host plants. In Medicago–*Sinorhizobium* interaction, root volatiles of Medicago, especially dimethylsulfide, were shown to attract nematodes which bring *Sinorhizobium meliloti* to the proximity of plant roots (Horiuchi et al. 2005). In a similar way, it may also be possible that other volatile organic compounds attract rhizobia from a distance.

In the 1980s, many genes and proteins involved in the recognition of plant signal molecules were identified in rhizobia: such as NodD (a LysR-type regulator which acts as a transcriptional activator for the *nod* operon), NodA (*N*-acetyltransferase), NodB (de-*N*-acetylase), and NodC (UDP-Gluc *N*-acetyltransferase), which together synthesize the backbone of a lipochitooligosaccharide called Nod factor (Rossen et al. 1984; Egelhoff et al. 1985; Egelhoff and Long 1985; Mulligan and Long 1985; Rossen et al. 1985; Fisher and Long 1992). To identify the signaling molecules from plant roots, a nod promoter–*LacZ* fusion reporter system was employed. With this reporter system, signaling molecules from legume plants were identified to be luteolin from alfalfa (*M. sativa*), 7,4′-dihydroxyflavone and geraldone from white clover (*Trifolium repens*), and daidzein and genistein from soybean (*G. max*) (Peters et al. 1986; Redmond et al. 1986; Djordjevic et al. 1987; Kosslak et al. 1987) (Table 1), whereas formononetin and umbelliferone were shown to exhibit inhibitory effects on the *nod* gene expression (Djordjevic et al. 1987). Beside these flavones and isoflavonoids, a chalcone (4,4′-dihydroxy-2′-methoxychalcone) from alfalfa (Maxwell et al. 1989), anthocyanidins (petunidin and malvidin) from the common bean (*P. vulgaris*) (Hungria et al. 1991), betains (trigonelline and stachydrine) from alfalfa (Phillips et al. 1992), and aldonic acids (erythronic acid and tetronic acid) from white lupine (*Lupinus albus*) (Gagnon and Ibrahim 1998) were also reported to be *nod* gene inducers (Table 1). These various data suggest that a structurally diverse variety of phytochemicals can function as signal molecules. Signaling molecules from a legume tree species, black locust (*Robinia pseudoacacia*), were also identified to be flavonoids (7,4′-dihydroxyflavone,

Table 1 Nod-gene-inducing compounds of root exudates of legume plants

Plant	Compound	Reference
Alfalfa (*Medicago sativa*)	Luteolin	Peters et al. (1986)
	7,4′-Dihydroxyflavone	Maxwell et al. (1989)
	7,4′-Dihydroxyflavanone	
	4,4′-Dihydroxy-2′-methoxychalcone	
	Chrysoeriol	Hartwig et al. (1990)
	Trigonelline	Phillips et al. (1992)
	Stachydrine	
Barrel Medic (*Medicago truncatula*)	7,4′-Dihydroxyflavone	Zhang et al. (2006)
Black Locust (*Robinia pseudoacacia*)	7,4′-Dihydroxyflavone	Scheidemann and Wetzel (1997)
	Apigenin	
	Naringenin	
	Chrysoeriol	
	Isoliquiritigenin	
Common bean (*Phaseolus vulgaris*)	Eriodictyol	Hungria et al. (1991a)
	Naringenin	
	Genistein 7-*O*-glycoside	
	Delphinidin	Hungria et al. (1991b)
	Petunidin	
	Malvidin	
	Myricetin	
	Quercetin	
	Kaempferol	
Common vetch (*Vicia sativa*)	3,5,7,3′-Tetrahydroxy-4′-methoxyflavanone	Zaat et al. (1989)
	7,3′-Dihydroxy-4′-methoxyflavanone	
Cowpea (*Vigna unguiculata*)	Daidzein	Dakora et al. (2000)
	Genistein	
	Coumestrol	
Miyakogusa (*Lotus japonicus*)	Unknown	
Pea (*Pisum sativum*)	Apigenin	Firmin et al. (1986)
	Eriodictyol	
Rostrate sesbania (*Sesbania rostrata*)	7,4′-Dihydroxyflavanone	Messens et al. (1991)
Soybean (*Glycine max*)	Daidzein	Kosslak et al (1987)
	Genistein	
	Coumestrol	Bassam et al. (1988)
White clover (*Trifolium repens*)	7,4′-Dihydroxyflavone	Redmond et al. (1986)
	Geraldone	
White lupine (*Lupinus albus*)	Erythronic acid	Gagnon and Ibrahim (1998)
	Tetronic acid	

apigenin, naringenin, chrysoeriol, isoliquiritigenin) (Scheidemann and Wetzel 1997). The synthesis of these signaling molecules was induced under nitrogen deficiency (Cho and Harper 1991), and the application of signaling molecules to the legume plants was shown to increase the number of nodules in the roots (Begum et al. 2001; Novak et al. 2002).

These signaling compounds were secreted from root tissues into the rhizosphere using an energy-dependent transport system, and therefore diffuse around plant roots. They bind to the NodD receptor in the rhizobial cell surface and induce the expression of *nod* genes leading to the synthesis of Nod factors in a species-specific manner. These signaling flavonoids were also shown to induce the type III secretion systems (TTSS) and the excretion of proteins (Viprey et al. 1998; Fauvart and Michiels 2008). TTSS, which is composed of ca. 20 proteins, is a molecular machine found in both symbiotic (such as *Bradyrhizobium japonicum*, *Mesorhizobium loti*, *Rhizobium fredii*) and pathogenic bacteria to deliver effector, translocator, and regulator proteins to eukaryotic cells. It was reported that a mutation in TTSS affects symbiosis in a host-specific manner, i.e., the mutation in TTSS of *Rhizobium* sp. NGR234 resulted in increased number of nodules in yam bean (*Pachyrhizus tuberosus*), less number of nodules in fish bean (*Tephrosia vogelii*), and had no effect on nodulation in cowpea (*Vigna unguiculata*) and white leadtree (*Leucaena leucocephala*) (Viprey et al. 1998). It was also shown that the TTSS mutant of *Rhizobium* sp. NGR234 formed small and nonnitrogen fixing nodules that seemed to be devoid of meristematic cells in sunhemp (*Crotalaria juncea*) (Marie et al. 2003). The precise function of TTSS and its secreted proteins (called Nops; nodulation outer proteins) in the nodulation process is not yet well understood, but three Nops (NopL, NopP, and NopT) are proposed to be involved in modulating the host signaling pathways such as the down regulation of host plant defenses, and two other Nops (NopD and NopM) may be involved in the interference of host protein regulation in the nuclei (Krishnan 2002; Krishnan et al. 2003; Ausmees et al. 2004; Fauvart and Michiels 2008). It is noteworthy that TTSS is not common machinery conserved in all *Rhizobia*, for instance, *S. meliloti* does not contain TTSS genes in its genome (Galibert et al. 2001).

3.2 Signaling Molecules from Rhizobia to Legume Plants

Upon the recognition of plant-derived signal molecules, rhizobia induce the expression of *nod* genes, which are responsible for the synthesis of a lipochitooligosaccharide, Nod factor. Each *Rhizobium* species contains both common *nod* genes such as *nod A*, *nod B*, and *nod C*, as well as other species-specific *nod* genes. The common nod genes are responsible for the synthesis of the backbone structure of Nod factors, while species-specific *nod* genes are responsible for the modification of that structure, especially the side chain of the oligosaccharide. Functions and properties of species-specific *nod* genes were summarized previously (Fisher and Long 1992; Werner and Muller 2002). These species-specific modifications are necessary for the divergence of the structures of various Nod factors, which is indispensable in the specific host–*Rhizobium* interactions. The secretion of Nod factors from rhizobia is an energy-dependent process, and several genes involved in the secretion of the Nod factor have been identified; *nodI*, *nodJ*, *nodT*, *nodFGHI* (Rivilla et al. 1995; Spaink et al. 1995; Cardenas et al. 1996; Fernandez-Lopez et al. 1996). The products of these genes comprise a bacterial type ATP-binding cassette (ABC) transporter which consists of two transmembrane proteins and two nucleotide binding proteins.

Nod factors secreted into the rhizosphere are received by the Nod receptors located at the plasma membrane of the host legume root cells. Upon the recognition of Nod factors, plant roots undergo a series of drastic physiological changes: (1) Nod factors induce the formation of a transient subcellular gradient of chloride, potassium, and calcium ions as well as pH, followed by sharp oscillations of the cytoplastic calcium ion concentration, called "calcium spiking" in root hair cells; (2) Nod factors induce the curling of root hairs, which entrap rhizobia; (3) rhizobia begin to penetrate plant roots by forming an infection thread; (4) cortical cells of the roots begin to divide to form the nodule primordia; (5) rhizobia eventually enter the plant cell via endocytosis, resulting in the formation of symbiosomes in which rhizobia are surrounded by the plasma-membrane-derived peribacteroid membrane (Verma and Hong 1996).

It is also noteworthy that the backbone structure of Nod factors is chitin (chitooligosaccharides), which is a major component of fungal cell walls and induces various defense responses in plants. Chitooligosaccharides have been shown to induce various defense responses in plant cells, such as the induction of defense-related genes, synthesis of phytoalexin, and the production of reactive oxygen species (Baier et al. 1999; Stergiopoulos and De Wit 2009).

Screening of legume mutants defective in the nodulation identified genes involved in the recognition of Nod factors. LysM domain-containing receptor-like kinases (NFR1, NFR5, SYM2) in legumes were shown to be crucial for the perception of Nod factors in the legume–*Rhizobium* symbiosis (Limpens et al. 2003; Madsen et al. 2003; Radutoiu et al. 2003). As Nod factors are structurally similar to chitooligosaccharides, it has been presumed that plants possess a specific receptor to perceive chitooligosaccharides leading to defense gene induction. In fact, Arabidopsis LysM type receptor-like kinase, which has an amino acid similarity to Nod factor receptors, was shown to be involved in the recognition of chitin-oligosaccharides (Miya et al. 2007; Wan et al. 2008). Interestingly, it was very recently shown that Arabidopsis LysM receptor kinase can transmit signals leading to the symbiosis in *L. japonicus* when a part of the kinase domain was modified to match the sequence of *L. japonicus* (Nakagawa et al. 2011).

When *L. japonicus* was inoculated with *M. loti*, many genes involved in the defense mechanism, such as those responsible for phenylalanine ammonia-lyases, 4-coumarate, CoA ligase, chalcone reductase, chitinase, β-1,3-glucanase, and peroxidase, were induced in the early stage of nodulation and then suppressed to normal levels (Kouchi et al. 2004). Considering the similarity between the Nod factor receptor and chitin receptor, this phenomenon indicates that when legume plants recognize rhizobia, plants first treat them as "potential pathogens" with the induction of defense genes and the production of phytoalexins and do not automatically accept the foreign bacteria into root cells until the plant further recognizes their symbiotic partners with different machineries such as extracellular lipo-polysaccharides (EPS) and lipo-polysaccharides (LPS), as well as effector proteins secreted by TTSS.

3.3 Transporters Involved in the Secretion of Root Exudates

The mechanism by which plant roots secrete phytochemicals was thought to be a passive process mediated by diffusion and channels. However, recent advances suggest that both primary and secondary transporters are involved in the secretion of phytochemicals into the rhizosphere. In plants, there are two major transporter families which are able to transport organic substrates, the ABC transporter family and multidrug and toxic compound extrusion (MATE) transporter family, which consist of 123 and 56 genes in Arabidopsis, respectively (Omote et al. 2006; Verrier et al. 2008; Yazaki et al. 2008). It has been shown that some transporters in these families are also involved in the secretion of root exudates (Loyola-Vargas et al. 2007; Sugiyama et al. 2007; Badri et al. 2008; Yazaki et al. 2008, 2009). For example, a MATE transporter of barley was shown to secrete citrate into the rhizosphere, protecting plant roots from aluminum toxicity (Furukawa et al. 2007; Wang et al. 2007).

In legume–*Rhizobium* interaction, an ABC-type transporter was shown to be involved in the secretion of the isoflavonoid genistein, a signal molecule from soybean to *B. japonicum* (Sugiyama et al. 2007) (Fig. 1). The genistein transport activity of the soybean root plasma membrane was inhibited by sodium orthovanadate, which is a general inhibitor for ABC transporters and P-type ATPases, but not by other inhibitors of electrochemical gradients such as nigericin, valinomycin, or gramicidin D which act as ionophores. It has been known that the synthesis and secretion of genistein is up-regulated under nitrogen deficiency, but the genistein transport activity of the soybean root plasma membrane remained unchanged even under nitrogen deficiency. This suggests that transport activity is constitutively active while the genistein secretion is regulated by the control of the biosynthesis level (Sugiyama et al. 2007). Furthermore, a pharmacological approach using subfamily-specific inhibitors such as verapamil and cyclosporine A, which are often used as general inhibitors of ABCB (P-glycoprotein)-type ABC transporters, and glybenclamide, a sulfonylurea derivative that acts as an effective inhibitor of several ABCC-type members, suggested that these transporter members are unlikely to be responsible for the genistein secretion in soybean. Bioinformatic analysis was indicative that a PDR (pleiotropic drug resistance)-type (full-size ABCG-type) ABC transporter is a primary candidate in the secretion of genistein (Sugiyama et al. 2008).

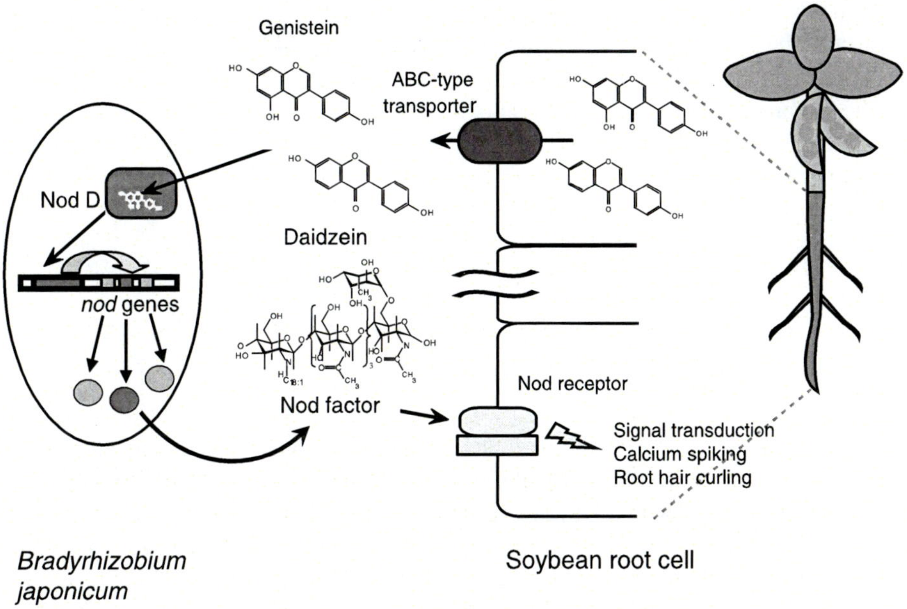

Fig. 1 A model of flavonoid secretion from soybean roots and the interrecognition between soybean and *B. japonicum*

4 Functions of Root Exudates in Symbiosis with Mycorrhiza

4.1 Symbiosis with Arbuscular Mycorrhizal Fungi

Mycorrhizal fungi are major components of the soil microbial community, aiding in the transfer of nutrients from the soil to the plants. They are divided into two groups: endomycorrhiza, such as arbuscular, ericoid, and orchid mycorrhiza, and ectomycorrhiza, which are found especially in temperate forests. These heterogeneous fungi colonize the roots of about 240,000 plant species in a wide range of terrestrial ecosystems. Among the mycorrhiza, the arbuscular mycorrhiza fungi symbiotically interact with more than 80% of plant species widely distributed throughout the plant kingdom (Parniske 2008), but some plant families such as Brassicaceae and Chenopodiaceae are nonhost plants (Smith and Read 1997). Interestingly, lupins are also a nonhost for arbuscular mycorrhizal fungi, although they belong to Leguminosae. Symbiosis with plants results in the formation of tree-shaped subcellular structures, called arbuscules, within the plant cells. These structures are thought to be the main site of nutrient exchange between the fungi and plants. Because the arbuscular mycorrhizal hyphal network reaches to more than 100 m per cubic centimeter of soil and is positioned to efficiently take up various nutrients and water from the surrounding soil (Miller et al. 1995), plants can obtain nutrients such as phosphate and various micronutrients, as well as water, by utilizing this hyphal network. Fossil records have revealed that the origin of arbuscular mycorrhizal symbiosis occurred at least 420–460 million years ago, which coincides with the appearance of the first terrestrial plants, suggesting that the colonization of land by plants from the water was assisted by ancestral arbuscular mycorrhizal fungi (Simon et al. 1993; Remy et al. 1994; Redecker et al. 2000).

4.2 *Signal Molecules Between Legume Plants and Arbuscular Mycorrhizal Fungi*

Because arbuscular mycorrhizal fungi are obligate biotrophs and depend on a living photoautotrophic host to complete their life cycle, a critical developmental step is hyphal branching which enables them to make contact with the host's roots and establish the symbiosis. In a similar way that flavonoids function as a signaling molecule for rhizobia, the branching factor is hypothesized to be a plant signal molecule to trigger hyphal branching (Buee et al. 2000), and these authors have ruled out flavonoids as candidates for the branching factor because root exudates of maize mutants deficient in chalcone synthase show branching activity similar to those of the wild type (Buee et al. 2000). It was also reported that root exudates of plants grown under phosphate deficient conditions have higher activity than those under sufficient phosphate nutrition (Tawaraya et al. 1995, 1998). Using root exudates of *L. japonicas*, the chemical structure of the branching factor was identified to be a strigolactone (Akiyama et al. 2005). Strigolactones have been previously isolated from the root exudates of a variety of plants (Bouwmeester et al. 2003) as a seed germination factor for parasitic weeds such as Striga and Orobanche (see Sect. 4.3).

It was shown that strigolactones were derived from a carotenoid pathway that is also induced under phosphate deficiency (Lopez-Raez et al. 2008). Strigolactones are found in the root exudates of tomato, sorghum, and pea, as well as *L. japonicus*, but not in carrot, tobacco, or alfalfa (Garcia-Garrido et al. 2009), suggesting the presence of other compounds in root exudates that activate hyphal branching. It should also be mentioned that strigolactones were identified in the root exudates of both Arabidopsis and lupin which are nonhost plants to arbuscular mycorrhizal fungi (Goldwasser et al. 2008; Yoneyama et al. 2008). There will be divergent phytochemicals that function as signals to arbuscular mycorrhizal fungi depending on the plant species, as is the signals to rhizobia. It was recently shown that lupin secretes pyranoisoflavones which inhibit hyphal development in arbuscular mycorrhizal fungi (Akiyama et al. 2010).

There is a large amount of interest in the molecular identification of signaling molecules from fungi to plants that induce symbiosis-specific responses in the host root. These hypothetical compounds are called Myc factors, and the existence of such factors became evident by using an ENOD11-promoter GUS (β-glucuronidase) reporter gene fusion in the roots of *M. truncatula* (Kosuta et al. 2003), because hyphae from germinating spores produced a diffusible factor that was perceived by Medicago roots separated by a physical barrier which prevented direct physical contact. Calcium spiking, which occurs in root hairs within a few minutes after the Nod factor application, is also observed following the recognition of Myc factors by plant roots (Navazio et al. 2007). Rapid and transient elevations in cytosolic calcium ions were observed in response to the culture medium of spores of *Gigaspora margarita*, indicating that diffusible molecules released by arbuscular mycorrhizal fungi are perceived by plant cells. The fungal molecules were found to be heat stable with a molecular weight of less than 3,000 and partially lipophilic (Navazio et al. 2007). It is still unknown whether the production of Myc factor is induced by strigolactones or other factors of plants.

4.3 Diverse Functions of Strigolactones

Strigolactones are a group of apocarotenoids and at least nine strigolactones (strigol, strigyl acetate, 5-deoxystrigol, orobanchol, orobanchyl acetate, sorgolactone, epi-orobanchol, solanacol, and sorgomol) have been structurally characterized from the root exudates of various plant species (Rani et al. 2008). The function of strigolactones in the rhizosphere had been identified to stimulate seed germination of harmful parasitic weeds such as witchweed (Striga) and broomrape (Orobanche) before the identification as a hyphal branching factor for arbuscular mycorrhizal fungi. Parasitic weeds are noxious root parasites on many crop species and cause devastating losses of crop yield in many parts of the world including Africa, India, and the Middle East (Rani et al. 2008). The first step in the life cycle is the germination of the seeds of these parasitic plants, which resemble the hyphal branching induced by strigolactones, and both parasitic weeds and arbuscular mycorrhizal fungi are obligate biotrophs. Strigolactones are short-lived in the rhizosphere because of a labile ether bond which spontaneously hydrolyses in water. The fragility of these compounds forms a steep concentration gradient from the plant roots, and this concentration can be a conscientious indicator for the distance to the plant roots (Parniske 2005), thereby enabling these obligate biotrophs to find living plant roots. Considering the fact that the origin of arbuscular mycorrhizal symbiosis occurred far before the appearance of parasitic weeds, it can be concluded that they have evolved to utilize this ancient signal of living plants to parasitize their roots (Fig. 2).

Besides being the signal molecule to arbuscular mycorrhizal fungi and parasitic weeds, strigolactones have been shown to function as an endogenous phytohormone. It was shown that strigolactone levels were reduced in rice mutants that have enhanced shoot branching and that application of strigolactones inhibited shoot branching (Umehara et al. 2008). Another group reported that pea mutants which have reduced levels of strigolactones and altered axial bud growth have root exudates that were shown to exhibit a significantly reduced amount of activity in

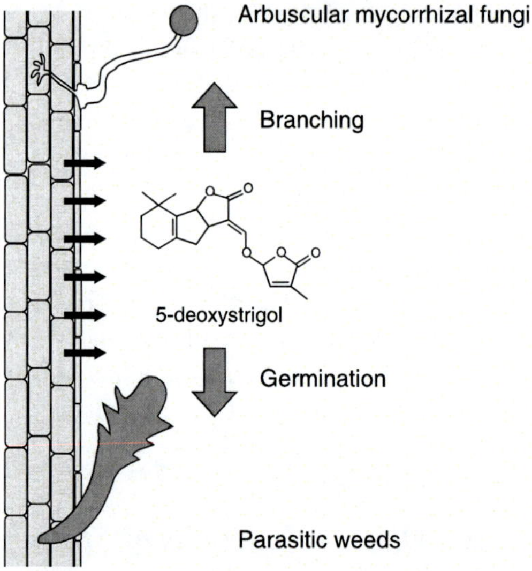

Fig. 2 Strigolactone stimulates both hyphal branching of arbuscular mycorrhizal fungi and germination of parasitic weeds

fungal hyphae branching (Gomez-Roldan et al. 2008). There is no information on how strigolactone acts as a phytohormone or on the receptor for strigolactone, but arbuscular mycorrhizal fungi may use the ancestral receptor for strigolactones to perceive signals of strigolactones in the rhizosphere. Identification of receptors for strigolactones in plants, as well as in fungi, is an interesting topic for future research. There also remains an open question on the original physiological role of strigolactones, i.e., whether strigolactones evolved first as an endogenous phytohormone or as a signal for arbuscular mycorrhizal fungi.

5 Functions of Root Exudates in Mineral Acquisition

5.1 Mineral Requirements for Legume Plants

Plants require 17 essential elements to complete their life cycles, and beside C, H, and O, they absorb mineral nutrients from the soil through their roots. Legume plants are unique in that they form symbiotic interaction with rhizobia. Symbiosis with nitrogen-fixing rhizobia enables legume plants to grow under low nitrogen conditions; however, legume plants require some micronutrients in greater quantities to maintain the symbiosis. One of the examples of these micronutrients is molybdenum (Mo). Mo is a cofactor of a few, but important, enzymes such as nitrate reductase and sulfite oxidase which are involved in nitrogen assimilation and sulfur metabolism, respectively (Hansch and Mendel 2009). Aldehyde oxidase and xanthine dehydrogenase are also Mo-containing enzymes. Legume plants need Mo not only for these enzymes but also for nitrogenase, the most important rhizobial enzyme for the fixation of nitrogen. Nitrogenase has a heterometal complex (FeMoCo) in its active site, and transport of Mo to the bacteroid is required for the proper function of nitrogenase. Application of molybdenum was shown to increase the yield and nitrogen content in legume crops in both laboratory and field conditions (Weeraratna 1980; Yanni 1992; Vieira et al. 1998). It is also reported that a *B. japonicum* strain deficient in molybdenum transport showed impaired nitrogen fixation activity when inoculated to soybean roots (Delgado et al. 2006).

During the reduction of atmospheric N_2 by nitrogenase, gaseous H_2 is produced. In order to utilize H_2 as an energy source, a limited number of rhizobia have hydrogenase that oxidizes H_2 and generates ATP. This hydrogen recycling was shown to increase productivity in symbiotic systems such as soybean. Hydrogenase is a nickel (Ni)-containing enzyme and consists of large and small subunits (HupL and HupS, respectively). It was reported that the exogenous application of Ni to pea roots increased the mature Hup proteins and hydrogenase activity in nodules (Brito et al. 1994), but it is not yet clear how Ni is transported from plant roots into bacteroids.

Iron (Fe) is an important element in photosynthesis because up to 80% of the cellular iron is found in the chloroplasts, but in legume plants iron plays an essential role for leghemoglobin formation, which is the most abundant protein in the nodules. Leghemoglobin is a hemoprotein that has a high affinity for oxygen leading to low oxygen content in the bacteroids to protect the oxygen-sensitive nitrogenase in the nodules (Johnston et al. 2001). Fe is also required for nitrogenase (Fe–S cluster and FeMoCo active site) and nitrogenase reductase (Fe–S cluster) as well as cytochromes and other electron donors that have Fe centers. Legume plants, therefore, need to absorb Fe from soils to meet this high demand, but soluble Fe is very limited in soils because most of the Fe exists as insoluble Fe^{3+}. In fact, Fe deficiency causes a drastic effect on nodule development (O'Hara et al. 1988). Gramineae plants, for instance, secrete phytosiderophores into the rhizosphere for iron acquisition (Takagi 1976), and in legume nodules, rhizobia produce siderophores to aid the iron acquisition as well. Recently, it was reported that red clover promotes the growth of siderophore-producing microbes in the rhizosphere under iron deficient conditions, where phenolic exudates from the red clover roots were responsible for this phenomenon (Jin et al. 2010). Rhizobial genes involved in the biosynthesis of siderophores and related Fe uptake have been reported, but the functions of these genes were studied only in a free-living state (Johnston et al. 2001). In *R. lequminosarum*, it was shown that the mutation of genes involved in the synthesis of siderophores has no influence in the symbiotic nitrogen fixation in pea in which these genes are actually not expressed in the mature bacteroids (Carter et al. 2002). In contrast, *B. japonicum* mutant defective in the uptake of the siderophore have a drastically divergent phenotype in planta, i.e., soybean nodules without siderophore do not fix nitrogen (Benson et al. 2005), suggesting that rhizobial siderophores have functions in Fe uptake at least in the soybean–*B. japonicum* interaction, while the functions of siderophores in nodules are still to be elucidated. It would be the future target of molecular breeding that genes of rhizobia are modified so that they produce siderophores to help legume plants to acquire more Fe from soils.

Contrary to Mo, Ni, and Fe that have symbiosis-specific functions in nodules, the specific relevance of zinc (Zn) for symbiosis is not known; although legume plants do contain a higher concentration of Zn. Zn is an important trace element for humans as well as plants, necessary for DNA replication, protein synthesis, and oxidative stress reduction. In fact, Zn malnutrition affects more than one third of world's population (Hess et al. 2009; Hirschi 2009). To overcome this huge problem, the legume crop lentil has been used for biofortification to increase Zn concentration in planta (Thavarajah et al. 2009).

5.2 *Root Exudates for Phosphorus Acquisition*

Phosphorus (P) is one of the major macronutrients in plants, but its availability is very limited because most P in soil is in insoluble forms such as organic phosphate or insoluble mineral phosphate. Organic phosphate, which can account for ca. 80% of the total phosphorus in soil (Li et al. 1997), has to be mineralized to inorganic phosphate in order to be absorbed by plant roots. As a strategy to acquire phosphate, plants secrete phosphohydrolases into the rhizosphere, which convert organic phosphate into inorganic soluble phosphate, e.g., acid phosphatase is secreted in response to phosphorus deficiency (Lefebvre et al. 1990; Duff et al. 1994). Another strategy is that plants also secrete protons to acidify the rhizosphere which increases the phosphates solubility (Staunton and Leprince 1996).

There are some plant species that can survive on infertile soil containing only limited amounts of available phosphorus: one example of these species is white lupin, a legume plant that has been extensively studied. In response to phosphorus starvation, white lupin develops special bottlebrush-like root clusters, called cluster roots, and secrete protons and a large amount of carboxylates, such as malate and citrate, which solubilizes the phosphates (Neumann and Martinoia 2002). Secretion of organic acids follows a spatial and temporal release pattern in the cluster roots of white lupin: cluster roots secrete low amounts of organic acids, mainly malate in early stage, while the roots secrete a larger amount of organic acids, mainly citrate in the mature stage, which is then accompanied by acidification of the rhizosphere (Neumann et al. 2000). It was also shown that the cluster roots of white lupin secrete isoflavonoids such as genistein and hydroxygenistein (Weisskopf et al. 2006b), which may act as antimicrobials by inhibiting the growth of soil microbes to suppress the biodegradation of citrate. In addition, cluster roots secrete antifungal enzymes, such as glucanase and chitinase, which is thought to prevent the biodegradation of the carboxylates by fungi (Weisskopf et al. 2006a).

6 Conclusions

This chapter provides an overview of literatures on the functions of root exudates in plant–microbe interactions and the acquisition of mineral nutrients from the soil. We have focused on the mutualistic interactions with rhizobia and arbuscular mycorrhizal fungi because recent research has identified genes and metabolites involved in these sophisticated interactions; however, there still remains many plant–microbe interactions in the soil to be analyzed in detail. For example, how plants recruit PGPR, how plants protect themselves from pathogens using phytoalexins, and how plants interfere with quorum sensing using *N*-acyl homoserine lactone mimics (Teplitski et al. 2000; Keshavan et al. 2005). There will also be underground multitrophic interactions, which involve both soluble and volatile root exudates. In maize, (*E*)-β-caryophyllene is emitted from insect-damaged roots that recruit entomopathogenic nematodes (Rasmann et al. 2005). It is possible that legume plants have a tri-interaction system mediated by root exudates including volatiles. Most studies of plant–microbe interactions so far have been focused on a particular microbe such as *Rhizobium* and arbuscular mycorrhizal fungi; however, in nature, legume plants have interactions with many microbes in the soil. Therefore, the interaction of legume plants with entire microbial communities is of particular interest for future research as we need to expand our understanding of the plant microbe interactions to the level that could help develop sustainable agriculture using the various functions of microbes on the soil. There is no argument that plants influence the soil microbial communities, but root exudates are not the only key player to the functions of those interactions. For instance, border cells and other rhizodeposits are nutrient sources and signals for microbes in soils as well (Dennis et al. 2010). The nature of root exudation enables plants to actively regulate the rhizosphere microbial communities, and further research on legume root exudates could open the door to the possibilities of sustainable agriculture practices utilizing legume crops that actively secrete metabolites to recruit beneficial microbes and prevent pathogens.

References

Akiyama K, Matsuzaki K, Hayashi H (2005) Plant sesquiterpenes induce hyphal branching in arbuscular mycorrhizal fungi. Nature 435:824–827

Akiyama K, Tanigawa F, Kashihara T, Hayashi H (2010) Lupin pyranoisoflavones inhibiting hyphal development in arbuscular mycorrhizal fungi. Phytochemistry 71:1865–1871

Ausmees N, Kobayashi H, Deakin WJ, Marie C, Krishnan HB, Broughton WJ, Perret X (2004) Characterization of NopP, a type III secreted effector of Rhizobium sp. strain NGR234. J Bacteriol 186:4774–4780

Badri DV, Vivanco JM (2009) Regulation and function of root exudates. Plant Cell Environ 32:666–681

Badri DV, Loyola-Vargas VM, Broeckling CD, De-la-Pena C, Jasinski M, Santelia D, Martinoia E, Sumner LW, Banta LM, Stermitz F, Vivanco JM (2008) Altered profile of secondary metabolites in the root exudates of Arabidopsis ATP-binding cassette transporter mutants. Plant Physiol 146:762–771

Badri DV, Weir TL, van der Lelie D, Vivanco JM (2009) Rhizosphere chemical dialogues: plant-microbe interactions. Curr Opin Biotechnol 20:642–650

Baier R, Schiene K, Kohring B, Flaschel E, Niehaus K (1999) Alfalfa and tobacco cells react differently to chitin oligosaccharides and *Sinorhizobium meliloti* nodulation factors. Planta 210:157–164

Bassam BJ, Djordjevic MA, Redmond JW, Batley M, Rolfe BG (1988) Identification of a nodD-dependent locus in the *Rhizobium* strain NGR234 activated by phenolic factors secreted by soybeans and other legumes. Mol Plant-Microbe Interact. 1:161–168

Batten K, Scow K, Davies K, Harrison S (2006) Two invasive plants alter soil microbial community composition in serpentine grasslands. Biol Invasions 8:217–230

Begum AA, Leibovitch S, Migner P, Zhang F (2001) Specific flavonoids induced nod gene expression and pre-activated nod genes of *Rhizobium leguminosarum* increased pea (*Pisum sativum* L.) and lentil (*Lens culinaris* L.) nodulation in controlled growth chamber environments. J Exp Bot 52:1537–1543

Benson HP, Boncompagni E, Guerinot ML (2005) An iron uptake operon required for proper nodule development in the *Bradyrhizobium japonicum*-soybean symbiosis. Mol Plant Microbe Interact 18:950–959

Bouwmeester HJ, Matusova R, Zhongkui S, Beale MH (2003) Secondary metabolite signalling in host-parasitic plant interactions. Curr Opin Plant Biol 6:358–364

Brechenmacher L, Lei Z, Libault M, Findley S, Sugawara M, Sadowsky MJ, Sumner LW, Stacey G (2010) Soybean metabolites regulated in root hairs in response to the symbiotic bacterium *Bradyrhizobium japonicum*. Plant Physiol 153:1808–1822

Brito B, Palacios JM, Hidalgo E, Imperial J, Ruiz-Argueso T (1994) Nickel availability to pea (*Pisum sativum* L.) plants limits hydrogenase activity of *Rhizobium leguminosarum* bv. viciae bacteroids by affecting the processing of the hydrogenase structural subunits. J Bacteriol 176: 5297–5303

Buee M, Rossignol M, Jauneau A, Ranjeva R, Becard G (2000) The pre-symbiotic growth of arbuscular mycorrhizal fungi is induced by a branching factor partially purified from plant root exudates. Mol Plant Microbe Interact 13:693–698

Caetano-Anolles G, Crist-Estes DK, Bauer WD (1988) Chemotaxis of *Rhizobium meliloti* to the plant flavone luteolin requires functional nodulation genes. J Bacteriol 170:3164–3169

Cardenas L, Dominguez J, Santana O, Quinto C (1996) The role of the nodI and nodJ genes in the transport of Nod metabolites in *Rhizobium etli*. Gene 173:183–187

Carter RA, Worsley PS, Sawers G, Challis GL, Dilworth MJ, Carson KC, Lawrence JA, Wexler M, Johnston AW, Yeoman KH (2002) The vbs genes that direct synthesis of the siderophore vicibactin in *Rhizobium leguminosarum*: their expression in other genera requires ECF sigma factor RpoI. Mol Microbiol 44:1153–1166

Cho MJ, Harper JE (1991) Effect of inoculation and nitrogen on isoflavonoid concentration in wild-type and nodulation-mutant soybean roots. Plant Physiol 95:435–442

Christie RM (2007) Why is indigo blue? Biotech Histochem 82:51–56

Currier WW, Strobel GA (1976) Chemotaxis of *Rhizobium* spp. to plant root exudates. Plant Physiol 57:820–823

Dakora FD (2000) Commonality of root nodulation signals and nitrogen assimilation in tropical grain legumes belonging to the tribe Phaseoleae. Australian Journal of Plant Physiology 27: 885–892

De-la-Pena C, Lei Z, Watson BS, Sumner LW, Vivanco JM (2008) Root-microbe communication through protein secretion. J Biol Chem 283:25247–25255

De-la-Pena C, Badri DV, Lei Z, Watson BS, Brandao MM, Silva-Filho MC, Sumner LW, Vivanco JM (2010) Root secretion of defense-related proteins is development-dependent and correlated with flowering time. J Biol Chem 285:30654–30665

Delgado MJ, Tresierra-Ayala A, Talbi C, Bedmar EJ (2006) Functional characterization of the *Bradyrhizobium japonicum* modA and modB genes involved in molybdenum transport. Microbiology 152:199–207

Dennis PG, Miller AJ, Hirsch PR (2010) Are root exudates more important than other sources of rhizodeposits in structuring rhizosphere bacterial communities? FEMS Microbiol Ecol 72: 313–327

Desbrosses GG, Kopka J, Udvardi MK (2005) *Lotus japonicus* metabolic profiling. Development of gas chromatography-mass spectrometry resources for the study of plant-microbe interactions. Plant Physiol 137:1302–1318

Dharmatilake AJ, Bauer WD (1992) Chemotaxis of *Rhizobium meliloti* towards nodulation gene-inducing compounds from Alfalfa roots. Appl Environ Microbiol 58:1153–1158

Dixon RA, Sumner LW (2003) Legume natural products: understanding and manipulating complex pathways for human and animal health. Plant Physiol 131:878–885

Djordjevic MA, Redmond JW, Batley M, Rolfe BG (1987) Clovers secrete specific phenolic compounds which either stimulate or repress nod gene expression in *Rhizobium trifolii*. EMBO J 6:173–1179

Downs CT, McDonald PM, Brown K, Ward D (2003) Effects of Acacia condensed tannins on urinary parameters, body mass, and diet choice of an Acacia specialist rodent, *Thallomys nigricauda*. J Chem Ecol 29:845–858

Doyle JJ, Luckow MA (2003) The rest of the iceberg. Legume diversity and evolution in a phylogenetic context. Plant Physiol 131:900–910

Duff SMG, Sarath G, Plaxton WC (1994) The role of acid phosphatases in plant phosphorus metabolism. Physiol Plant 90:791–800

Egelhoff TT, Long SR (1985) *Rhizobium meliloti* nodulation genes: identification of nodDABC gene products, purification of nodA protein, and expression of nodA in *Rhizobium meliloti*. J Bacteriol 164:591–599

Egelhoff TT, Fisher RF, Jacobs TW, Mulligan JT, Long SR (1985) Nucleotide sequence of *Rhizobium meliloti* 1021 nodulation genes: *nodD* is read divergently from *nodABC*. DNA 4: 241–248

Farag MA, Huhman DV, Dixon RA, Sumner LW (2008) Metabolomics reveals novel pathways and differential mechanistic and elicitor-specific responses in phenylpropanoid and isoflavonoid biosynthesis in *Medicago truncatula* cell cultures. Plant Physiol 146:387–402

Farag MA, Deavours BE, de Fatima A, Naoumkina M, Dixon RA, Sumner LW (2009) Integrated metabolite and transcript profiling identify a biosynthetic mechanism for hispidol in *Medicago truncatula* cell cultures. Plant Physiol 151:1096–1113

Fauvart M, Michiels J (2008) Rhizobial secreted proteins as determinants of host specificity in the rhizobium-legume symbiosis. FEMS Microbiol Lett 285:1–9

Fernandez-Lopez M, D'Haeze W, Mergaert P, Verplancke C, Prome JC, Van Montagu M, Holsters M (1996) Role of nodl and nodJ in lipo-chitooligosaccharide secretion in *Azorhizobium caulinodans* and *Escherichia coli*. Mol Microbiol 20:993–1000

Fisher RF, Long SR (1992) Rhizobium–plant signal exchange. Nature 357:655–660

Firmin JL, Wilson KE, Rossen L, Johnston AWB (1986) Flavonoid activation of nodulation genes in *Rhizobium* reversed by other compounds present in plants. Nature 324:90–92

Furukawa J, Yamaji N, Wang H, Mitani N, Murata Y, Sato K, Katsuhara M, Takeda K, Ma JF (2007) An aluminum-activated citrate transporter in barley. Plant Cell Physiol 48:1081–1091

Gagnon H, Ibrahim RK (1998) Aldonic Acids: A Novel Family of *nod* Gene Inducers of *Mesorhizobium loti*, *Rhizobium lupini*, and *Sinorhizobium meliloti*. Mol Plant Microbe Interact 11:988–998

Galibert F, Finan TM, Long SR, Puhler A, Abola P, Ampe F, Barloy-Hubler F, Barnett MJ, Becker A, Boistard P, Bothe G, Boutry M, Bowser L, Buhrmester J, Cadieu E, Capela D, Chain P, Cowie A, Davis RW, Dreano S, Federspiel NA, Fisher RF, Gloux S, Godrie T, Goffeau A, Golding B, Gouzy J, Gurjal M, Hernandez-Lucas I, Hong A, Huizar L, Hyman RW, Jones T, Kahn D, Kahn ML, Kalman S, Keating DH, Kiss E, Komp C, Lelaure V, Masuy D, Palm C, Peck MC, Pohl TM, Portetelle D, Purnelle B, Ramsperger U, Surzycki R, Thebault P, Vandenbol M, Vorholter FJ, Weidner S, Wells DH, Wong K, Yeh KC, Batut J (2001) The composite genome of the legume symbiont *Sinorhizobium meliloti*. Science 293:668–672

Garcia-Garrido JM, Lendzemo V, Castellanos-Morales V, Steinkellner S, Vierheilig H (2009) Strigolactones, signals for parasitic plants and arbuscular mycorrhizal fungi. Mycorrhiza 19:449–459

Goldwasser Y, Yoneyama K, Xie X, Yoneyama K (2008) Production of Strigolactones by *Arabidopsis thaliana* responsible for *Orobanche aegyptiaca* seed germination. Plant Growth Regul 55:21–28

Gomez-Roldan V, Fermas S, Brewer PB, Puech-Pages V, Dun EA, Pillot JP, Letisse F, Matusova R, Danoun S, Portais JC, Bouwmeester H, Becard G, Beveridge CA, Rameau C, Rochange SF (2008) Strigolactone inhibition of shoot branching. Nature 455:189–194

Graham PH, Vance CP (2003) Legumes: importance and constraints to greater use. Plant Physiol 131:872–877

Hansch R, Mendel RR (2009) Physiological functions of mineral micronutrients (Cu, Zn, Mn, Fe, Ni, Mo, B, Cl). Curr Opin Plant Biol 12:259–266

Hartwig UA, Maxwell CA, Joseph CM, Phillips DA (1990) Chrysoeriol and Luteolin Released from Alfalfa Seeds Induce nod Genes in *Rhizobium meliloti*. Plant Physiol 92:116–122

Hernandez G, Valdes-Lopez O, Ramirez M, Goffard N, Weiller G, Aparicio-Fabre R, Fuentes SI, Erban A, Kopka J, Udvardi MK, Vance CP (2009) Global changes in the transcript and metabolic profiles during symbiotic nitrogen fixation in phosphorus-stressed common bean plants. Plant Physiol 151:1221–1238

Hess SY, Lonnerdal B, Hotz C, Rivera JA, Brown KH (2009) Recent advances in knowledge of zinc nutrition and human health. Food Nutr Bull 30:S5–S11

Hirschi KD (2009) Nutrient biofortification of food crops. Annu Rev Nutr 29:401–421

Horiuchi J, Prithiviraj B, Bais HP, Kimball BA, Vivanco JM (2005) Soil nematodes mediate positive interactions between legume plants and *rhizobium* bacteria. Planta 222:848–857

Hungria M, Joseph CM, Phillips DA (1991) Anthocyanidins and flavonols, major *nod* gene Inducers from seeds of a black-seeded common bean (*Phaseolus vulgaris* L.). Plant Physiol 97:751–758

Ikeda S, Rallos LE, Okubo T, Eda S, Inaba S, Mitsui H, Minamisawa K (2008) Microbial community analysis of field-grown soybeans with different nodulation phenotypes. Appl Environ Microbiol 74:5704–5709

Innes L, Hobbs PJ, Bardgett RD (2004) The impacts of individual plant species on rhizosphere microbial communities in soils of different fertility. Biol Fertil Soils 40:7–13

Jin CW, Li GX, Yu XH, Zheng SJ (2010) Plant Fe status affects the composition of siderophore-secreting microbes in the rhizosphere. Ann Bot 105:835–841

Johnston AW, Yeoman KH, Wexler M (2001) Metals and the rhizobial-legume symbiosis–uptake, utilization and signalling. Adv Microb Physiol 45:113–156

Keshavan ND, Chowdhary PK, Haines DC, Gonzalez JE (2005) L-Canavanine made by Medicago sativa interferes with quorum sensing in *Sinorhizobium meliloti*. J Bacteriol 187:8427–8436

Kosslak RM, Bookland R, Barkei J, Paaren HE, Appelbaum ER (1987) Induction of *Bradyrhizobium japonicum* common nod genes by isoflavones isolated from *Glycine max*. Proc Natl Acad Sci USA 84:7428–7432

Kosuta S, Chabaud M, Lougnon G, Gough C, Denarie J, Barker DG, Becard G (2003) A diffusible factor from arbuscular mycorrhizal fungi induces symbiosis-specific MtENOD11 expression in roots of *Medicago truncatula*. Plant Physiol 131:952–962

Kouchi H, Shimomura K, Hata S, Hirota A, Wu GJ, Kumagai H, Tajima S, Suganuma N, Suzuki A, Aoki T, Hayashi M, Yokoyama T, Ohyama T, Asamizu E, Kuwata C, Shibata D, Tabata S (2004) Large-scale analysis of gene expression profiles during early stages of root nodule formation in a model legume, *Lotus japonicus*. DNA Res 11:263–274

Kowalchuka GA, Hola WHG, Van Veen JA (2006) Rhizosphere fungal communities are influenced by Senecio jacobaea pyrrolizidine alkaloid content and composition. Soil Biol Biochem 38:2852–2859

Krishnan HB (2002) NolX of *Sinorhizobium fredii* USDA257, a type III-secreted protein involved in host range determination, Iis localized in the infection threads of cowpea (*Vigna unguiculata* [L.] Walp) and soybean (*Glycine max* [L.] Merr.) nodules. J Bacteriol 184:831–839

Krishnan HB, Lorio J, Kim WS, Jiang G, Kim KY, DeBoer M, Pueppke SG (2003) Extracellular proteins involved in soybean cultivar-specific nodulation are associated with pilus-like surface appendages and exported by a type III protein secretion system in *Sinorhizobium fredii* USDA257. Mol Plant Microbe Interact 16:617–625

Lefebvre DD, Duff SM, Fife CA, Julien-Inalsingh C, Plaxton WC (1990) Response to phosphate deprivation in Brassica nigra suspension cells: enhancement of intracellular, cell surface, and secreted phosphatase activities compared to increases in Pi-absorption rate. Plant Physiol 93: 504–511

Li M, Osaki M, Rao IM, Tadano T (1997) Secretion of phytase from the roots of several plant species under phosphorus-deficient conditions. Plant Soil 195:161–169

Limpens E, Franken C, Smit P, Willemse J, Bisseling T, Geurts R (2003) LysM domain receptor kinases regulating rhizobial Nod factor-induced infection. Science 302:630–633

Lopez-Raez JA, Charnikhova T, Gomez-Roldan V, Matusova R, Kohlen W, De Vos R, Verstappen F, Puech-Pages V, Becard G, Mulder P, Bouwmeester H (2008) Tomato strigolactones are derived from carotenoids and their biosynthesis is promoted by phosphate starvation. New Phytol 178:863–874

Loyola-Vargas VM, Broeckling CD, Badri D, Vivanco JM (2007) Effect of transporters on the secretion of phytochemicals by the roots of *Arabidopsis thaliana*. Planta 225:301–310

Lucas Garcia JA, Barbas C, Probanza A, Barrientos ML, Gutierrez Manero FJ (2001) Low molecular weight organic acids and fatty acids in root exudates of two Lupinus cultivars at flowering and fruiting stages. Phytochem Anal 12:305–311

Madsen EB, Madsen LH, Radutoiu S, Olbryt M, Rakwalska M, Szczyglowski K, Sato S, Kaneko T, Tabata S, Sandal N, Stougaard J (2003) A receptor kinase gene of the LysM type is involved in legume perception of rhizobial signals. Nature 425:637–640

Marie C, Deakin WJ, Viprey V, Kopcinska J, Golinowski W, Krishnan HB, Perret X, Broughton WJ (2003) Characterization of Nops, nodulation outer proteins, secreted via the type III secretion system of NGR234. Mol Plant Microbe Interact 16:743–751

Maxwell CA, Hartwig UA, Joseph CM, Phillips DA (1989) A chalcone and two related flavonoids released from alfalfa roots induce nod genes of *Rhizobium meliloti*. Plant Physiol 91:842–847

Mazzola M, Funnell DL, Raaijmakers JM (2004) Wheat cultivar-specific selection of 2,4-diacetylphloroglucinol-producing fluorescent Pseudomonas species from resident soil populations. Microb Ecol 48:338–348

Messens E, Geelen D, van Montagu M, Holsters M (1991) 7,4-Dihydroxyflavanone is the major *Azorhizobium* nod gene-inducing factor present in *Sesbania rostrata* seedling exudate. Mol Plant-Microbe Interact. 4:262–267

Micallef SA, Shiaris MP, Colon-Carmona A (2009) Influence of Arabidopsis thaliana accessions on rhizobacterial communities and natural variation in root exudates. J Exp Bot 60:1729–1742

Miller RM, Reinhardt DR, Jastrow JD (1995) External hyphal production of vesicular-arbuscular mycorrhizal fungi in pasture and tallgrass prairie communities. Oecologia 103:17–23

Miya A, Albert P, Shinya T, Desaki Y, Ichimura K, Shirasu K, Narusaka Y, Kawakami N, Kaku H, Shibuya N (2007) CERK1, a LysM receptor kinase, is essential for chitin elicitor signaling in Arabidopsis. Proc Natl Acad Sci USA 104:19613–19618

Mougel C, Offre P, Ranjard L, Corberand T, Gamalero E, Robin C, Lemanceau P (2006) Dynamic of the genetic structure of bacterial and fungal communities at different developmental stages of *Medicago truncatula* Gaertn. cv. Jemalong line J5. New Phytol 170:165–175

Mulligan JT, Long SR (1985) Induction of *Rhizobium meliloti* nodC expression by plant exudate requires nodD. Proc Natl Acad Sci USA 82:6609–6613

Nakagawa T, Kaku H, Shimoda Y, Sugiyama A, Shimamura M, Takanashi K, Yazaki K, Aoki T, Shibuya N, Kouchi H (2011) From defense to symbiosis: limited alterations in the kinase domain of LysM receptor-like kinases are crucial for evolution of legume-*Rhizobium* symbiosis. Plant Journal 65:169–180

Narasimhan K, Basheer C, Bajic VB, Swarup S (2003) Enhancement of plant-microbe interactions using a rhizosphere metabolomics-driven approach and its application in the removal of polychlorinated biphenyls. Plant Physiol 132:146–153

Navazio L, Moscatiello R, Genre A, Novero M, Baldan B, Bonfante P, Mariani P (2007) A diffusible signal from arbuscular mycorrhizal fungi elicits a transient cytosolic calcium elevation in host plant cells. Plant Physiol 144:673–681

Neumann G, Martinoia E (2002) Cluster roots – an underground adaptation for survival in extreme environments. Trends Plant Sci 7:162–167

Neumann G, Massonneau A, Langlade N, Dinkelaker B, Hengeler C, Römheld V, Martinoia E (2000) Physiological aspect of cluster root function and development in phosphorus-deficient White Lupin (*Lupinus albus* L.). Ann Bot 85:909–919

Novak K, Chovanec P, Skrdleta V, Kropacova M, Lisa L, Nemcova M (2002) Effect of exogenous flavonoids on nodulation of pea (*Pisum sativum* L.). J Exp Bot 53:1735–1745

O'Hara GW, Dilworth MJ, Boonkerd N, Parkpian P (1988) Iron-deficiency specifically limits nodule development in peanut inoculated with *Bradyrhizobium* sp. New Phytol 108:51–57

Offre P, Pivato B, Siblot S, Gamalero E, Corberand T, Lemanceau P, Mougel C (2007) Identification of bacterial groups preferentially associated with mycorrhizal roots of *Medicago truncatula*. Appl Environ Microbiol 73:913–921

Omote H, Hiasa M, Matsumoto T, Otsuka M, Moriyama Y (2006) The MATE proteins as fundamental transporters of metabolic and xenobiotic organic cations. Trends Pharmacol Sci 27:587–593

Parke D, Rivelli M, Ornston LN (1985) Chemotaxis to aromatic and hydroaromatic acids: comparison of *Bradyrhizobium japonicum* and *Rhizobium trifolii*. J Bacteriol 163:417–422

Parniske M (2005) Plant-fungal associations: cue for the branching connection. Nature 435: 750–751

Parniske M (2008) Arbuscular mycorrhiza: the mother of plant root endosymbioses. Nat Rev Microbiol 6:763–775

Peters NK, Frost JW, Long SR (1986) A plant flavone, luteolin, induces expression of *Rhizobium meliloti* nodulation genes. Science 233:977–980

Phillips DA, Joseph CM, Maxwell CA (1992) Trigonelline and stachydrine released from Alfalfa seeds activate NodD2 protein in *Rhizobium meliloti*. Plant Physiol 99:1526–1531

Priha O, Grayston SJ, Pennanen T, Smolander A (1999) Microbial activities related to C and N cycling and microbial community structure in the rhizospheres of *Pinus sylvestris*, *Picea abies* and *Betula pendula* seedlings in an organic and mineral soil. FEMS Microbiol Ecol 30: 187–199

Radutoiu S, Madsen LH, Madsen EB, Felle HH, Umehara Y, Gronlund M, Sato S, Nakamura Y, Tabata S, Sandal N, Stougaard J (2003) Plant recognition of symbiotic bacteria requires two LysM receptor-like kinases. Nature 425:585–592

Rani K, Zwanenburg B, Sugimoto Y, Yoneyama K, Bouwmeester HJ (2008) Biosynthetic considerations could assist the structure elucidation of host plant produced rhizosphere signalling compounds (strigolactones) for arbuscular mycorrhizal fungi and parasitic plants. Plant Physiol Biochem 46:617–626

Rasmann S, Kollner TG, Degenhardt J, Hiltpold I, Toepfer S, Kuhlmann U, Gershenzon J, Turlings TC (2005) Recruitment of entomopathogenic nematodes by insect-damaged maize roots. Nature 434:732–737

Redecker D, Kodner R, Graham LE (2000) Glomalean fungi from the Ordovician. Science 289: 1920–1921

Redmond J, Batley M, Djordjevic M, Innes R, Kuempel P, Rolfe B (1986) Flavones induce expression of nodulation genes in *Rhizobium*. Nature 323:632–635

Remy W, Taylor TN, Hass H, Kerp H (1994) Four hundred-million-year-old vesicular arbuscular mycorrhizae. Proc Natl Acad Sci USA 91:11841–11843

Rispail N, Hauck B, Bartholomew B, Watson AA, Nash RJ, Webb KJ (2010) Secondary metabolite profiling of the model legume *Lotus japonicus* during its symbiotic interaction with *Mesorhizobium loti*. Symbiosis 50:119–128

Rivilla R, Sutton JM, Downie JA (1995) *Rhizobium leguminosarum* NodT is related to a family of outer-membrane transport proteins that includes TolC, PrtF, CyaE and AprF. Gene 161:27–31

Rossen L, Johnston AW, Downie JA (1984) DNA sequence of the *Rhizobium leguminosarum* nodulation genes nodAB and C required for root hair curling. Nucleic Acids Res 12:9497–9508

Rossen L, Shearman CA, Johnston AWB, Downie JA (1985) The nodD gene of Rhizobium leguminosarum is autoregulatory and in the presence of plant exudate induces the nodA, B, C genes. EMBO J 4:3369–3373

Scheidemann P, Wetzel A (1997) Identification and characterization of flavonoids in the root exudate of *Robinia pseudoacacia*. Trees 11:316–321

Simon L, Bousquet J, Levesque RC, Lalonde M (1993) Origin and diversification of endomycorrhizal fungi and coincidence with vascular land plants. Nature 363:67–69

Smil V (1999) Nitrogen in crop production. Global Biogeochem Cycles 13:647–662

Smit G, Puvanesarajah V, Carlson RW, Barbour WM, Stacey G (1992) *Bradyrhizobium japonicum* nodD1 can be specifically induced by soybean flavonoids that do not induce the nodYABCSUIJ operon. J Biol Chem 267:310–318

Smith SE, Read DJ (1997) Mycorrhizal symbiosis. Academic, San Diego

Spaink HP, Wijfjes AH, Lugtenberg BJ (1995) Rhizobium NodI and NodJ proteins play a role in the efficiency of secretion of lipochitin oligosaccharides. J Bacteriol 177:6276–6281

Staunton S, Leprince F (1996) Effect of pH and some organic anions on the solubility of soil phosphate: implications for P bioavailability. Eur J Soil Sci 47:231–239

Stergiopoulos I, De Wit PJ (2009) Fungal effector proteins. Annu Rev Phytopathol 47:233–263

Sugiyama A, Shitan N, Yazaki K (2007) Involvement of a soybean ATP-binding cassette-type transporter in the secretion of genistein, a signal flavonoid in legume-*Rhizobium* symbiosis. Plant Physiol 144:2000–2008

Sugiyama A, Shitan N, Yazaki K (2008) Signaling from soybean roots to *rhizobium*: An ATP-binding cassette-type transporter mediates genistein secretion. Plant Signal Behav 3:38–40

Suzuki H, Sasaki R, Ogata Y, Nakamura Y, Sakurai N, Kitajima M, Takayama H, Kanaya S, Aoki K, Shibata D, Saito K (2008) Metabolic profiling of flavonoids in Lotus japonicus using liquid chromatography Fourier transform ion cyclotron resonance mass spectrometry. Phytochemistry 69:99–111

Takagi S (1976) Naturally occurring iron-chelating compounds in oat and rice root-washings. Soil Sci Plant Nutr 22:423–433

Tawaraya K, Watanabe S, Yoshida E, Wagatsuma T (1995) Effect of onion (*Allium cepa*) root exudates on the hyphal growth of *Gigaspora margarita*. Mycorrhiza 6:57–59

Tawaraya K, Hashimoto K, Wagatsuma T (1998) Effect of root exudate fractions from P-deficient and P-sufficient onion plants on root colonisation by the arbuscular mycorrhizal fungus *Gigaspora margarita*. Mycorrhiza 8:67–70

Teplitski M, Robinson JB, Bauer WD (2000) Plants secrete substances that mimic bacterial N-acyl homoserine lactone signal activities and affect population density-dependent behaviors in associated bacteria. Mol Plant Microbe Interact 13:637–648

Thavarajah D, Thavarajah P, Sarker A, Vandenberg A (2009) Lentils (*Lens culinaris* Medikus Subspecies culinaris): a whole food for increased iron and zinc intake. J Agric Food Chem 57: 5413–5419

Ueda H, Sugimoto Y (2010) Vestitol as a chemical barrier against intrusion of parasitic plant *Striga hermonthica* into *Lotus japonicus* roots. Biosci Biotechnol Biochem 74:1662–1667

Umehara M, Hanada A, Yoshida S, Akiyama K, Arite T, Takeda-Kamiya N, Magome H, Kamiya Y, Shirasu K, Yoneyama K, Kyozuka J, Yamaguchi S (2008) Inhibition of shoot branching by new terpenoid plant hormones. Nature 455:195–200

Uren NC (2007) Types, amounts and possible functions of compounds released into the rhizosphere of soil-grown plants. In: Pinton R, Varanini Z, Nannipicro P (eds) The rhizosphere: biochemistry and organic substances at the soil-plant interface. CRC, New York, pp 1–22

Verma D, Hong Z (1996) Biogenesis of the peribacteroid membrane in root nodules. Trends Microbiol 4:364–368

Verrier PJ, Bird D, Burla B, Dassa E, Forestier C, Geisler M, Klein M, Kolukisaoglu U, Lee Y, Martinoia E, Murphy A, Rea PA, Samuels L, Schulz B, Spalding EJ, Yazaki K, Theodoulou FL

(2008) Plant ABC proteins – a unified nomenclature and updated inventory. Trends Plant Sci 13:151–159

Vieira RF, Cardoso EJBN, Vieira C, Cassini STA (1998) Foliar application of molybdenum in common beans. I. Nitrogenase and reductase activities in a soil of high fertility. J Plant Nutr 21:169–180

Viprey V, Greco AD, Golinowski W, Broughton WJ, Perret X (1998) Symbiotic implications of type III protein secretion machinery in *Rhizobium*. Mol Microbiol 28:1381–1389

Wan J, Zhang XC, Neece D, Ramonell KM, Clough S, Kim SY, Stacey MG, Stacey G (2008) A LysM receptor-like kinase plays a critical role in chitin signaling and fungal resistance in Arabidopsis. Plant Cell 20:471–481

Wang J, Raman H, Zhou M, Ryan PR, Delhaize E, Hebb DM, Coombes N, Mendham N (2007) High-resolution mapping of the Alp locus and identification of a candidate gene HvMATE controlling aluminium tolerance in barley (*Hordeum vulgare* L.). Theor Appl Genet 115: 265–276

Weeraratna CS (1980) Studies on the molybdenum application to soybean. Beitr Trop Landwirtsch Veterinarmed 18:131–134

Weisskopf L, Abou-Mansour E, Fromin N, Tomasi N, Santelia D, Edelkott I, Neumann G, Aragno M, Tabacchi R, Martinoia E (2006a) White lupin has developed a complex strategy to limit microbial degradation of secreted citrate required for phosphate acquisition. Plant Cell Environ 29:919–927

Weisskopf L, Tomasi N, Santelia D, Martinoia E, Langlade NB, Tabacchi R, Abou-Mansour E (2006b) Isoflavonoid exudation from white lupin roots is influenced by phosphate supply, root type and cluster-root stage. New Phytol 171:657–668

Werner D (2007) Molecular biology and ecology of the *rhizobia*-legume symbiosis. In: Pinton R, Varanini Z, Nannipiero P (eds) The rhizosphere: biochemistry and organic substances at the soil-plant interface. CRC, New York, pp 237–266

Werner D, Muller P (2002) Communication and efficiency in the symbiotic signal exchange. In: Heldmaier G, Werner D (eds) Environmental signal processing and adaptation. Springer, Heidelberg

Yang CH, Crowley DE (2000) Rhizosphere microbial community structure in relation to root location and plant iron nutritional status. Appl Environ Microbiol 66:345–351

Yanni YG (1992) Performance of chickpea, lentil and lupin nodulated with indigenous or inoculated rhizobia micropartners under nitrogen, boron, cobalt and molybdenum fertilization schedules. World J Microbiol Biotechnol 8:607–613

Yazaki K, Sugiyama A, Morita M, Shitan N (2008) Secondary transport as an efficient membrane transport mechanism for plant secondary metabolites. Phytochem Rev 7:513–524

Yazaki K, Shitan N, Sugiyama A, Takanashi K (2009) Cell and molecular biology of ATP-binding cassette proteins in plants. Int Rev Cell Mol Biol 276:263–299

Yoneyama K, Xie X, Sekimoto H, Takeuchi Y, Ogasawara S, Akiyama K, Hayashi H, Yoneyama K (2008) Strigolactones, host recognition signals for root parasitic plants and arbuscular mycorrhizal fungi, from Fabaceae plants. New Phytol 179:484–494

Yost CK, Rochepeau P, Hynes MF (1998) Rhizobium leguminosarum contains a group of genes that appear to code for methyl-accepting chemotaxis proteins. Microbiology 144:1945–1956

Zaat SAJ, Schripsema J, Wijffelman CA, Brussel AAN, Lugtenberg BJJ (1989) Analysis of the major inducers of the *Rhizobium* nodA promoter from *Vicia sativa* root exudate and their activity with different nodD genes. Plant Mol Biol 13:175–188.

Zhang J, Subramanian S, Zhang Y, Yu O (2007) Flavone synthases from *Medicago truncatula* are flavanone-2-hydroxylases and are important for nodulation. Plant Physiol 144:741–751

Strigolactones in Root Exudates as a Signal in Symbiotic and Parasitic Interactions

Hinanit Koltai, Radoslava Matusova, and Yoram Kapulnik

Abstract Plants produce numerous secondary metabolites, many of which have a role in their development. The presence of such compounds in the rhizosphere led other organisms in the course of evolution to recognize these root exudates as signals for the presence of a host plant. Strigolactones (SLs) were recently identified as a new plant hormone. However, they were first identified, more than 40 years ago, as germination stimulants of the parasitic plants *Striga* and *Orobanche*, and later as stimulants of hyphal branching of the symbiotic arbuscular mycorrhizal fungi. In this chapter, we focus on SLs in root exudates as a signal in these parasitic and symbiotic interactions. The possible evolution of the biological role(s) of SLs, their essentialness to, and their involvement in determining host recognition by parasitic plants and symbiotic fungi will be discussed.

1 Introduction

Plants produce numerous secondary metabolites, many of which play a role in their development. In some cases, these plant products can also be found outside the plant, by active or passive secretion. In the course of evolution, the presence of such compounds in the rhizosphere led other organisms to recognize these root exudates as signals for the presence of a host plant. Moreover, these plant-derived signals were not only recognized, but in some cases became triggers for physiological responses in the plant-interacting organisms.

H. Koltai (✉) • Y. Kapulnik
Institute of Plant Sciences, ARO, Volcani Center, Bet-Dagan 50250, Israel
e-mail: hkoltai@agri.gov.il

R. Matusova
Institute of Plant Genetics and Biotechnology, Slovak Academy of Sciences, 39 A, Akademicka 2, 950 07 Nitra, Slovak Republic

J.M. Vivanco and F. Baluška (eds.), *Secretions and Exudates in Biological Systems*, Signaling and Communication in Plants 12, DOI 10.1007/978-3-642-23047-9_3,

Strigolactones (SLs) are carotenoid-derived terpenoid lactones (Matusova et al. 2005) which were recently identified as a new plant hormone, and demonstrated to have a role in plant development (reviewed by Dun et al. 2009; Leyser 2009; Beveridge and Kyozuka 2010; Waldie et al. 2010). However, they were first identified, more than 40 years ago as germination stimulants of the parasitic plants *Striga* and *Orobanche* and later, as stimulants of hyphal branching of the symbiotic arbuscular mycorrhizal fungi (AMF) (reviewed by Xie et al. 2010). These functions suggest that SLs are important components of plant communication with the soil biota, which leads to the establishment of plant interactions: either as a harmful association with parasitic plants or as a beneficial one with symbiotic fungi.

In this chapter, we focus on SLs in root exudates as a signal in these parasitic and symbiotic interactions. We will also outline SLs' role in plant development and their associated pathways. In addition, the evolutionary implications of SL exploitation for plant–host recognition and interactions by parasitic plants and symbiotic fungi will be discussed.

2 Roles of Strigolactones in Plant Development

2.1 Shoot Development

A particular class of mutants was identified some years ago that displayed a specific increase in bud outgrowth; this increase could not be entirely explained by the plant hormones known back then, including cytokinin and auxin. Indeed, using grafting studies and hormone measurements, it was shown that this class of mutants is defective in synthesizing or responding to a novel branch-inhibiting signal, SMS, acting as a long-distance cue. SMS was suggested to be involved in branching control and to be regulated by auxin and another feedback signal (reviewed by Dun et al. 2009).

SLs, first identified as germinators of parasitic plant seeds (Cook et al. 1972), were found to fit the characteristics of SMS in their ability to act as long-distance branching factors that suppress growth of preformed axillary buds (Gomez-Roldan et al. 2008; Umehara et al. 2008). Hence, SLs were defined as a new group of plant hormones or their derivatives (e.g., Gomez-Roldan et al. 2008; Umehara et al. 2008; Brewer et al. 2009; Ferguson and Beveridge 2009; reviewed by Dun et al. 2009).

The presence of SLs has been demonstrated in several plant species, including cotton (*Gossypium hirsutum* L.), sorghum [*Sorghum bicolor* (L.) Moench], maize (*Zea mays* L.), common millet (*Panicum miliaceum* L.), cowpea [*Vigna unguiculata* (L.) Walp.], red clover (*Trifolium pratense* L.), garden pea (*Pisum sativum* L.), rice (*Oryza sativa* cv. Nipponbare), tobacco (*Nicotiana tabacum* L.), and tomato (*Solanum lycopersicum* L.) (reviewed by Xie et al. 2010). Moreover, in each of the examined species, a mixture of several SL compounds was identified; the components of this mixture were shown to vary in ratio with plant age, growth conditions and cultivars, and to overlap between plant species (reviewed by Yoneyama et al. 2008, 2009).

SLs were suggested to be derived from carotenoids (Matusova et al. 2005), and hence to be possible products of carotenoid cleavage dioxygenase (CCD) enzymes (Umehara et al. 2008). Indeed, several of the hyperbranching mutants identified to date in several plant species, including *Arabidopsis thaliana*, rice (*O. sativa*), petunia (*Petunia hybrida*), pea (*P. sativum*), tomato (*S. lycopersicum*), and chrysanthemum (*Dendranthema grandiflorum*) (e.g., Drummond et al. 2009; Liang et al. 2010; Vogel et al. 2010; reviewed by Dun et al. 2009; Leyser 2009), have been found to be flawed in CCDs. These include CCD7, also known as MAX3, RMS5 and HTD1/D17 (Booker et al. 2004; Johnson et al. 2006; Zou et al. 2006), and CCD8, also known as MAX4, RMS1, D10, and DAD1 (Sorefan et al. 2003; Bainbridge et al. 2005; Snowden et al. 2005; Arite et al. 2007). CCD7 and CCD8 have been suggested, although not conclusively determined, to sequentially catalyze carotenoid cleavage reactions (Schwartz et al. 2004; Auldridge et al. 2006). A cytochrome P450 monooxygenase, MAX1, was also suggested to be involved in a later SL biosynthetic step (Booker et al. 2005; Gomez-Roldan et al. 2008). Additional steps in SL synthesis were anticipated but are still unknown (Matusova et al. 2005).

SLs were shown to be synthesized mainly in the roots. Grafting of wild-type (WT) roots to a SL-synthesis-mutant shoot was shown, in a variety of species, to be sufficient to reverse the mutant phenotype to that of the WT. Moreover, interstock grafting of the WT lower shoot part between mutant roots and shoot was also sufficient to reverse the SL-synthesis-mutant shoot phenotype to the WT, suggesting that SLs are synthesized in the lower part of the shoot as well (Napoli 1996; Foo et al. 2001; Booker et al. 2004; Gomez-Roldan et al. 2008; Umehara et al. 2008; Koltai et al. 2010a; reviewed by Dun et al. 2009). It was also suggested that SLs, their metabolites, or other unknown secondary messengers move in the root-to-shoot direction (Napoli 1996; Beveridge et al. 1997, 2000; Koltai et al. 2010a; reviewed by Dun et al. 2009) to confer a significant reduction in shoot branching (Foo et al. 2001; Brewer et al. 2009; Ferguson and Beveridge 2009).

Although a receptor for SLs has not yet been identified, MAX2, the overshooting mutant of which is flawed in an F-box protein, was suggested to be a component of SL signaling. It was further suggested that it might act as a receptor for SLs and function in ubiquitin-mediated degradation of as yet unknown protein targets (Stirnberg et al. 2007; Umehara et al. 2008).

Several lines of evidence suggest a connection between SL pathways and light (Waldie et al. 2010). The SL-signaling mutant *max2* was found to be hyposensitive to both red (R) and far-red (FR) light-induced seed germination, whereas several light-harvesting-associated genes were found to exhibit a slower rate of induction in *max2* mutants relative to the WT upon R light exposure (Shen et al. 2007).

SLs were suggested to be potentially positive regulators of plant light-harvesting components, based on SL-induced gene transcription profiling and a reduced level of chlorophyll in *Sl-ORT1* (Mayzlish-Gati et al. 2010); *Sl-ORT1* is a tomato mutant deficient in SL biosynthesis (Dor et al. 2010; Koltai et al. 2010a). Furthermore, a recent study has demonstrated that some light-signaling genes are positive regulators of SLs, whereas SLs were suggested to regulate the nuclear localization of COP1 ubiquitin ligase, which partially controls the level of light regulators, including HY5, thereby mimicking light-adapted seedling growth (Tsuchiya et al. 2010).

A large number of studies suggested a cross talk between SLs and other regulators of shoot branching. For auxin, it was suggested that SL is an auxin-promoted secondary messenger that moves up into the buds to repress their outgrowth (Brewer et al. 2009; Ferguson and Beveridge 2009; reviewed by Dun et al. 2009). Alternatively, it was suggested that restrained bud outgrowth results from a SL-mediated reduction in the shoot of the capacity for polar auxin transport from the apical meristem, leading to inhibition of polar auxin transport from the buds (e.g., Bennett et al. 2006; Mouchel and Leyser 2007; Ongaro and Leyser 2008; Leyser 2009, 2010). It was suggested that for the coordinated control of axillary branching, both auxin and SLs have the ability to change each other's levels and distribution in a dynamic feedback loop (Hayward et al. 2009).

2.2 *Root Development*

SLs have also been shown to affect root development. Based on analysis of mutants flawed in SL biosynthesis or signaling, and the treatment of seedlings with GR24 (a bioactive, synthetic SL; Johnson et al. 1981), SLs were suggested to control lateral root formation and to have a positive effect on root-hair elongation; both root responses to GR24 were shown to be mediated via the MAX2 F-box (Kapulnik et al. 2011; Ruyter-Spira et al. 2011). Moreover, auxin-efflux carriers were shown to be involved in SLs' effect on root growth and root-hair elongation (Koltai et al. 2010b), suggesting, in agreement with the findings of SLs-auxin cross talk in shoot (reviewed by Dun et al. 2009), possible cross-talk junctures between SLs and auxin in controlling root development.

2.3 *Seed Germination*

At the basis of host recognition in several parasitic plant species (e.g., within the Orobanchaceae) is the ability to respond to SLs as seed-germination factors (reviewed by Westwood et al. 2010 and discussed further on). Although an extensive role for SLs in the regulation of seed dormancy and germination in plants has yet to be established, their ability to break seed dormancy was also demonstrated in nonparasitic plant species, including lettuce (*Lactuca sativa*), wild oats (*Avena fatua*) (Bradow et al. 1988, 1990), and *A. thaliana*, in which the involvement of SLs in enhancement of seed germination was demonstrated (Nelson et al., 2011; Tsuchiya et al. 2010).

3 Strigolactones as Signals for Plant Interactions

Even before the identification of SLs as plant hormones, they were identified as germination stimulants of the parasitic plants *Striga* and *Orobanche* (e.g., Cook et al. 1972; Yokota et al. 1998; Matusova et al. 2005; Akiyama and Hayashi 2006; Xie et al. 2007, 2008a, b, 2009a, b; Goldwasser et al. 2008; Gomez-Roldan et al. 2008; recently reviewed by Xie et al. 2010). SLs were also identified as stimulants of hyphal branching in AMF (e.g., Akiyama et al. 2005; Besserer et al. 2006, 2008; Gomez-Roldan et al. 2008; Yoneyama et al. 2008; reviewed by Akiyama and Hayashi 2006; Bouwmeester et al. 2007; García-Garrido et al. 2009; Xie et al. 2010). Hence, two seemingly unrelated biological systems of plant–fungus and plant–plant associations emerged to include SLs as signals for host recognition, while taking advantage of a plant hormone produced by the plant roots. Below, we present each of these plant–fungus and plant–plant biological systems, and discuss SLs as important communication signals in these plant interactions.

3.1 Mycorrhizal Symbiosis

The arbuscular mycorrhizal (AM) symbiosis is an association between the roots of higher plants and the soil AMF. AMF are members of the fungal phylum Glomeromycota (Redecker and Raab 2006), which contains 10 genera of AMF: *Glomus*, *Gigaspora*, *Scutellospora*, *Acaulospora*, *Entrophospora*, *Pacispora*, *Diversispora*, *Archaeospora*, *Geosiphon*, and *Paraglomus* (Redecker and Raab 2006 and references within). AM symbiosis has been suggested to be the most prevalent symbiosis on earth; under suitable conditions, symbiotic associations are formed with most terrestrial vascular flowering plants (Smith and Read 1997). AM symbiosis is suggested to be an ancient plant association: arbuscules were discovered in an early Devonian land plant, suggesting that mycorrhizae were present by the Early Devonian (about 400 million years ago), and probably much earlier (Remy et al. 1994). It is suggested that this symbiosis played an important role in plants' ability to colonize the land (Simon et al. 1993).

Two stages can be discerned during the AMF–host association. The first is the "presymbiotic stage," in which the fungal spore germinates in the soil and the hyphae develop but exhibit limited growth. Then, in the presence of a host root, the hypha starts to branch (Mosse and Hepper 1975; Gianinazzi-Pearson et al. 1989; Giovannetti et al. 1996; Buée et al. 2000; reviewed by Bécard et al. 2004; Requena et al. 2007). Once the hypha comes into contact with a host root, the fungus forms an appressorium, which is a contact structure with the root epidermis, and a prepenetration apparatus is formed by the host root in a cell-layer-controlled fashion (Genre et al. 2005; Siciliano et al. 2007). These morphological changes are likely to necessitate a reciprocal exchange of signals between the fungus and the plant, and are necessary to proceed to the second stage of the symbiosis process, the "symbiotic stage."

In the symbiotic stage, the fungus penetrates the root epidermis and grows within the root cortex, where it forms morphologically distinct, specialized structures. Two morphological types of mycorrhization may be distinguished. One is the *Arum*-type, in which short side branches migrate within the root cortex, penetrate the cortical cell walls, and branch dichotomously to produce characteristic arbuscules (Fig. 1). Despite penetration through the cell wall, the arbuscular hypha remains in the apoplastic compartment of the root cortex, surrounded by an interfacial matrix and the peri-arbuscular membrane. In the second morphological type of mycorrhization, the *Paris*-type, colonization is characterized by the development of extensive intracellular coiled hyphae, with relatively little intracellular growth (Smith and Read 1997).

The association between the fungi and the plant host is considered beneficial. During the symbiosis, bidirectional exchange between the two organisms is established; arbuscules are thought to be the site for nutrient exchange between symbionts (reviewed by, e.g., Harrison 2005; Paszkowski 2006). The main benefit to the host is thought to be associated with the ability of AMF to enhance the plant's ability to bridge the "depletion zone." This zone is established in the soil once mineral concentrations decrease in close proximity to root surfaces; it might be bridged by AMF hyphae, which potentially provide a greater root surface area to exploit a larger volume of soil, thereby leading to an increase in the amount of nutrients available to the plant (Rausch and Bucher 2002). Specifically, the availability and absorbance of phosphate (Pi) is promoted by the mycorrhitic symbiosis, not only by the greater root surface provided by the hyphae, but also by their active import, accumulation, and storage in the fungus vacuoles (Harrison and van Buuren 1995; Maldonado-Mendoza et al. 2001; Rausch and Bucher 2002; Smith et al. 2003, 2004; Benedetto et al. 2005; reviewed by Javot et al. 2007).

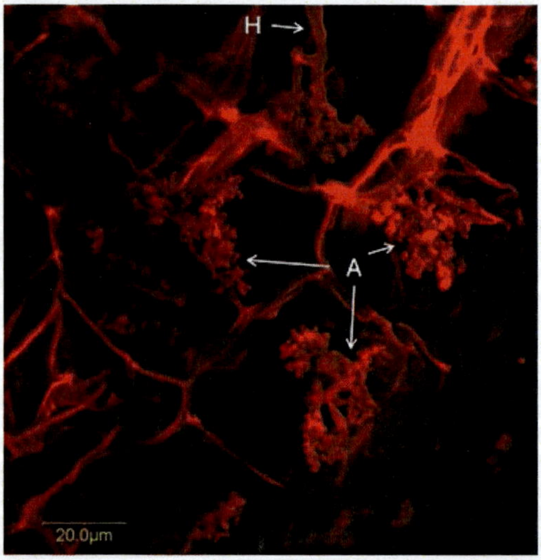

Fig. 1 Structure of arbuscular mycorrhizal fungus *Glomus intraradices* in the root tissues of snapdragon (*Antirrhinum majus*). *A* arbuscules; *H* hyphae. Scale bar: 20 μm

In return, the fungus receives fixed carbon from the host. It has been found that the fungi may take up glucose, whereas this uptake is estimated at between 4 and 20% of the plant's total photosynthetic products, and takes place either in the intracellular hyphae or the arbuscules (reviewed by Douds et al. 2000). It was suggested that the plant can control the flux of carbon to the root and to the fungi via jasmonic acid-related pathways (reviewed by Hause et al. 2007). Either glucose (reviewed by Douds et al. 2000), hexose (Shachar-Hill et al. 1995; Solaiman and Saito 1997), or sucrose was suggested to be delivered to the mycorrhized root. Sucrose was suggested to be cleaved in mycorrhitic roots by either symbiosis-induced sucrose synthases (Hohnjec et al. 2003) or invertases (Schaarschmidt et al. 2006).

The fungal carbon demand has been shown to affect carbon partitioning in the plant. The fungi affect the carbon source, e.g., by increasing host leaf area. They also reduce the root-to-shoot ratio, whereas within the root, mycorrhization affects the level of root carbon and its distribution between soluble and insoluble forms (reviewed by Douds et al. 2000). Toward completion of the fungal life cycle, the fungus produces an extraradical mycelium from which spores are eventually formed (Smith and Read 1997).

3.2 *Strigolactones as Signal Molecules of AM Symbiosis*

It has been suggested that the evolutionary success of AMF in entering into symbiotic associations indicates that they evolved strategies to increase the probability of meeting their host's roots (Harrison 2005). One of these strategies might be AMF hyphal branching in response to root exudates, reflecting communication between the host and fungi to enhance successful mycorrhization (Koske and Gemma 1992).

Despite spore germination and limited development of the fungal hyphae in the absence of a host, AM hyphae only undergo extensive branching in the presence of host roots (Buée et al. 2000). Moreover, root exudates were shown to be sufficient to induce hyphal branching (Giovannetti et al. 1996). These branching factors were first found to be a group of lipophilic compounds (Nagahashi and Douds 2000), and later specified to be SLs; it was shown that both SLs purified from root exudates of *Lotus japonicus* and synthetic SLs are capable of inducing hyphal branching of *Gigaspora margarita* at subnanogram levels (Akiyama and Hayashi 2006; Akiyama et al. 2005). The synthetic SL GR24 was shown to effectively induce AMF branching at 10^{-8} M (Gomez-Roldan et al. 2008). In addition, reduced AMF hyphal branching was found in the presence of root exudates of SL-deficient mutants of pea and tomato compared to the WT root exudates (Gomez-Roldan et al. 2008; Koltai et al. 2010a).

Moreover, SLs were shown to induce mitosis in the AMF *Gigaspora rosea*, as well as mitochondrial activation; the latter was also demonstrated in *Glomus intraradices* (Besserer et al. 2006, 2008, 2009). SLs were shown to induce rapid changes in the shape, density, and motility of the mitochondria (within an hour), as well as in NADH concentrations and dehydrogenase activity, and ATP content (within minutes) in *G. rosea* hyphae, suggesting that SLs rapidly enhance the fungus' energy metabolism; 5 days later, they also led to gene-expression activation in this fungus (Besserer et al. 2006, 2008).

Interestingly, it was suggested that AMF negatively interfere with the production of SLs: the mycorrhized plants were suggested to produce and/or exude less SLs, since they showed reduced susceptibility to *Striga* or *Orobanche* due to a lower level of induction of seed germination (Lendzemo et al. 2007; Fernández-Aparicio et al. 2010). Moreover, strigolactone production was shown to be significantly reduced upon AM symbiosis in tomato (López-Ráez et al. 2011). Hence, SLs may be negatively regulated by AMF via a feedback loop: SLs induce AMF colonization, and AMF colonization then leads to a reduction in SL synthesis or exudation.

It is not yet clear if SLs have any role during the symbiotic stages of AM symbiosis. Moreover, it is not clear whether SLs are essential for the AMF interaction or whether they specify it. These issues are further discussed later.

3.3 Parasitic Weeds

Parasitic plants belonging to the family Orobanchaceae are the most economically important parasitic weeds, as they have a devastating impact on the production of many crop species. Parasitic witchweed (*Striga* spp.) and broomrape (*Orobanche* and *Phelipanche* spp.) are obligate root parasites. These plant pathogens attack their hosts underground and most of the damage is done before the parasites become visible aboveground. Several *Striga*, *Orobanche*, and *Phelipanche* species negatively affect crop production on many millions of hectares, with an estimated billion dollars (US) in crop losses each year (Parker 2009). According to the Food and Agricultural Organization of the United Nations (FAO) (http://www.fao.org/), *Striga* infests the cereal-producing areas of sub-Saharan Africa and causes a 50% yield loss in maize and sorghum. Distribution and estimated crop losses caused by individual parasitic weed species around the globe were recently summarized by Parker (2009).

The broomrapes (*Orobanche* and *Phelipanche*) include about 170 species worldwide (Rumsey and Jury 1991). The root-holoparasitic angiosperm *Phelipanche ramosa* (L.) Pomel (syn. *Orobanche ramosa* L.) attacks economically important crops such as the rapeseeds, tomato, tobacco, potato, carrot, and hemp. *Phelipanche aegyptiaca* (Pers.) Pomel (syn. *Orobanche aegyptiaca* Pers.) parasitizes members of the Solanaceae, Cruciferae, Umbelliferae, Compositae, and Papilionaceae (Eizenberg and Joel 2001). The parasitic plant *Orobanche crenata* Forsk. damages faba bean, pea, lentil, vetches, grass, pea, and other legumes in the Mediterranean region (Rubiales 2001). Some legumes are attacked by *Orobanche foetida* Poir. as well. *Orobanche minor* Smith parasitizes a broad range of plant families, with preference for members of the Fabaceae and Asteraceae (Rumsey and Jury 1991).

Orobanche cumana Wallr. is known for its high host specificity, parasitizing sunflower exclusively (Eizenberg and Joel 2001). All of these species are considered serious threats to crop production, mainly in the Mediterranean region, and warm temperate regions of Europe, North America, and India.

Parasitism not only causes quantitative crop losses, it also influences crop quality. For example, parasitism of *Phelipanche ramosa* causes reductions in tomato fruit weight, mesocarp thickness, sugar and ascorbic acid contents and firmness, as well as changes in fruit color (Longo et al. 2010). The hemiparasitic genus *Striga* (witchweed) comprises 28 species and 6 subspecies in Africa (Mohamed et al. 2001). Of these species, *S. hermonthica* (Del.) Benth., *S. asiatica* (L.) Kuntze, *S. aspera* (Willd.) Benth., and *S. gesnerioides* (Willd.) Vatke cause significant agricultural yield losses in sub-Saharan Africa. *S. hermonthica* has the largest geographical distribution and causes the greatest crop damage. *S. hermonthica, S. asiatica,* and *S. aspera* parasitize grasses and cereals such as maize, sorghum, rice, and proso millet. *S. gesnerioides* parasitizes broad-leaf plants, including cowpea and tobacco (Mohamed et al. 2001; for review Joel et al. 2007 and Parker 2009).

The life cycles of *Striga*, *Phelipanche,* and *Orobanche* species are very similar and tightly linked to the life cycles of their host plants. This coordinated development ensures that parasitic weeds do not germinate at the end of their host's growing season without conditions to finish their life cycle. The communication between parasites and host plants strongly depends on signaling molecules exuded from the roots of host (and nonhost) plants at different stages of development. Among them, the compounds inducing germination of the parasitic plants have been the most studied due to their crucial role in parasite germination. The tiny seeds of weedy parasitic plants contain only a small amount of energy reserves, for the parasitic plant's short autonomous growth period (Joel et al. 1995); within a few days they must attach to a host to acquire nutrients or they die.

Parasitic plants have therefore developed safety mechanisms to prevent germination in the absence of a host plant. Seeds start to germinate upon recognition of some compound(s) (i.e., germination stimulant) exuded from the roots of the host plant. Moreover, the seeds are able to perceive these signaling molecules after incubation for a specified period of time in a warm and moist environment to release dormancy. This period is called conditioning or preconditioning. Optimal temperature during the conditioning period varies among species, corresponding to the temperature at the beginning of their host plant's growing season. During the first few days of conditioning, metabolic activity involving intense respiration, protein synthesis (Bar-Nun and Mayer 1993), carbohydrate metabolism (Bar-Nun and Mayer 2002), and gibberellin synthesis (Zehhar et al. 2002) was observed in *Phelipanche* seeds. In *S. hermonthica,* there was evidence of ethylene biosynthesis during conditioning (Sugimoto et al. 2003). Respiration activity and protein synthesis decreased when the seeds became responsive to the germination stimulant (Bar-Nun and Mayer 1993).

Dormancy release and reinduction in parasitic weed seeds are highly dependent on the conditioning temperature and conditioning period. Upon longer conditioning or in the absence of germination stimulant, secondary dormancy develops and seeds lose their responsiveness to the germination stimulant. This safety mechanism prevents germination of seeds in the absence of their host plant or at the end of its growing season (Matusova et al. 2004).

Once parasitic weed seeds germinate, they must attach to the root of a host plant or die. Germinated seeds attach to and invade the roots of the host plant via a specialized organ called haustorium. The haustorium forms the morphological and physiological bridge between the parasite and the host plant root. It develops from the radicle tip (Joel et al. 2007) to form a globular structure covered by hair-like structures. The latter are more obvious in *Striga* spp. than in *Orobanche* or *Phelipanche* spp. Haustorium formation is induced by specific secondary metabolites that are released into the rhizosphere by the roots of the host plant. Haustorium-initiation factors have been well studied in the facultative parasite *Triphysaria versicolor* (Albrecht et al. 1999; Tomilov et al. 2004) and include simple phenolics, flavonoids, or quinones. Similar molecules have been identified for *Striga* and *Agalinis* (Riopel and Timko 1995). Haustorium formation in germinating *S. asiatica* seeds was induced by 2,6-dimethoxy-*p*-benzoquinone (2,6-DMBQ) (Wolf and Timko 1991) or ethylene (Rich and Ejeta 2007). Attachment of the parasite to the host root does not appear to be specific. *P. ramosa* attaches to host or nonhost roots (Zehhar et al. 2003). After attachment to a host, the haustorium penetrates the host root and establishes connections with the host vascular system, from which it obtains the water and nutrients needed for growth (Parker and Riches 1993). Outside the root the parasite forms a tubercle – a swollen structure that accumulates nutrients (Joel 2000). The tubercle develops underground for several weeks and then shoot outgrowth is initiated. Shoots emerge from the soil, flower, and produce tens of thousands of seeds. Seeds enter into primary dormancy and the next season's life cycle of parasitic plants is launched.

3.4 Strigolactones as Signal Molecules of Parasitic Weeds

Many plants have evolved safety mechanisms (e.g., sensitivity to light, temperature, CO_2 level) to ensure suitable conditions for their germination and development. The angiosperm parasite's strategy involves tight coordination of early developmental stages with chemical signals from the host plant. Induction of germination by compounds exuded by the host plant's roots ensures that the seeds of the parasitic plant will germinate near the host root and that the parasite will be able to reach the host root within several days, attach to it and develop further. Several compounds have been found to stimulate germination of parasitic plant seeds under laboratory conditions. Chemicals found in smoke and coumarin promote germination of the parasitic weed *P. aegyptiaca* (Bar-Nun and Mayer 2005) or ethylene for *S. hermonthica* (Eplee 1975).

In nature, parasitic plants have learned to sense specific chemical molecules exuded by the roots of potential host plants at extremely low concentrations. It is interesting that the parasitic *Striga*, *Orobanche*, and *Phelipanche* spp. require structurally similar signaling molecules to induce germination. Three types of isoprenoid compounds have been described that stimulate germination of root-parasitic plants: dihydrosorgoleone, the sesquiterpene lactones, and the SLs (Bouwmeester et al. 2003). Most of the germination stimulants identified to date have been SLs.

Strigol and strigyl acetate were the first naturally occurring germination stimulants for *Striga* isolated from the false host cotton (Cook et al. 1966, 1972). In 1993, Siame et al. (1993) detected strigol in root exudates of the true *Striga* hosts maize, sorghum, and proso millet. Hauck et al. (1992) described sorgolactone as a major *Striga* germination stimulant from sorghum root exudates and alectrol from root exudates of cowpea (Műller et al. 1992). The name "strigolactones" was proposed for these strigol-related compounds (Butler 1995). In the following years, the SL orobanchol was isolated from red clover (Yokota et al. 1998), sorgomol from sorghum (Xie et al. 2008b), and 5-deoxystrigol from *L. japonicus* (Akiyama et al. 2005), sorghum, maize, pearl millet (Awad et al. 2006), and several Fabaceae species (Yoneyama et al. 2008). Many other SLs from host plants and root exudates of the nonhost plant white lupin (Yoneyama et al. 2008) or root culture of *Menispermum dauricum* (Yasuda et al. 2003) were reported. According to the proposed SL biosynthetic pathway (Matusova et al. 2005), 5-deoxystrigol might be a common precursor for SL. Indeed, conversion of 5-deoxystrigol to sorgolactone with the intermediate precursor sorgomol was suggested (Xie et al. 2008b).

The growing evidence of SLs' presence in diverse plant species in recent years indicates their wide distribution in the plant kingdom (Bouwmeester et al. 2003; Yoneyama et al. 2006). Today, more than 14 SLs have been identified and characterized in root exudates of many host and nonhost plant species (Yoneyama et al. 2009). The SLs identified so far contain a tricyclic SL backbone (ABC part) coupled to butenolide (D-ring) via an enol ether bridge (Yoneyama et al. 2010). The D-ring may have a key function in SL's biological activity (Wigchert and Zwanenburg 1999). The SLs are active at extremely low concentrations (on the order of 10^{-7}–10^{-15} M; Joel 2000). A variety of natural SLs have been able to induce germination of *O. minor* from 10 pM SL (for orobanchol, 2′-epiorobanchol, and sorgomol) to 10 nM for 7-oxoorobanchol. The synthetic analog GR24 is 100-fold less active than natural SLs (Kim et al. 2010). Advances in chromatography and mass spectometry are enabling the discovery and characterization of novel SLs.

3.5 *Other Rhizosphere Organisms*

It is quite surprising that among the multitude of organisms in the rhizosphere that interact with plants, only two (i.e., AMF and parasitic weeds) have acquired the ability to recognize their host through SLs. Hence, a search is being conducted for other soilborne fungi that may recognize and respond to SLs. So far, no effect of SLs on hyphal branching has been recorded in soilborne fungi, including *Rhizoctonia solani*, *Fusarium oxysporum*, and *Verticillium dahlia* (Steinkellner et al. 2007). However, a role was discovered for SLs in rhizobia nodulation (Foo and Davis, 2011; Soto et al. 2010).

4 Strigolactone Secretion and Stability in the Soil

Root-secreted secondary metabolites are used to regulate the rhizosphere. They are either used to the detriment of neighboring plants through allelopathy or are exploited by other plants and microorganisms to initiate their development. However, despite the ecophysiological significance of plant-secreted compounds and the large number of compounds produced by plant roots, very little is known about the molecular mechanisms involved in the regulation of root exudation.

Several reports suggest a higher secretion of SLs when mycotrophic plants are exposed to low Pi levels relative to plants exposed to higher Pi levels (Yoneyama et al. 2007a, b; López-Ráez and Bouwmeester 2008; López-Ráez et al. 2008). Yoneyama et al (2007a) showed that under limited supply of Pi (and N), SL contents in both sorghum root tissues and root exudates increase, suggesting that low Pi and N conditions increase both SL production and secretion; once produced in the roots, SLs appear to be rapidly secreted. In addition, under Pi deficiency, the exudation of SLs in red clover was significantly stimulated (Yoneyama et al. 2007b).

However, reports suggest that root exudation of amino acids and reducing sugars is also greater for plants grown under Pi deficiency vs. those grown under high Pi conditions. This increase in root exudation was suggested to be a result of changes in the membrane permeability of Pi-deficient roots, rather than of changes in the content of these compounds in the root (Graham et al. 1981).

Moreover, the mechanism for Pi inhibition of AM formation was associated with a membrane-mediated decrease in root exudation. Furthermore, it was demonstrated that subsequent mycorrhizal infection was highly correlated with initial differences in root exudation (Ratnayake et al. 1978). These biological findings can now be explained, at least in part, by SL availability in the rhizosphere: the suggested higher AMF infection under low Pi conditions may correlate with an increase in SL availability in the rhizosphere, due to higher membrane permeability of the root cells under these conditions. Hence, one cannot exclude the possibility that the higher SL exudation to the rhizosphere under low Pi conditions is the result of increased membrane permeability, rather than of a well-controlled and specific mechanism for increased SL secretion.

It is not yet known where exactly along the longitudinal root axis SLs are exuded. It is widely recognized that the gradual maturation of root tissues along the root axis is not the only variation in metabolic activity. Yet, from available information, it can be concluded that the pattern of exudation is not homogeneous along the root axis. For example, release of phytosiderophores in response to Fe deficiencies appears to concentrate in the apical root zone (Marschner 1995), while release of organic anions follows a heterogeneous pattern along the roots (Hoffland et al. 1989).

To fulfill a role in plant interactions, following root exudation, SLs have to have a reasonable level of stability to be recognized by microorganisms or other plants in the rhizosphere; a critical factor in host location for AMF and parasitic weeds is the lifetime of individual SLs in the rhizosphere. Although little is known about SL stability in soil, some indication of it was revealed by water-degradation experiments, suggesting that SL stability differs considerably between natural and synthetic SLs (Akiyama et al. 2010).

5 Evolutionary Aspects of Strigolactones

5.1 Evolution of Strigolactones' Biological Role

Whether SLs were originally produced by the plant to control organogenesis or to promote symbiosis is still not known. It seems reasonable to assume that they were not produced by the plant to promote plant–parasitic interactions. On the one hand, SLs appear to have a pivotal role in determining plant architecture. This role is crucial for plant survival and productivity. For example, for plants to survive and produce, their stems must have the ability to respond to damage by growing lateral shoot branches, which may be induced by alterations in SL levels. Another example is associated with the plant's response to light quality and intensity. These two factors have been shown to affect multiple processes in plants, including the shade-avoidance response, which includes enhanced apical dominance (Franklin 2008; Pierik et al. 2009). This relationship between light quality and branching was suggested to be mediated by effects of light on auxin fluxes and auxin response (reviewed by Leyser 2009). It might be that SLs, which have been identified as shoot-branching inhibitors and as mediators of auxin flux (e.g., see review by Leyser 2010), are involved in this response. Such pivotal roles for SLs may have been a major force in their evolution, suggesting that the primary role of SLs is plant morphogenesis.

On the other hand, the involvement of SLs in AMF symbiosis may suggest an ancient role for SLs as signals in plant–fungus interactions (Wilkinson 2001). AMF have been dated as far back as 400 million years: an arbuscular fossil was found in *Aglaophytes major*, a Devonian plant showing features of both vascular plants and bryophytes from the early Devonian (Remy et al. 1994). Moreover, fossilized fungal hyphae and spores that strongly resemble those of the modern genus *Glomus* were found in Ordovician rocks in Wisconsin dating back 460 million years; these fossilized fungi might have been free-living or perhaps they formed mycorrhizal-like relationships with the bryophytes, the first terrestrial plants, predating the first vascular plants on earth (Redecker et al. 2000).

It has been suggested that the first plants were so closely associated with fungi that they could, in fact, be considered as partly composed of fungus: a blueprint for terrestrial plants was drawn once an endosymbiotic relationship between a fungus and an alga was established (Pirozynski and Malloch 1975), implying that the mycorrhizal mutualism was vital in allowing plants to colonize the land. This places the symbiotic relationship with AMF as a crucial step in evolution. Hence, it might be that SLs were produced by primitive land plants very early in evolution to promote the establishment of a symbiosis that was essential for terrestrial life. Undoubtedly, characterization of SL production and its role in gymnosperms and nonvascular plants will shed light on the evolutionary history of SLs, and on their primary role as either plant morphogen or promoter of an essential symbiosis.

By the same token, understanding the mechanism of SL secretion may provide more clues to their primary role. Active secretion from the roots by a controlled transport mechanism may support a primary role for SLs in the establishment of symbioses; passive SL dispersal, i.e., via diffusion or leakage from the plant root, may suggest that their primary role is in plant morphogenesis, while the plant interactors simply exploited a leaking substance from the roots for selection of their plant host.

The dependence of parasitic weeds on SLs for their germination provides a case in which a plant substance dedicated to a certain function (i.e., plant development or symbiosis establishment) may be exploited by another organism, to serve an alternative function. It is suggested that because SL-mediated regulation of branching appears to be a fundamental and thus ancient mechanism in angiosperms, both nonparasitic and parasitic plants shared this signaling pathway. The ability of SLs to positively affect seed germination was also demonstrated for some nonparasitic species (discussed earlier). Hence, the SL-mediated regulator of germination in parasitic plants may be derived from a preexisting mechanism of self-regulation of plant germination and development. However, adaptation of this pathway for host detection by the parasitic plant species may have necessitated modifications in their biology. These modifications may have included parasitic species stopping their SL biosynthesis, either permanently or temporarily, or specification of SL responses such that the parasitic plants could distinguish endogenous SLs from exogenous ones, for recognition of a plant host (reviewed by Westwood et al. 2010).

5.2 *Does Strigolactone Confer Plant-Interaction Specificity?*

Different plant species and even cultivars produce and secrete different combinations of SLs (reviewed by Xie et al. 2010; Yoneyama et al. 2008). Hence, SLs may be associated with determination of interaction specificity between the plant host and the AMF or parasitic weed.

Although many land plants have established symbiosis with AMF, several families have been reported as non-AMF–host plants; these include the Chenopodiaceae, the Brassicaceae, and lupins [*Lupinus albus*, which is an exception in the mycorrhizal host family of the Fabaceae (or Leguminosae)] (Smith and Read 1997). The basis for plants being nonhosts was suggested to be either the existence of inhibitors of AMF symbiosis in those plants or the lack of a substance important for AMF symbiosis (reviewed by Giovannetti and Sbrana 1998; Vierheilig and Bago 2005). Accordingly, root exudates of non-AMF–host plants were not able to induce AMF hyphal branching (Buée et al. 2000). Since SLs are known to be promoters of AMF symbiosis, it might be that they are at least partially responsible for the non-AMF–host phenotype.

Nevertheless, similar SLs were identified in root exudates of AMF host plants and two non-AMF–host plants – *A. thaliana* (Brassicaceae) and lupin (Fabaceae) (Goldwasser et al. 2008; Yoneyama et al. 2008). However, neither phosphorus nor nitrogen deficiency increased exudation of these SLs in *L. albus*. This is in contrast to the mycotrophic Fabaceae plant *T. pratense*, in which phosphorus deficiency promoted SL exudation (Yoneyama et al. 2008). Hence, it might be that the regulation of SL production and/or exudation, rather than their composition, plays a role in host/nonhost recognition. However, more studies are needed to determine whether SLs indeed confer host capacity for AM symbiosis.

Similarly, there is no clear evidence for specificity of SLs for parasitic plants. Many parasitic plants of the Orobanchaceae have a broad host range. For example, *P. ramosa* parasitizes tobacco in which (+)-orobanchol, 2′-epi-orobanchol, solanacol (a tetradehydrostrigol isomer), and a didehydrostrigol isomer were identified (Xie et al. 2007). Tomato is producing orobanchol, solanacol and two or three didehydroorobanchol isomers (López-Ráez et al. 2008), and *Arabidopsis* is producing orobanchol (Goldwasser et al. 2008). These host plants differ in the composition and quantity of SLs produced, with no proof of different SL functions in parasitic weed germination and/or development. The difficulty is caused by the complex mixture of molecules exuded by plant roots and other factors in the rhizosphere that influence the germination of parasitic plants and cannot be omitted in evaluating SLs' effects on parasitic plant development underground. Another problem is the extremely low production of SLs by individual plants, making their isolation difficult for studies of natural SLs. Eleven naturally occurring SLs induced germination of *O. minor* seeds, but the number of germinated seeds depended on lipophilicity and stability of the SL molecules (Kim et al. 2010). Lower efficiency of sorgolactone and 5-deoxystrigol in germination-stimulating activity on *O. minor* seeds could be explained by a lack of oxygen-containing substituents on the A/B ring moiety (Kim et al. 2010). Similar experiments using several parasitic weed seeds collected from several host plants with known SL production might be very helpful in determining the potential specificity of different SLs in germination of parasitic weed seeds.

More data suggest nonspecific behavior of parasitic plants at early stages of their development. Seed germination can be induced by root exudates of host and nonhost plants and seeds are able to connect to the roots of host and nonhost plants (e.g., Zehhar et al. 2003). At a later stage, multiple layers of compatibility/incompatibility may contribute to host specificity in parasitic plants (Yoshida and Shirasu 2009; Thorogood and Hiscock 2010).

Interestingly, determination of differential SL activity on the AMF *G. margarita*'s hyphal branching suggested structural requirements for induction of hyphal branching which are similar, but not identical, to those required for root-parasitic weeds (Akiyama et al. 2010). This may suggest that each organism, i.e., parasitic plants and AMF, developed the ability to exploit the host SLs independently, leading to similar but not identical requirements for their interaction.

5.3 *Are SLs Essential Components of Plant Interactions?*

SLs may not be an essential signal for AMF. Both spore germination and some degree of hyphal development and branching occur even without the host or its root exudates in the vicinity (reviewed by Giovannetti et al. 2010). Chance encounters between the developing hyphae and the host root, even in the absence of SLs, occur. If SLs are associated only with the presymbiosis steps of AM symbiosis (and this remains to be determined), they may not be essential for mycorrhization.

On the other hand, SL is an essential signal for parasitic weeds. Other compounds can induce germination of parasitic plants (Eplee 1975; Bar-Nun and Mayer 2005; Daws et al. 2008), but these compounds do not ensure their germination in the presence of host plants, which is a prerequisite for the existence of plants that are fully dependent on nutrients acquired from the host. Perception of SLs, compounds produced by plants in very low quantities, guarantees germination in the presence of, and close to only host roots.

6 Concluding Remarks

SLs fulfill several key roles in plant development and interactions; these are summarized in Fig. 2. SLs' role as regulators of shoot morphology and root development suggest a pivotal role for these hormones in the regulation of an array of plant responses, some of which are essential for plant survival and reproduction. It is likely that as SL-related research progresses, more developmental roles will be unveiled for SLs, as well as their mode of integration into the hormonal network for coordinated regulation of plant development.

Despite the suggested essential nature of AM symbiosis to the evolution of terrestrial life, it appears that SLs are not essential for this symbiosis. Rather, they may be considered enhancers of this interaction, which may take place even without them. Whether SLs evolved to serve as enhancers of AM symbiosis or as regulators of plant development remains to be determined. On the other hand, it is likely that SLs' pivotal role in weed plant parasitism reflects exploitation of a preexisting plant substance which was either actively or passively exuded from the host roots, and used by the parasite to select its host. It is possible that SLs affect additional soil microorganisms; this, however, remains to be determined.

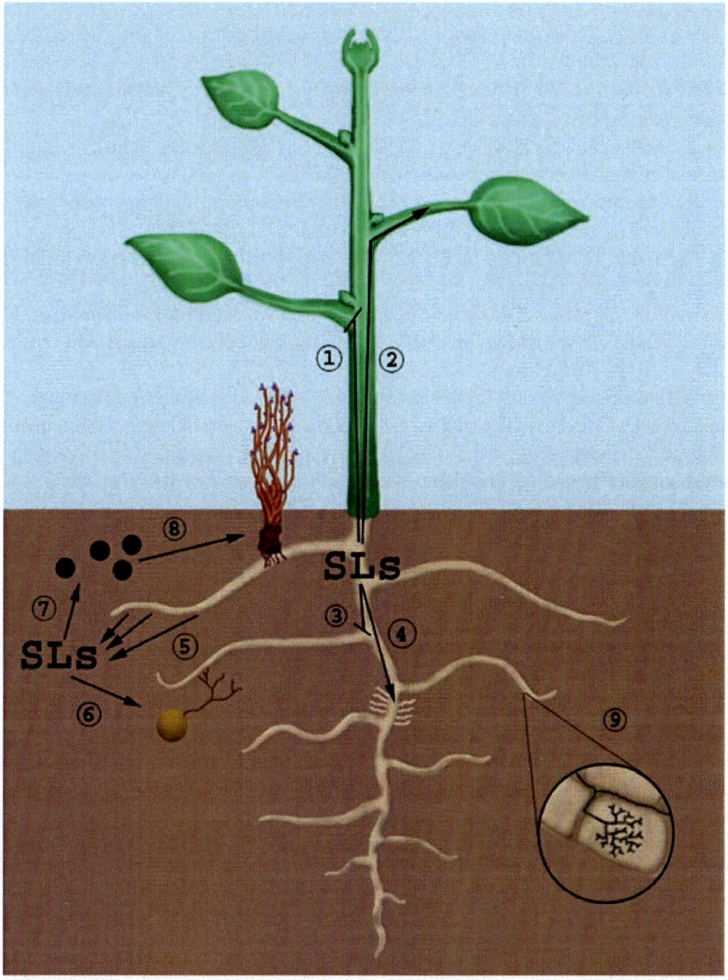

Fig. 2 An illustration of the presently known biological functions of strigolactones (SLs). SLs play a role in plant development: Aboveground, SLs repress shoot branching (1) but potentially enhance plant light harvesting (2). In roots, SLs are suggested to control lateral root formation (3) and have a positive effect on root-hair elongation (4). SLs are released from the roots to the rhizosphere (5), where they induce arbuscular mycorrhiza hyphal branching (6) and parasitic weed seed germination (7); the latter is an essential step in the establishment of parasitic weed–host associations (8). Whether SLs have a role in the symbiotic phase of mycorrhizal associations, however, remains to be determined (9). Figure illustrated by Omer Koltai

Today, additional roles and functions for SLs are being revealed. It is expected that we will soon know more on SLs' biological role, on the components involved in SL signaling and synthesis, and on SL evolution.

References

Akiyama K, Hayashi H (2006) Strigolactones: chemical signals in fungal symbionts and parasitic weeds in plant roots. Ann Bot 97:925–931

Akiyama K, Matsuzaki K, Hayashi H (2005) Plant sesquiterpenes induce hyphal branching in arbuscular mycorrhizal fungi. Nature 435:824–827

Akiyama K, Ogasawara S, Ito S, Hayashi H (2010) Structural requirements of strigolactones for hyphal branching in AM fungi. Plant Cell Physiol 51:1104–1117

Albrecht H, Yoder JY, Philips DA (1999) Flavonoids promote haustoria formation in the root parasite *Triphysaria versicolor*. Plant Physiol 119:585–591

Arite T, Iwata H, Ohshima K, Maekawa M, Nakajima M, Kojima M, Sakakibara H, Kyozuka J (2007) DWARF10, an RMS1/MAX4/DAD1 ortholog, controls lateral bud outgrowth in rice. Plant J 51:1019–1029

Auldridge ME, Block A, Vogel JT, Dabney-Smith C, Mila I, Bouzayen M, Magallanes-Lundback M, DellaPenna D, McCarty DR, Klee HJ (2006) Characterization of three members of the Arabidopsis carotenoid cleavage dioxygenase family demonstrates the divergent roles of this multifunctional enzyme family. Plant J 45:982–993

Awad AA, Sato D, Kusumoto D, Kamioka H, Takeuchi Y, Yoneyama K (2006) Characterization of strigolactones, germination stimulants for the root parasitic plants Striga and Orobanche, produced by maize, millet and sorghum. Plant Growth Regul 48:221–227

Bainbridge K, Sorefan K, Ward S, Leyser O (2005) Hormonally controlled expression of the Arabidopsis MAX4 shoot branching regulatory gene. Plant J 44:569–580

Bar-Nun N, Mayer AM (1993) Preconditioning and germination of *Orobanche* seeds: respiration and protein synthesis. Phytochemistry 34:39–45

Bar-Nun N, Mayer AM (2002) Composition of and changes in storage compounds in *Orobanche aegyptiaca* seeds during preconditioning. Isr J Plant Sci 50:277–279

Bar-Nun N, Mayer AM (2005) Smoke chemicals and coumarin promote the germination of the parasitic weed *Orobanche aegyptiaca*. Isr J Plant Sci 53:97–101

Bécard G, Kosuta S, Tamasloukht M, Séjalon-Delmas N, Roux C (2004) Partner communication in the arbuscular mycorrhizal interaction. Can J Bot 82:1186–1197

Benedetto A, Magurno F, Bonfante P, Lanfranco L (2005) Expression profiles of a phosphate transporter gene (GmosPT) from the endomycorrhizal fungus *Glomus mosseae*. Mycorrhiza 15:620–627

Bennett T, Sieberer T, Willett B, Booker J, Luschnig C, Leyser O (2006) The Arabidopsis MAX pathway controls shoot branching by regulating auxin transport. Curr Biol 16:553–563

Besserer A, Puech-Pages V, Kiefer P, Gomez-Roldan V, Jauneau A, Roy S, Portais J, Roux C, Becard G, Sejalon-Delmas N (2006) Strigolactones stimulate arbuscular mycorrhizal fungi by activating mitochondria. PLoS Biol 4:1239–1247

Besserer A, Bécard G, Jauneau A, Roux C, Séjalon-Delmas N (2008) GR24, a synthetic analog of strigolactones, stimulates the mitosis and growth of the arbuscular mycorrhizal fungus *Gigaspora rosea* by boosting its energy metabolism. Plant Physiol 148:402–413

Besserer A, Bécard G, Roux C, Séjalon-Delmas N (2009) Role of mitochondria in the response of arbuscular mycorrhizal fungi to strigolactones. Plant Signal Behav 4:75–77

Beveridge CA, Kyozuka J (2010) New genes in the strigolactone-related shoot branching pathway. Curr Opin Plant Biol 13:34–39

Beveridge CA, Symons GM, Murfet IC, Ross JJ, Rameau C (1997) The *rms1* mutant of pea has elevated indole-3-acetic acid levels and reduced root sap zeatin riboside content but increased branching controlled by graft transmissible signal(s). Plant Physiol 15:1251–1258

Beveridge CA, Symons GM, Turnbull CG (2000) Auxin inhibition of decapitation-induced branching is dependent on graft-transmissible signals regulated by genes Rms1 and Rms2. Plant Physiol 123:689–698

Booker J, Auldridge M, Wills S, McCarty D, Klee H, Leyser O (2004) MAX3/CCD7 is a carotenoid cleavage dioxygenase required for the synthesis of a novel plant signaling molecule. Curr Biol 14:1232–1238

Booker J, Sieberer T, Wright W, Williamson L, Willett B, Stirnberg P, Turnbull C, Srinivasan M, Goddard P, Leyser O (2005) MAX1 encodes a cytochrome P450 family member that acts downstream of MAX3/4 to produce a carotenoid-derived branch-inhibiting hormone. Dev Cell 8:443–449

Bouwmeester HJ, Matusova R, Zhongkui S, Beale MH (2003) Secondary metabolite signalling in host-parasitic plant interactions. Curr Opin Plant Biol 6:358–364

Bouwmeester HJ, Roux C, Lopez-Raez JA, Becard G (2007) Rhizosphere communication of plants, parasitic plants and AM fungi. Trends Plant Sci 12:224–230

Bradow JM, Connick WJ, Pepperman AB (1988) Comparison of the seed germination effects of synthetic analogs of strigol, gibberellic acid, cytokinins, and other plant growth regulators. J Plant Growth Regul 7:227–239

Bradow JM, Connick W, Pepperman AB, Wartelle L (1990) Germination stimulation in wild oats (*Avena fatua* L.) by synthetic strigol analogs and gibberellic acid. J Plant Growth Regul 9:35–41

Brewer PB, Dun EA, Ferguson BJ, Rameau C, Beveridge CA (2009) Strigolactone acts downstream of auxin to regulate bud outgrowth in pea and Arabidopsis. Plant Physiol 150:482–493

Buée M, Rossignol M, Jauneau A, Ranjeva R, Bécard G (2000) The pre-symbiotic growth of arbuscular mycorrhizal fungi is induced by a branching factor partially purified from plant root exudates. Mol Plant Microbe Interact 13:693–698

Butler LG (1995) Chemical communication between the parasitic weed Striga and its crop host, vol 582nd edn, ACS symposium series. American Chemical Society, Washington, DC, pp 158–168

Cook CE, Whichard LP, Turner B, Wall ME, Egley GH (1966) Germination of witchweed (*Striga lutea* Lour.): isolation and properties of a potent stimulant. Science 154:189–1190

Cook CE, Whichard LP, Wall ME, Egley GH, Coggon P, Luhan PA, McPhail AT (1972) Germination stimulants. 2. The structure of strigol – a potent seed germination stimulant for witchweed (*Striga lutea* Lour.). J Am Chem Soc 94:6198–6199

Daws MI, Pritchard HW, van Staden J (2008) Butenolide from plant-derived smoke functions as a strigolactone analogue: evidence from parasitic weed seed germination. S Afr J Bot 74:116–120

Dor E, Alperin B, Wininger S, Ben-Dor B, Somvanshi VS, Koltai H, Kapulnik Y, Hershenhorn J (2010) Characterization of a novel tomato mutant resistant to *Orobanche* and *Phelipanche* spp. weedy parasites. Euphytica 171:371–380

Douds DDJ, Pfeffer PE, Shachar-Hill Y (2000) Carbon partitioning, cost, and metabolism of arbuscular mycorrhizas. In: Kapulnik Y, Douds DDJ (eds) Arbuscular mycorrhizas: physiology and function. Kluwer Academic, Dordrecht, pp 107–129

Drummond RS, Martínez-Sánchez NM, Janssen BJ, Templeton KR, Simons JL, Quinn BD, Karunairetnam S, Snowden KC (2009) *Petunia hybrida* carotenoid cleavage dioxygenase7 is involved in the production of negative and positive branching signals in petunia. Plant Physiol 151:1867–1877

Dun EA, Brewer PB, Beveridge CA (2009) Strigolactones: discovery of the elusive shoot branching hormone. Trends Plant Sci 14:364–372

Eizenberg H, Joel DM (2001) *Orobanche* species in Israeli agriculture. in: COST Action 849, Parasitic plant management in sustainable agriculture. Workshop 'State of the art in *Orobanche* control', Bari, Italy, 18–20 Oct 2001

Eplee RE (1975) Ethylene: a witchweed seed germination stimulant. Weed Sci 23:433–436

Ferguson BJ, Beveridge CA (2009) Roles for auxin, cytokinin and strigolactone in regulating shoot branching. Plant Physiol 149:1929–1944

Fernández-Aparicio M, García-Garrido JM, Ocampo JA, Rubiales D (2010) Colonisation of field pea roots by arbuscular mycorrhizal fungi reduces *Orobanche* and *Phelipanche* species seed germination. Weed Res 50:262–268

Foo E, Turnbull CG, Beveridge CA (2001) Long-distance signaling and the control of branching in the rms1 mutant of pea. Plant Physiol 126:203–209

Foo E, Davies NW. 2011. Strigolactones promote nodulation in pea. Planta, in press. DOI 10.1007/s00425-011-1516-7

Franklin KA (2008) Shade avoidance. New Phytol 179:930–944

García-Garrido JM, Lendzemo V, Castellanos-Morales V, Steinkellner S, Vierheilig H (2009) Strigolactones, signals for parasitic plants and arbuscular mycorrhizal fungi. Mycorrhiza 19:449–459

Genre A, Chabaud M, Timmers T, Bonfante P, Barker DG (2005) Arbuscular mycorrhizal fungi elicit a novel intracellular apparatus in *Medicago truncatula* root epidermal cells before infection. Plant Cell 17:3489–3499

Gianinazzi-Pearson V, Branzanti B, Gianinazzi S (1989) In vitro enhancement of spore germination and early hyphal growth of a vesicular-arbuscular mycorrhizal fungus by host root exudates and plant flavonoids. Symbiosis 7:243–255

Giovannetti M, Sbrana C (1998) Meeting a non-host: the behaviour of AM fungi. Mycorrhiza 8:123–130

Giovannetti M, Sbrana C, Citernesi AS, Avio L (1996) Analysis of factors involved in fungal recognition responses to host-derived signals by arbuscular mycorrhizal fungi. New Phytol 133:65–71

Giovannetti M, Avio L, Sbrana C (2010) Fungal spore germination and pre-symbiotic mycelial growth – physiological and genetic aspects. In: Koltai H, Kapulnik Y (eds) Arbuscular mycorrhizas: physiology and function, 2nd edn. Springer, Dorderecht, pp 3–32

Goldwasser Y, Yoneyama K, Xie X, Yoneyama K (2008) Production of strigolactones by *Arabidopsis thaliana* responsible for *Orobanche aegyptiaca* seed germination. Plant Growth Regul 55:21–28

Gomez-Roldan V, Fermas S, Brewer PB, Puech-Pagès V, Dun EA, Pillot JP, Letisse F, Matusova R, Danoun S, Portais JC, Bouwmeester H, Bécard G, Beveridge CA, Rameau C, Rochange SF (2008) Strigolactone inhibition of shoot branching. Nature 455:189–194

Graham JH, Leonard RT, Menge JA (1981) Membrane-mediated decrease of root-exudation responsible for phosphorus inhibition of vesicular-arbuscular mycorrhiza formation. Plant Physiol 6:548–552

Harrison MJ (2005) Signaling in the arbuscular mycorrhizal symbiosis. Annu Rev Microbiol 59:19–42

Harrison MJ, van Buuren ML (1995) A phosphate transporter from the mycorrhizal fungus *Glomus versiforme*. Nature 378:626–629

Hauck C, Műller S, Schildknecht H (1992) A germination stimulant produced for parasitic flowering plants from *Sorghum bicolor*, a genuine host plant. J Plant Physiol 139:474–478

Hause B, Mrosk C, Isayenkov S, Strack D (2007) Jasmonates in arbuscular mycorrhizal interactions. Phytochemistry 68:101–110

Hayward A, Stirnberg P, Beveridge C, Leyser O (2009) Interactions between auxin and strigolactone in shoot branching control. Plant Physiol 151:400–412

Hoffland E, Findenegg GR, Nelemans JA (1989) Solubilization of rock phosphate by rape. II. Local root exudation of organic acids as a response to P-starvation. Plant Soil 113:161–165

Hohnjec N, Perlick AM, Pühler A, Küster H (2003) The *Medicago truncatula* sucrose synthase gene *MtSucS1* is activated both in the infected region of root nodules and in the cortex of roots colonized by arbuscular mycorrhizal fungi. Mol Plant Microbe Interact 16:903–915

Javot H, Pumplin N, Harrison MJ (2007) Phosphate in the arbuscular mycorrhizal symbiosis: transport properties and regulatory roles. Plant Cell Environ 30:310–322

Joel DM (2000) The long-term approach to parasitic weeds control: manipulation of specifc developmental mechanisms of the parasite. Crop Prot 19:753–758

Joel DM, Steffens JC, Matthews DM (1995) Germination of weedy root parasites. In: Kigel J, Galili G (eds) Seed development and germination. Marcel Dekker, New York, pp 567–597

Joel DM, Hershenhorn J, Eizenberg H, Aly R, Ejeta G, Rich P, Ransom J, Sauerborn J, Rubiales D (2007) Biology and management of weedy root parasites. In: Janick J (ed) Horticultural reviews, vol 33. Wiley, New Jersey, pp 267–349

Johnson AW, Gowda G, Hassanali A, Knox J, Monaco S, Razavi Z, Rosebery G (1981) The preparation of synthetic analogues of strigol. J Chem Soc Perkin Trans 1:1734–1743

Johnson X, Brcich T, Dun EA, Goussot M, Haurogné K, Beveridge CA, Rameau C (2006) Branching genes are conserved across species. Genes controlling a novel signal in pea are coregulated by other long-distance signals. Plant Physiol 142:1014–1026

Kapulnik Y, Delaux P-M, Resnick N, Mayzlish-Gati E, Wininger S, Bhattacharya C, Séjalon-Delmas N, Combier J-P, Bécard G, Belausov E, Beeckman T, Dor E, Hershenhorn J, Koltai H (2011) Strigolactones affect lateral root formation and root hair elongation in Arabidopsis. Planta 233(1):209–216

Kim HI, Xie X, Kim HS, Chun JC, Yoneyama K, Nomura T, Takeuchi Y, Yoneyama K (2010) Structure-activity relationship of naturally occurring strigolactones in *Orobanche minor* seed germination stimulation. J Pestic Sci 35:344–347

Koltai H, LekKala SP, Bhattacharya C, Mayzlish-Gati E, Resnick N, Wininger S, Dor E, Yoneyama K, Yoneyama K, Hershenhorn J, Joel DM, Kapulnik Y (2010a) A tomato strigolactone-impaired mutant displays aberrant shoot morphology and plant interactions. J Exp Bot 61:1739–1749

Koltai H, Dor E, Hershenhorn J, Joel DM, Weininger S, Lekalla S, Shealtiel H, Bahattacharya C, Eliahu E, Resnick N, Barg R, Kapulnik Y (2010b) Strigolactones' effect on root growth and root-hair elongation may be mediated by auxin-efflux carriers. J Plant Growth Regul 29:129–136

Koske RE, Gemma JN (1992) Fungal reactions to plants prior to mycorrhizal formation. In: Allen MF (ed) Mycorrhizal functioning: an integrative plant-fungal process. Chapman & Hall, New York, pp 3–36

Lendzemo VW, Kuyper TW, Matusova R, Bouwmeester HJ, Van Ast A (2007) Colonization by arbuscular mycorrhizal fungi of sorghum leads to reduced germination and subsequent attachment and emergence of *Striga hermonthica*. Plant Signal Behav 2:58–62

Leyser O (2009) The control of shoot branching: an example of plant information processing. Plant Cell Environ 32:694–703

Leyser O (2010) The power of auxin in plants. Plant Physiol 154:501–505

Liang J, Zhao L, Challis R, Leyser O (2010) Strigolactone regulation of shoot branching in chrysanthemum (*Dendranthema grandiflorum*). J Exp Bot 61:3069–3078

Longo AMG, Lo Monaco A, Mauromicale G (2010) The effect of *Phelipanche ramosa* infection on the quality of tomato fruit. Weed Res 50:58–66

López-Ráez JA, Bouwmeester H (2008) Fine-tuning regulation of strigolactone biosynthesis under phosphate starvation. Plant Signal Behav 3:963–965

López-Ráez JA, Charnikhova T, Gómez-Roldán V, Matusova R, Kohlen W, De Vos R, Verstappen F, Puech-Pages V, Bécard G, Mulder P, Bouwmeester H (2008) Tomato strigolactones are derived from carotenoids and their biosynthesis is promoted by phosphate starvation. New Phytol 178:863–874

López-Ráez JA, Charnikhova T, Fernandez I, Bouwmeester H, Pozo MJ (2011) Arbuscular mycorrhizal symbiosis decreases strigolactone production in tomato. J Plant Physiol 168(3):294–297

Maldonado-Mendoza IE, Dewbre GR, Harrison MJ (2001) A phosphate transporter gene from the extra-radical mycelium of an arbuscular mycorrhizal fungus *Glomus intraradices* is regulated in response to phosphate in the environment. Mol Plant Microbe Interact 14:1140–1148

Marschner H (1995) Mineral nutrition of higher plants, 2nd edn. Academic, San Diego

Matusova R, van Mourik T, Bouwmeester HJ (2004) Changes in the sensitivity of parasitic weed seeds to germination stimulants. Seed Sci Res 14:335–344

Matusova R, Rani K, Verstappen FW, Franssen MC, Beale MH, Bouwmeester HJ (2005) The strigolactone germination stimulants of the plant-parasitic *Striga* and *Orobanche* spp. are derived from the carotenoid pathway. Plant Physiol 139:920–934

Mayzlish-Gati E, LekKala SP, Resnick N, Wininger S, Bhattacharya C, Lemcoff JH, Kapulnik Y, Koltai H (2010) Strigolactones are positive regulators of light-harvesting genes in tomato. J Exp Bot 61:3129–3136

Mohamed KI, Musselman LJ, Riches CR (2001) The genus *Striga* (Scrophulariaceae) in Africa. Ann MI Bot Gard 88:60–103

Mosse B, Hepper C (1975) Vesicular-arbuscular mycorrhizal infections in root organ cultures. Physiol Plant Pathol 5:215–223

Mouchel CF, Leyser O (2007) Novel phytohormones involved in long-range signaling. Curr Opin Plant Biol 10:473–476

Műller S, Hauck C, Schildknecht H (1992) Germination stimulants produced by *Vigna unguiculata* Walp cv Saunders Upright. J Plant Growth Regul 11:77–84

Napoli CA (1996) Highly branched phenotype of the petunia dad1-1 mutant is reversed by grafting. Plant Physiol 111:27–37

Nagahashi G, Douds Jr DD (2000) Partial separation of root exudate components and their effects upon the growth of germinated spores of AM fungi. Mycological Research 104:1453–64.

Nelson DC, Scaffidi A, Dun EA, Waters MT, Flematti GR, Dixon KW, Beveridge CA, Ghisalberti EL, Smith SM 2011 F-box protein MAX2 has dual roles in karrikin and strigolactone signaling in Arabidopsis thaliana. Proc Natl Acad Sci USA 108:8897–902

Ongaro V, Leyser O (2008) Hormonal control of shoot branching. J Exp Bot 59:67–74

Parker C (2009) Observations on the current status of *Orobanche* and *Striga* problems worldwide. Pest Manag Sci 65:453–459

Parker C, Riches CR (1993) Parasitic weeds of the world: biology and control. CAB International, Wallingford

Paszkowski U (2006) A journey through signaling in arbuscular mycorrhizal symbioses. New Phytol 172:35–46

Pierik R, Keuskamp DH, Sasidharan R, Djakovic-Petrovic T, de Wit M, Voesenek LA (2009) Light quality controls shoot elongation through regulation of multiple hormones. Plant Signal Behav 4:755–756

Pirozynski KA, Malloch DW (1975) The origin of land plants: a matter of mycotrophism. Biosystems 6:153–164

Ratnayake M, Leonard RT, Menge JA (1978) Root exudation in relation to supply of phosphorus and its possible relevance to mycorrhizal formation. New Phytol 81:543–552

Rausch C, Bucher M (2002) Molecular mechanisms of phosphate transport in plants. Planta 216:23–37

Redecker D, Kodner R, Graham LE (2000) Glomalean fungi from the Ordovician. Science 289:1920–1921

Redecker D, Raab P (2006) Phylogeny of the glomeromycota (arbuscular mycorrhizal fungi): recent developments and new gene markers. Mycologia 98:885–895

Remy W, Taylor TN, Hass H, Kerp H (1994) Four hundred-million-year-old vesicular arbuscular mycorrhizae. Proc Natl Acad Sci USA 91:11841–11843

Requena N, Serrano E, Ocon A, Breuninger M (2007) Plant signals and fungal perception during arbuscular mycorrhiza establishment. Phytochemistry 68:33–40

Rich PJ, Ejeta G (2007) Biology of host-parasite interactions in *Striga* species. In: Ejeta G, Gressel J (eds) Integrating new technologies for *Striga* control. Towards ending the witch-hunt. World Scientific, Singapore, pp 19–32

Riopel JL, Timko MP (1995) Haustorial initiation and differentiation. In: Press MC, Graves JD (eds) Parasitic plants. Chapman and Hall, London, pp 39–79

Rubiales D (2001) Parasitic plants: an increasing threat. Grain Legumes 33:10–11

Rumsey FJ, Jury SL (1991) An account of *Orobanche* L. in Britain and Ireland. Watsonia 18:257–295

Ruyter-Spira C, Kohlen W, Charnikhova T, et al. 2011 Physiological effects of the synthetic strigolactone analog GR24 on root system architecture in Arabidopsis: Another below-ground role for strigolactones? Plant Physiology, 155:721–34.

Schaarschmidt S, Roitsch T, Hause B (2006) Arbuscular mycorrhiza induces gene expression of the apoplastic invertase LIN6 in tomato (*Lycopersicon esculentum*) roots. J Exp Bot 57:4015–4023
Schwartz SH, Qin X, Loewen MC (2004) The biochemical characterization of two carotenoid cleavage enzymes from *Arabidopsis* indicates that a carotenoid-derived compound inhibits lateral branching. J Biol Chem 279:46940–46945
Shachar-Hill Y, Pfeffer PE, Douds D, Osman SF, Doner LW, Ratcliffe RG (1995) Partitioning of intermediate carbon metabolism in VAM colonized leek. Plant Physiol 108:7–15
Shen H, Luong P, Huq E (2007) The F-box protein MAX2 functions as a positive regulator of photomorphogenesis in Arabidopsis. Plant Physiol 145:1471–1483
Siame BA, Weerasuriya Y, Wood K, Ejeta G, Hitler LG (1993) Isolation of strigol, a germination stimulant for *Striga asiatica*, from host plants. J Agric Food Chem 41:1488–1491
Siciliano V, Genre A, Balestrini R, Cappellazzo G, deWit PJ, Bonfante P (2007) Transcriptome analysis of arbuscular mycorrhizal roots during development of the prepenetration apparatus. Plant Physiol 144:1455–1466
Simon L, Bousquet J, Levesque RC, Lalonde M (1993) Origin and diversification of endomycorrhizal fungi and coincidence with vascular land plants. Nature 363:67–69
Smith SE, Read DJ (1997) Mycorrhizal symbiosis. Academic, San Diego
Smith SE, Smith FA, Jakobsen I (2003) Mycorrhizal fungi can dominate phosphate supply to plants irrespective of growth responses. Plant Physiol 133:16–20
Smith SE, Smith FA, Jakobsen I (2004) Functional diversity in arbuscular mycorrhizal (AM) symbioses: the contribution of the mycorrhizal P uptake pathway is not correlated with mycorrhizal responses in growth or total P uptake. New Phytol 162:511–524
Snowden KC, Simkin AJ, Janssen BJ, Templeton KR, Loucas HM, Simons JL, Karunairetnam S, Gleave AP, Clark DG, Klee HJ (2005) The *Decreased apical dominance1/Petunia hybrida carotenoid cleavage dioxygenase8* gene affects branch production and plays a role in leaf senescence, root growth, and flower development. Plant Cell 17:746–759
Solaiman MD, Saito M (1997) Use of sugars by intraradical hyphae of arbuscular mycorrhizal fungi revealed by radiorespirometry. New Phytol 136:533–538
Sorefan K, Booker J, Haurogné K, Goussot M, Bainbridge K, Foo E, Chatfield S, Ward S, Beveridge C, Rameau C, Leyser O (2003) *MAX4* and *RMS1* are orthologous dioxygenase-like genes that regulate shoot branching in *Arabidopsis* and pea. Genes Dev 17:1469–1474
Soto MJ, Fernández-Aparicio M, Castellanos-Morales V, García-Garrido JM, Ocampo JA, Delgado MJ, Vierheilig H. 2010. First indications for the involvement of strigolactones on nodule formation in alfalfa (Medicago sativa). Soil Biology and Biochemistry 42: 383–385.
Steinkellner S, Lendzemo V, Langer I, Schweiger P, Khaosaad T, Toussaint JP, Vierheilig H (2007) Flavonoids and strigolactones in root exudates as signals in symbiotic and pathogenic plant-fungus interactions. Molecules 12:1290–1306
Stirnberg P, Furner IJ, Leyser O (2007) MAX2 participates in an SCF complex which acts locally at the node to suppress shoot branching. Plant J 50:80–94
Sugimoto Y, Ali AM, Yabuta S, Kinoshita H, Inanaga S, Itai A (2003) Germination strategy of *Striga hermonthica* involves regulation of ethylene biosynthesis. Physiol Planta 119:137–145
Thorogood CJ, Hiscock SJ (2010) Compatibility interactions at the cellular level provide the basis for host specificity in the parasitic plant Orobanche. New Phytol 186:571–575
Tomilov A, Tomilova N, Yoder JI (2004) In vitro haustorium development in roots and root cultures of the hemiparasitic plant *Triphysaria versicolor*. Plant Cell Tissue Organ Cult 77:257–265
Tsuchiya Y, Vidaurre D, Toh S, Hanada A, Nambara E, Kamiya Y, Yamaguchi S, McCourt P (2010) A small-molecule screen identifies new functions for the plant hormone strigolactone. Nat Chem Biol 6:741–749
Umehara M, Hanada A, Yoshida S, Akiyama K, Arite T, Takeda-Kamiya N, Magome H, Kamiya Y, Shirasu K, Yoneyama K, Kyozuka J, Yamaguchi S (2008) Inhibition of shoot branching by new terpenoid plant hormones. Nature 455:195–200

Vierheilig H, Bago B (2005) Host and non-host impact on the physiology of the symbiosis. In: Declerck S, Strullu S, Fortin A (eds) Root-organ cultures of mycorrhizal fungi. Springer, Heidelberg, pp 139–158

Vogel JT, Walter MH, Giavalisco P, Lytovchenko A, Kohlen W, Charnikhova T, Simkin AJ, Goulet C, Strack D, Bouwmeester HJ, Fernie AR, Klee HJ (2010) SlCCD7 controls strigolactone biosynthesis, shoot branching and mycorrhiza-induced apocarotenoid formation in tomato. Plant J 61:300–311

Waldie T, Hayward A, Beveridge CA (2010) Axillary bud outgrowth in herbaceous shoots: how do strigolactones fit into the picture? Plant Mol Biol 73:27–36

Westwood JH, Yoder JI, Timko MP, dePamphilis CV (2010) The evolution of parasitism in plants. Trends Plant Sci 15:227–235

Wigchert SCM, Zwanenburg B (1999) A critical account on the inception of Striga seed germination. J Agric Food Chem 47:1320–1325

Wilkinson DM (2001) Mycorrhizal evolution. Trends Ecol Evol 16:64–65

Wolf SJ, Timko MP (1991) In vitro root culture: a novel approach to study the obligate parasite *Striga asiatica* (L.) Kunte. Plant Sci 73:233–242

Xie X, Kusumoto D, Takeuchi Y, Yoneyama K, Yamada Y, Yoneyama K (2007) 2′-Epi-orobanchol and solanacol, two unique strigolactones, germination stimulants for root parasitic weeds, produced by tobacco. J Agric Food Chem 55:8067–8072

Xie X, Yoneyama K, Kusumoto D, Yamada Y, Takeuchi Y, Sugimoto Y, Yoneyama K (2008) Sorgomol, germination stimulant for root parasitic plants, produced by *Sorghum bicolor*. Tetrahedron Lett 49:2066–2068

Xie X, Yoneyama K, Harada Y, Fusegi N, Yamada Y, Ito S, Yokota T, Takeuchi Y, Yoneyama K (2009a) Fabacyl acetate, a germination stimulant for root parasitic plants from *Pisum sativum*. Phytochemistry 70:211–215

Xie X, Yoneyama K, Kurita J, Harada Y, Yamada Y, Takeuchi Y, Yoneyama K (2009b) 7-Oxoorobanchyl acetate and 7-oxoorobanchol as germination stimulants for root parasitic plants from flax (*Linum usitatissimum*). Biosci Biotechnol Biochem 73:1367–1370

Xie X, Yoneyama K, Yoneyama K (2010) The strigolactone story. Annu Rev Phytopathol 48:93–117

Yasuda N, Sugimoto Y, Kato M, Inanaga S, Yoneyama K (2003) (+)-Strigol, a witchweed seed germination stimulant, from *Menispermum dauricum* root culture. Phytochemistry 62:1115–1119

Yokota T, Sakai H, Okuno K, Yoneyama K, Takeuchi Y (1998) Alectrol and orobanchol, germination stimulants for *Orobanche minor*, from its host red clover. Phytochemisty 49:1967–1973

Yoneyama K, Sato D, Takeuchi Y, Sekimoto H, Yokota T, Sassa T (2006) Search for germination stimulants and inhibitors for root parasitic weeds. In: Duke SO, Rimando A (eds) Natural products for pest management. American Chemical Society, Washington, DC, pp 88–98

Yoneyama K, Xie X, Kusumoto D, Sekimoto H, Sugimoto Y, Takeuchi Y, Yoneyama K (2007a) Nitrogen deficiency as well as phosphorus deficiency in sorghum promotes the production and exudation of 5-deoxystrigol, the host recognition signal for arbuscular mycorrhizal fungi and root parasites. Planta 227:125–132

Yoneyama K, Yoneyama K, Takeuchi Y, Sekimoto H (2007b) Phosphorus deficiency in red clover promotes exudation of orobanchol, the signal for mycorrhizal symbionts and germination stimulant for root parasites. Planta 225:1031–1038

Yoneyama K, Xie X, Sekimoto H, Takeuchi Y, Ogasawara S, Akiyama K, Hayashi H, Yoneyama K (2008) Strigolactones, host recognition signals for root parasitic plants and arbuscular mycorrhizal fungi, from Fabaceae plants. New Phytol 179:484–494

Yoneyama K, Xie X, Yoneyama K, Takeuchi Y (2009) Strigolactones: structures and biological activities. Pest Manag Sci 65:467–470

Yoneyama K, Awad AA, Xie X, Yoneyama K, Takeuchi Y (2010) Strigolactones as germination stimulants for root parasitic plants. Plant Cell Physiol 51:1095–1103

Yoshida S, Shirasu K (2009) Multiple layers of incompatibility to the parasitic witchweed, *Striga hermonthica*. New Phytol 183:180–189

Zehhar N, Ingouff M, Bouya D, Fer A (2002) Possible involvement of gibberellins and ethylene in *Orobanche ramosa* germination. Weed Res 42:464–469

Zehhar N, Labrousse P, Arnaud M-C, Boulet C, Bouya D, Fer A (2003) Study of resistance to *Orobanche ramosa* in host (oilseed rape and carrot) and non-host (maize) plants. Eur J Plant Pathol 109:75–82

Zou J, Zhang S, Zhang W, Li G, Chen Z, Zhai W, Zhao X, Pan X, Xie Q, Zhu L (2006) The rice *high-tillering dwarf1* encoding an ortholog of *Arabidopsis MAX3* is required for negative regulation of the outgrowth of axillary buds. Plant J 48:687–698

Proteoid Roots and Exudation of Proteases by Plant Roots

Bartosz Adamczyk, Aino Smolander, Veikko Kitunen, and Mirosław Godlewski

Abstract Among the numerous strategies for improving P or N uptake by plants, development of proteoid roots and exudation of proteases can be found. Proteoid roots, which develop as a response to P deficiency during exudative burst secrete carboxylates and acid phosphatase, which improves P uptake. Proteoid root morphology and anatomy, factors that influence development and their role in plant nutrition, are described. In addition, this chapter summarizes our knowledge of the recently discovered phenomenon of protease secretion by intact plant roots, including their biochemical characterization and their potential role in the nitrogen nutrition of plants.

1 Proteoid Roots

Plants possess numerous strategies for improving their ability to survive in adverse conditions. One of the crucial points for survival is the need to obtain enough nutrients for growth and development. The number of plant adaptations for acquiring nutrients from the environment is impressive. The broad set of strategies for improving nutrient uptake includes formation of mycorrhiza, carnivory, cluster roots formation, formation of N_2-fixing nodules, and exudation of low and high molecular mass compounds. This section deals with proteoid roots (also called cluster roots) and their secretome.

B. Adamczyk (✉) • A. Smolander • V. Kitunen
Finnish Forest Research Institute, Vantaa Research Unit, 18, 01301 Vantaa, Finland
e-mail: bartek_adamczyk_79@o2.pl

M. Godlewski
Department of Plant Cytology and Cytochemistry, University of Łódź, 90-237 Łódź, Poland

J.M. Vivanco and F. Baluška (eds.), *Secretions and Exudates in Biological Systems*, Signaling and Communication in Plants 12, DOI 10.1007/978-3-642-23047-9_4,

1.1 History and Distribution of Proteoid Roots in the Plant Kingdom

In 1894, Adolf Engler observed atypical, clustered roots of plants in the family Proteaceae, but it was Helen M. Purnell who first called such specific roots "proteoid roots" (the term "proteoid roots" was made from the family name of these plants) (Purnell 1960). Plants in the genus *Persoonia* are an exception in that they do not create proteoid roots, as do the rest of the Proteaceae (Lamont 1982). Apart from the Proteaceae, proteoid roots were also found in some of the Betulaceae, Casurinaceae, Cucurbitaceae, Cyperaceae, Eleagnaceae, Leguminosae (Fabaceae), Mimosaceae, Moraceae, Myricaceae, and Restionaceae (Skene 1998; Neumann and Martinoia 2002; Shane and Lambers 2005). As these roots are not present only in the Proteaceae family, some authors called them cluster roots (Skene 1998). The geographical distribution of plants having cluster roots is very wide, covering all continents and even some Pacific islands. Although plants able to develop proteoid roots can be found worldwide, they usually occupy ecosystems with low levels of nutrients.

1.2 Morphology, Anatomy, and Development of Proteoid Roots

Proteoid roots can easily be recognized, because they are extensively branched and covered with densely grouped absorption hairs (Watt and Evans 1999). The morphology of proteoid roots is strongly dependent on the plant species. Clusters can form on roots singly (e.g., *Lupinus albus*) or be more complex, with a root within a cluster becoming the axis for another cluster. More complex, mat-like structures can also be observed (e.g., *Banksia* spp.) (Watt and Evans 1999) (Fig. 1a, b). It is striking that plants able to form cluster roots usually do not form mycorrhizal symbioses. However, plants from the family Casuarinaceae, for example, can form N_2-fixing nodules, develop mycorrhiza and develop cluster roots in response to nutrient deficiency (Diem et al. 2000).

Some plants in the family Cyperaceae develop dauciform roots (also called carrot-shaped) instead of cluster roots. The function of dauciform roots is exactly the same as that of cluster roots (see Sect. 1.3), but their structure is different (Fig. 1c). Dauciform roots are swollen lateral roots with an abundance of long dense root hairs (Davies et al. 1973), and they like cluster roots can exudate carboxylates and acid phosphatase (Playsted et al. 2006).

Proteoid roots increase the root surface area substantially, for example, *Hakea obliqua* proteoid roots extended the surface area of the roots by about 25 times (Dell et al. 1980). However, if the root-hair area is included in the comparison, this estimate increases to 140 times (Lamont 2003). Depending on plant species, mature cluster roots may be physiologically active for several days (Shane and Lambers 2005). During that period, they undergo "exudative burst," during which many

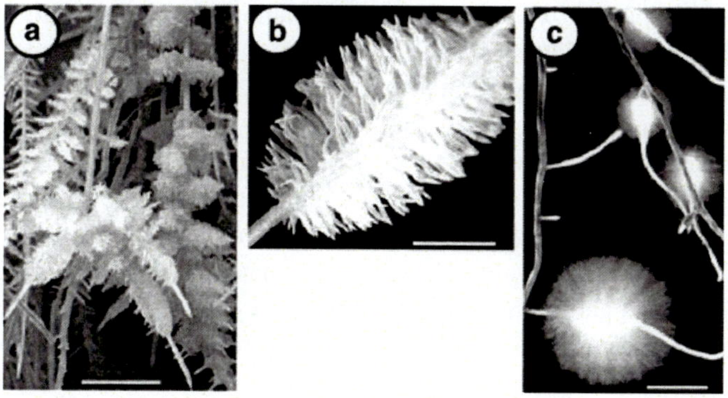

Fig. 1 Morphology of cluster roots and dauciform roots. (**a**) Proteoid roots (compound type) of *Banksia prionotes* (Proteaceae), scale bar 13 mm; (**b**) Proteoid root (simple type) of *Hakea prostrata*; (Proteaceae) scale bar 4 mm; (**c**) Dauciform roots of *Lepidosperma squamatum* (Cyperaceae) scale bar 2 mm (produced by Dr Michael W Shane, taken from Lambers et al. 2006)

compounds, including organic acids, phenolics, and acid phosphatases, are released (Watt and Evans 1999); after that the clusters senesce (Shane and Lambers 2005).

The meristems of rootlets arise from the pericycle cells located over the protoxylem poles, which is similar to the development of regular branch roots (Shane and Lambers 2005). Appearance of clusters is stimulated in conditions of P deficiency (Dinkelaker et al. 1995) and in some species in Fe deficiency (Arahou and Diem 1997), N deficiency (Lamont 1972), or K deficiency (Liang and Li 2003). Moreover, external factors are also involved. It was shown that cluster-root formation can be stimulated in nutrient-rich patches (Skene 1998) and in the soil organic layer (Lamont 1973). However, the influence of other factors such as the constituents of dissolved organic matter and microbial factors should be taken into consideration. Moreover, species-to-species differences in the pattern of cluster-root formation cannot be excluded (Neumann and Martinoia 2002).

Cluster-root formation is tightly controlled by auxin–cytokinin interactions. It has been shown that auxin promotes cluster formation, and that auxin antagonists and cytokinins inhibit cluster formation (Neumann and Martinoia 2002). Skene and James (2000) induced formation of cluster roots in *L. albus* with auxin, even when P was supplied at a level that inhibits their development. Watt and Evans (1999) suspected that ethylene may also be involved in the auxin signal for cluster-root formation. According to the studies of Zhou et al. (2008), sugar signaling mediates cluster-root formation in *L. albus*. In spite of multiple factors inducing the development of cluster roots and the wide array of their structures in different species, cluster roots seem to be devoted mainly to one function – improving nutrient uptake.

1.3 Role of Proteoid Roots in Plant Nutrition

According to the current stage of knowledge, proteoid roots improve the uptake of nutrients by plants. Studies were done on improving uptake of P by exudation of huge amounts of carboxylates, phenolics, mucilage, acid phosphatase, and phytase (e.g., Tarafdar and Claassen 2001; Lambers et al. 2006). Uptake of P can be significantly improved by formation of cluster roots. The uptake rate of P was 2–13 times higher in cluster roots than in ordinary roots (Lamont 1982). Carboxylates (e.g., acetate, *cis*- and *trans*-aconitate, citrate, fumarate, lactate, malate, maleate, malonate) mobilize inorganic and organic forms of phosphorus by complexing metal cations that bind to P (Gardner et al. 1982; Roelofs et al. 2001). Carboxylates can also displace phosphate from the soil matrix by ligand exchange (Shane and Lambers 2005). Tricarboxylates release inorganic P through ligand exchange or complexation of metal ions more effectively than mono- and di-carboxylates do (Dinkelaker et al. 1989), therefore citrate has a greater affinity for trivalent and divalent metals Fe^{3+} and Ca^{2+} than, e.g., malate does (Shane and Lambers 2005). Along with carboxylates via the plasma membrane, ATPase protons are exuded to maintain charge balance (Hinsinger 2001; Hinsinger et al. 2003). In P-deficient conditions, proteoid roots can also exude phenolic compounds (e.g., isoflavonoids) that can mobilize P by reducing phosphates that are bound to minerals (Shane and Lambers 2005). Moreover, mucilage can also act similarly to carboxylates (Neumann and Römheld 2001).

Roots are able to adjust exudation of certain carboxylates to the source of P in the soil. When P was supplied as aluminum phosphate, cluster roots of *Banksia grandis* were exuding di- and tricarboxylates, but when P was supplied as iron phosphate, mainly monocarboxylates were exuded (Lambers et al. 2002). Root clusters also modify their secretome depending on the type of deficiency – in conditions of Fe deficiency, cluster roots of lupin (*L. albus*) secreted mainly malate in conditions of P deficiency, mainly citrate (Liang and Li 2003).

After mobilization, inorganic phosphorus can easily be taken up by roots, but organic phosphorus needs to be hydrolyzed before that (George et al 2002). Acid phosphatases (EC 3.1.3.2) and phytases (hydrolyzing phytate) (EC. 3.1.3.26) are secreted by soil microbes (Tarafdar and Claassen 2001) but also by plant roots, with or without cluster roots (Adams and Pate 1992; Li et al. 1997; Tarafdar and Claassen 2001, 2005). The role of secreted enzymes in plant P nutrition could be highly relevant, because organic P accounts for 30–80% of the total P in soil (Tarafdar and Claassen 1988). Secretion of acid phosphatase from lupin roots is induced by P deficiency (Wasaki et al. 2008). Wasaki et al. (2003) showed that acid phosphatase is secreted from whole lupin roots, not only cluster roots; however, cluster roots secreted a vast amount of this enzyme in P-deficient conditions. Wasaki et al. (2003) proposed an interesting mechanism for the response of lupin roots to different levels of phosphorus: (1) in conditions of sufficient P, acid phosphatase is synthesized at low level and localized on the epidermis of the roots, (2) in conditions of slight P deficiency, acid phosphatase from the epidermis is secreted immediately, (3) in P deficient conditions, cluster roots are developed; and the expression and secretion of acid phosphatase to the soil increase substantially.

Acid phosphatase, secreted by lupin roots, was precisely characterized by Miller et al. (2001), who found that secreted acid phosphatase is a glycoprotein having a

31-amino acid presequence that targets it outside the cell. The molecular mass of the processed protein was about 49 kDa. Expression of genes encoding this enzyme is induced in conditions of P deficiency and is specific to cluster roots (Miller et al. 2001). Ozawa et al. (1995) showed that acid phosphatase secreted by cluster roots of lupin has a broad substrate specificity and stability at pH 4–9 (Ozawa et al. 1995). Acid phosphatase secreted by cluster roots seems to be very stable in the soil solution as its half-life is about 14 days (Tadano et al. 1993).

Cluster roots improve P uptake not only by increasing the amount of easily accessible P but also by improving P uptake directly, because they possess in the plasma membrane high density of high affinity P transporters (Neumann et al. 2000). According to the studies of Liu et al. (2001), one of the P transporters from lupin (*LaPT1*) is expressed only under P deficiency and mainly in cluster roots.

It has been suggested that proteoid roots also improve the ability of plants to use soil organic nitrogen as N source. Paungfoo-Lonhienne et al. (2009) showed that the roots of *H. actities* did not develop cluster roots in P-deficient conditions, but in conditions where there was lack of N, they observed numerous cluster roots. A growth-limiting supply of inorganic nitrogen or glycine resulted in production of a small number of small cluster roots, and supplementation with protein (bovine serum albumin) resulted in numerous small cluster roots. Such a morphological difference as a response to varying N supply indicates that cluster roots are also included in the plant strategy for obtaining N. For *L. albus* (Hawkins et al. 2005) and *H. actities* (Schmidt and Stewart 1997), uptake of glycine by cluster roots was higher than uptake by regular roots. Moreover, Schmidt and Stewart (1997) suggested that *H. actities* cluster roots can exude proteases, which was then proven by Paungfoo-Lonhienne et al. (2008). Cluster roots of *H. actities* possess amino acid transporters (Schmidt et al. 2003) and also peptide transporters (Paungfoo-Lonhienne et al. 2009), which point to their possible role in increasing N uptake. Wasaki et al. (2005) suggested that even chitinase (EC 3.2.1.14) may be secreted by cluster roots of white lupin. Such chitinase could potentially play a role in antifungal plant–pathogen interactions but also might improve N accessibility by digestion of chitin from the soil organic layer.

The proven ability of cluster roots to mobilize nutrients, especially P, can be used effectively in agriculture. Dinkelaker et al. (1995) showed that wheat intercropped with *L. albus* increased not only P uptake but also N and Mn uptake. Soybean intercropped with *L. albus* increased uptake of Cu, Fe, and Zn but not that of P (Braum and Helmke 1995). Plants of *Zea mays* grown after *Brassica napus* or *Beta vulgaris* took up more P in the presence of the previous crop's residue (Dessougi et al. 2003).

Cluster roots, extensively branched and covered with densely grouped absorption hairs, are developed in some plant families (e.g., Proteaceae) in response to nutrient deficiency. Morphology of proteoid roots is determined by plant species, but also depends on environmental conditions. A large proportion of plants able to develop cluster roots cannot produce mycorrhizal associations, and proteoid roots are somehow an alternative product. Improvement of P uptake by cluster roots is their best-known function. Cluster roots secrete a vast amount of carboxylates that mobilize both inorganic and organic forms of phosphorus. Moreover, acid phosphatase secreted by cluster roots digests organic phosphorus, thus creating an easily accessible pool of P. P-deficient conditions activate P transporters in cluster roots, making them effective P suppliers. According to the latest studies, cluster roots can also improve N uptake by taking up organic nitrogen (amino acids) more intensively than regular roots and by exudation of proteases. Cluster roots, a remarkable example of plants' ability to adapt to unfriendly conditions, require more study.

2 Secretion of Proteases by Plant Roots

In traditional soil science it was assumed that plants can take up only inorganic forms of N (i.e., ammonium and nitrate). However, studies from the last two decades clearly showed that certain plants can take up not only inorganic N forms but also amino acids. Effectiveness of the use of amino acids from the soil depends on many factors, e.g., plant species, soil concentration of amino acids, competition with microorganisms and mycorrhizal associations (reviewed by Lipson and Näsholm 2001; Näsholm et al. 2009). According to the latest findings, *H. actities* and *Arabidopsis thaliana* can take peptides up in intact form (Rentsh et al. 2007; Paungfoo-Lonhienne et al. 2009). However, the most substantial part of N is present in the soil in the form of proteins (Kaye and Hart 1997). To break down this rich source of N, proteolytic activity is needed. Proteases present in the soil are secreted by microorganisms (Elfstrand et al. 2007; Nannipieri et al. 2000). Secretion of proteases by plant roots was observed a few years ago by Professor Mirosław Godlewski (University of Łódź, Poland). In a study concerning a different problem, photographic film (exposed, fixed, and developed) was immersed in the sterile medium of hydroponically cultivated seedlings. After a few days of incubation, the film (containing gelatin) was digested. This effect was observed for several plant species (Adamczyk et al. 2004). In the following years, studies of the exudation of proteases by plant roots continued (Godlewski and Adamczyk 2007, Adamczyk et al. 2008a, 2009a, 2010a, b; Paungfoo-Lonhienne et al. 2008).

In this section of the chapter, we will present results showing that roots of several plants are also able to secrete proteases, and that they can improve their N budget by secretion of proteases as if they are improving the P budget using root-secreted acid phosphatase (see Sect. 1.3). Moreover, we will also present the biochemical characteristics of root-secreted proteases.

2.1 *Secretion of Proteases by Different Plant Species*

Secretion of proteases by roots seems to be prevalent in many plant species, both agricultural and wild species (Adamczyk et al. 2008b, 2010b; Godlewski and Adamczyk 2007; Paungfoo-Lonhienne et al. 2008); however, tree species have not been studied. Crucial differences were found in the level of proteolytic activity of such root-secreted proteases; thus this activity seems to be species dependent. Differences between two cultivars of *Cucumis sativus* (Hela vs. Julian) indicate that the level of activity of root-secreted proteases may also be cultivar dependent (Godlewski and Adamczyk 2007). In addition, environmental factors can also influence proteolytic activity of root-secreted proteases, e.g., availability of different sources of N (Adamczyk et al. 2008a, 2010a).

2.2 *Potential Role of Root-Secreted Proteases in Plant Nitrogen Nutrition*

2.2.1 Biochemical Characteristics of Root-Secreted Proteases

To understand the role of root-secreted proteases it is important to study their biochemical characteristics. Contrary to the well-known endogenic plant proteases relatively little information on the biochemistry of plant-secreted proteases is available. Until now we know only the pH preferences, the class to which these proteases belong and their mode of action on substrate.

Proteases secreted by roots of *Allium porrum*, *Helianthus annuus,* and *Z. mays* showed activity in the pH range 4–8, with maximum activity close to neutral pH (Godlewski and Adamczyk 2007). Such a pH range is characteristic for soil: from high pH, which can be found in agricultural soil (sometimes even 8 pH – as found by, e.g., Floch et al. 2009), to low pH, which is typical for boreal forest soils (sometimes even less than 4 pH; Tamminen and Derome 2005). Moreover, such a pH range for protease activity (pH 4–8) is typical for cysteine proteases (Nduwimana et al. 1995). In further studies, has been confirmed that root-secreted proteases belong to that protease family (Godlewski and Adamczyk 2007). Root-secreted proteases showed substantial inhibition of activity after preincubation with inhibitors of this class – iodoacetamide and E-64 (*N*-(*trans*-epoxysuccinyl)-L-leucine 4-guanidinobutylamide) and lack of inhibition, or poor inhibition after preincubation with inhibitors of other classes (Godlewski and Adamczyk 2007). Moreover other studies also suggest that this protease belongs to the cysteine protease family; identification of protease with LC-MS (liquid chromatography-mass spectrometry) and PEAKS online software showed that root-secreted protease of *A. porrum* has amino acid sequence similar to that of *A. thaliana* cysteine proteinase (Adamczyk et al. 2009a).

The ability of root-secreted proteases to digest proteins to a size that can be taken up by plant roots is a crucial factor for the role of root-secreted proteases in plant nitrogen nutrition. In LC-MS studies it was shown that the digestion products of bovine serum albumin and casein, which are mainly low molecular mass fragments, are released by *A. porrum* root-secreted proteases (Adamczyk et al. 2009a) (Fig. 2). Such a pattern of digestion can be explained by the endopeptidase activity of root-secreted proteases. Proteases, on the basis of their mode of action, are usually divided into two groups:

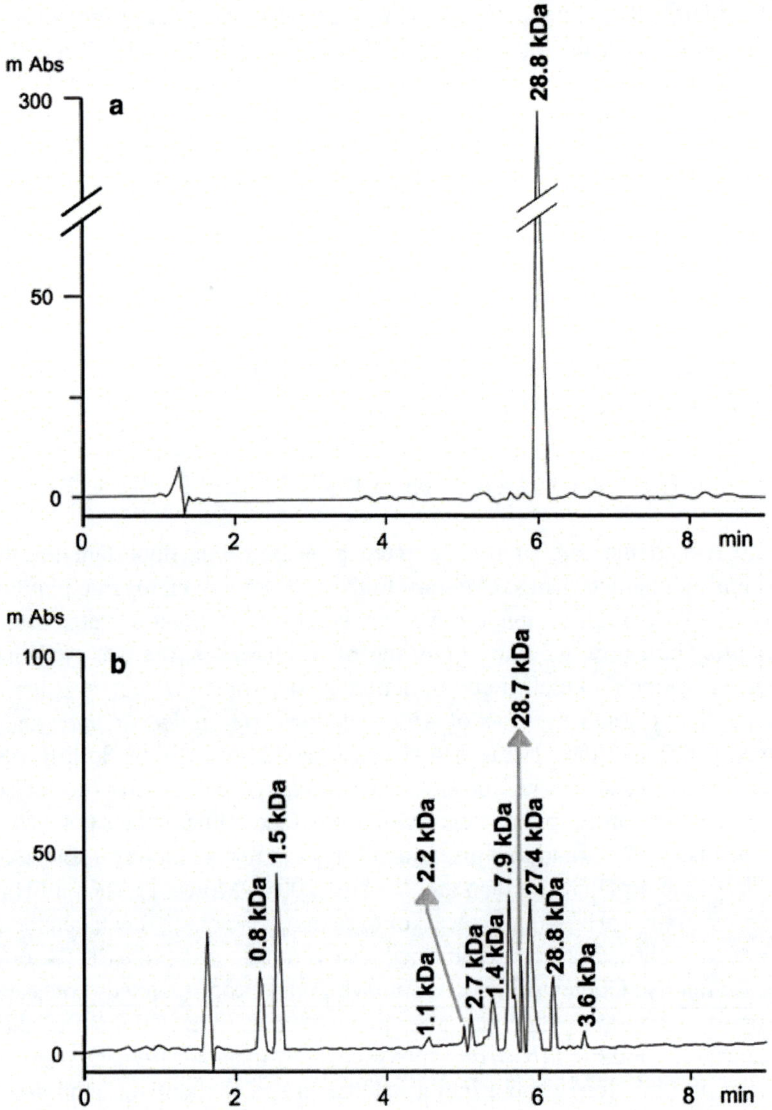

Fig. 2 Chromatograms of the degradation products of proteins (BSA, casein) by root-secreted proteases. (**a**) Casein incubated with denatured root-secreted protease (no digestion), (**b**) casein incubated for 1 h with root-secreted protease (taken from Adamczyk et al. 2010a). Above peaks there are molecular masses of peptide fragments

endopeptidases (digesting peptide bonds within the polypeptide chain) and exopeptidases (which remove the last or first amino acid residue from a polypeptide chain) (Nduwimana et al. 1995). Studies of the degradation products of the B-chain of insulin suggested that *A. porrum* root-secreted proteases possess mainly endopeptidase activity and preferentially cleave a peptide bond that has aromatic or hydrophobic amino acid at the P2 position and polar amino acid at the P1 position, which enhances the cleavage susceptibility of the peptide bond (Adamczyk et al. 2009a).

Such degradation products of proteins by root-secreted proteases can potentially act as an easily accessible pool of N for plants. It has already been shown that low molecular mass peptides and amino acids can be taken up by the roots of certain plant species with amino acids or peptide transporters (Delrot et al. 2001; Lee et al. 2007; Ortiz-Lopez et al. 2000; Paungfoo-Lonhienne et al. 2009).

2.2.2 Seedling Growth and Root-Protease Secretion with Different Sources of Nitrogen

According to the preceding section (see Sect. 2.2.1), plants can potentially use proteins as N source. Studies conducted on *Triticum aestivum* seedlings, obtained from embryos isolated from grains, indicate that plants can even compensate the lack of inorganic nitrogen with organic nitrogen, supplied as protein. Hydroponic sterile cultures of *T. aestivum*, cultivated on Murashige and Skoog medium in which inorganic nitrogen was replaced by protein (casein), reached an even higher biomass compared with seedlings cultivated on medium with inorganic nitrogen (Fig. 3). The improved growth was explained as being due to creation of an accessible pool of amino acids by digestion of proteins by root-secreted proteases. Such improvement in growth was also explained by concomitant uptake of organic

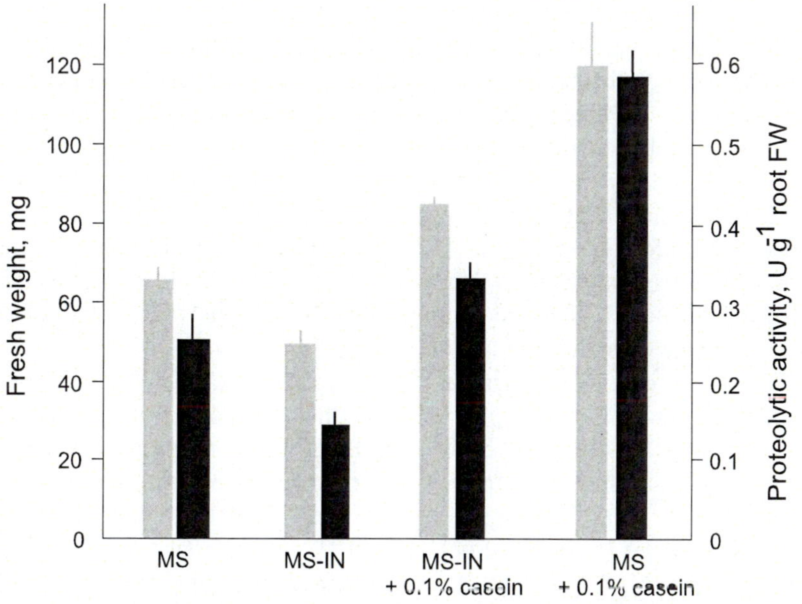

Fig. 3 Fresh weight of shoot (*gray columns*) and proteolytic activity (*black columns*) in the culture medium of seedlings cultivated on different media (means ± standard error of the mean, $n = 6$). Azocasein was used as a substrate for proteases. *MS* Murashige and Skoog medium, *MS-IN* Murashige and Skoog medium without inorganic nitrogen, *FW* fresh weight. This figure is a combination of two figures already published in Adamczyk et al. (2010a, b)

nitrogen and organic carbon originating from digestion of protein in the medium by root-secreted proteases (Adamczyk et al. 2008a). Moreover, higher concentrations of amino acids in the medium can induce expression of amino acid transporters in the rhizodermal cells of *A. thaliana*, improving uptake of amino acids (Hirner et al. 2006).

However, in natural conditions both forms of nitrogen, inorganic and organic (including amino acids and proteins) can be found (Kaye and Hart 1997). Studies conducted with hydroponic cultures of *T. aestivum*, growing on inorganic nitrogen, organic nitrogen (casein), or a mixture of both, suggested that the highest growth can be obtained with medium containing both sources of nitrogen in the proper proportion (Adamczyk et al. 2010a). In such medium, i.e., having both sources of nitrogen, the activity of root-secreted proteases was also found to be the highest (Adamczyk et al. 2010a) (Fig. 3). A similar result for growth of seedlings was obtained by Paungfoo-Lonhienne et al. (2008) for *A. thaliana*.

2.2.3 Role of Root-Secreted Proteases in Plant Nitrogen Nutrition in Field Conditions

All the above-mentioned studies, which were conducted in sterile conditions, mainly on agricultural species, showed the ability of certain plants to secrete proteases and, using them, to improve the N budget, but we know nothing about this phenomenon in field conditions. Moreover, no experiment has been able to properly quantify the extent to which plants utilize organic N in a specific ecosystem, even though the existence of such process has been shown in laboratory and field conditions (Gärdenäs et al. 2010). In field conditions, proteolysis may be driven mainly by microbial enzymes, as they are also secreting proteases (e.g., Nannipieri et al. 2000). However, microorganism communities depend on numerous factors (Wieland et al. 2001) and soil is not homogenous, which can strongly influence the proteolysis process. There are patches of organic nitrogen, originating from plants or animal residues (Hodge et al. 2000), but also from organic fertilizers. In such patches, rich in organic nitrogen (also as proteins), the amount of substrate can exceed the digestive abilities of microbial proteases.

The next problem in understanding the role of root-secreted proteases is the mycorrhizal status of the plant. Plants that do not create symbioses with ectomycorrhizal and ericoid mycorrhizal fungi (secreting proteases; Bending and Read 1996) could improve proteolysis in the soil by secreting proteases from the roots. In such a case, root-secreted proteases could substitute for mycorrhizal secreted proteases (Paungfoo-Lonhienne et al. 2008). However, plants which can develop mycorrhizal symbioses are also able to secrete proteases through the roots (Godlewski and Adamczyk 2007).

Last, but not least, is the problem of the form in which organic N is present in the soil. Some of the organic nitrogen in soil is in the form of recalcitrant complexes, e.g., with polyphenols (Bending and Read 1996). Moreover, tannins, an important group of phenolic compounds, can also decrease the activity of proteases directly, slowing proteolysis (Adamczyk et al. 2009b). Complexation of proteases and proteins with tannins may strongly influence soil proteolysis. There is a need to study proteolysis processes in the soil, keeping in mind that strong interactions between proteins (both substrates and enzymes) and soil constituents exist.

The ability of plants to support their nitrogen budget by the use of root-secreted proteases can be used in agricultural practice. In current fertilization practice, inorganic nitrogen fertilizers are used, commonly in excess. Inorganic nitrogen fertilizers, mainly nitrates, may undergo leaching because of their high mobility in soil (Jones et al. 2005), which can cause eutrophication of water reservoirs (Huang et al. 2003). Organic nitrogen seems to be more stable in the soil than inorganic nitrogen is (Jones et al. 2005), so organic nitrogen fertilizers seem to be more environment friendly than inorganic nitrogen fertilizers. Moreover, as was shown for sterile cultures of *T. aestivum*, the highest growth was obtained on medium with both inorganic nitrogen and organic nitrogen compared with inorganic nitrogen alone (Adamczyk et al. 2010a), so the best way to improve yield seems to be a mixture of both N sources. The ability to exude proteases and digest proteins from organic nitrogen fertilizers without microbial assistance could be of great importance in developing sustainable agriculture based on sources of organic nitrogen.

Secretion of proteases by roots, a newly discovered plant behavior, can change our view on nitrogen cycling in the soil – plants can potentially be involved directly in the turnover of organic N in the soil. Secretion of proteases seems to be common in the plant kingdom; however, the level of their activity is species dependent. These proteases are active in the pH range 4–8, they belong to the cysteine protease family, and they operate mainly as endopeptidases, digesting proteins to low molecular mass products, which can be taken up by plant roots. Plants, as was shown for wheat, can use these proteases to digest proteins in order to obtain nitrogen. The highest yield can be acquired after cultivation of a plant on a mixture of inorganic and organic sources of N. Protease secretion by roots needs more study, especially in field conditions, where factors like competition with soil microorganisms or accessibility of soil proteins are important.

References

Adamczyk B, Godlewski M, Domanska A, Bilecka A (2004) Can plant roots exude proteolytic enzymes into the culture medium? Acta Physiol Plant 26:45

Adamczyk B, Godlewski M, Zimny J, Zimny A (2008a) Wheat (*Triticum aestivum*) seedlings secrete proteases from roots and, after protein addition, grow well on medium without inorganic nitrogen. Plant Biol 10:718–724

Adamczyk B, Kitunen V, Smolander A (2008b) Protein precipitation by tannins in soil organic horizon and vegetation in relation to tree species. Biol Fertil Soils 45:55–64

Adamczyk B, Godlewski M, Smolander A, Kitunen V (2009a) Degradation of proteins by enzymes exuded by *Allium porrum* roots – a potentially important strategy for acquiring organic nitrogen by plants. Plant Physiol Biochem 47:919–925

Adamczyk B, Kitunen V, Smolander A (2009b) Polyphenol oxidase, tannase and proteolytic activity in relation to tannin concentration in the soil organic horizon under silver birch and Norway spruce. Soil Biol Biochem 41:2085–2093

Adamczyk B, Godlewski M, Zimny J, Zimny A (2010a) Growth and protease secretion from the roots of wheat (*Triticum aestivum* cv. Tacher) seedlings cultivated on different nitrogen sources. Indian J Plant Physiol 15:150–153

Adamczyk B, Smolander A, Kitunen V, Godlewski M (2010b) Proteins as nitrogen source for plants – a short story about exudation of proteases by plant roots. Plant Signal Behav 5:1–3

Adams MA, Pate JS (1992) Availability of organic and inorganic forms of phosphorus to lupins (*Lupinus* spp.). Plant Soil 145:107–113

Arahou M, Diem HG (1997) Iron deficiency induces cluster (proteoid) root formation in *Casuarina glauca*. Plant Soil 196:71–79

Bending GD, Read DJ (1996) Nitrogen mobilization from protein-polyphenol complex by ericoid and ectomycorrhizal fungi. Soil Biol Biochem 28:1603–1612

Braum SM, Helmke PA (1995) White lupin utilizes soil phosphorus that is unavailable to soybean. Plant Soil 176:95–100

Davies J, Briarty LG, Rieley JO (1973) Observations on the swollen lateral roots of the Cyperaceae. New Phytol 72:167–174

Dell B, Kuo J, Thompson GJ (1980) Development of proteoid roots in *Hakea obliqua* R.Br. (Proteaceae) grown in water culture. Aust J Bot 28:27–37

Delrot S, Atanassova R, Gomes E, Coutos-Thevenot P (2001) Plasma membrane transporters: a machinery for uptake of organic solutes and stress resistance. Plant Sci 161:391–404

Dessougi HI, Dreele zu, Claasen N (2003) Growth and phosphorus uptake of maize cultivated alone, in mixed culture with other crops or after incorporation of their residues. J Plant Nutr Soil Sci 166:254–261

Diem HG, Duchoux E, Zaid H, Arahou M (2000) Cluster roots in Casuarinaceae: role and relationship to soil nutrient factors. Ann Bot 85:929–936

Dinkelaker B, Römheld V, Marschner H (1989) Citric acid excretion and precipitation of calcium citrate in the rhizosphere of white lupin (*Lupinus albus* L). Plant Cell Environ 12:285–292

Dinkelaker B, Hengeler C, Bienfait HF (1995) Distribution and function of proteoid roots and other root clusters. Bot Acta 108:183–200

Elfstrand S, Bath B, Martensson A (2007) Influence of various forms of green manure amendment on soil microbial community composition, enzyme activity and nutrient levels in leek. Appl Soil Ecol 36:70–82

Floch C, Capowiez Y, Criquet S (2009) Enzyme activities in apple orchard agroecosystems: how are they affected by management strategy and soil properties. Soil Biol Biochem 41:61–68

Gärdenäs AI, Ågren GI, Bird JA, Clarholm M, Hallin S, Ineson P, Kätterer T, Knicker H, Nilsson I, Näsholm T, Ogle S, Paustian K, Persson T, Stendahl J (2010) Knowledge gaps in soil carbon and nitrogen interactions – from molecular to global scale. Soil Biol Biochem. doi:10.1016/j.soilbio.2010.04.006

Gardner WK, Parbery DG, Barber DA (1982) The acquisition of phosphorus by *Lupinus albus* L. I. Some characteristics of the soil/root interface. Plant Soil 68:19–32

George TS, Gregory PJ, Wood M, Read D, Buresh RJ (2002) Phosphatase activity and organic acids in the rhizosphere of potential agroforestry species and maize. Soil Biol Biochem 34:1487–1494

Godlewski M, Adamczyk B (2007) The ability of plants to secrete proteases by roots. Plant Physiol Biochem 45:657–664.

Hawkins H-J, Wolf G, Stock WD (2005) Cluster roots of *Leucadendron laureolum* (Proteaceae) and *Lupinus albus* (Fabaceae) take up glycine intact: an adaptive strategy to low mineral nitrogen in soils? Ann Bot 96:1275–1282.

Hinsinger P (2001) Bioavailability of soil inorganic P in the rhizosphere as affected by root-induced chemical changes: a review. Plant Soil 237:173–195

Hinsinger P, Plassard C, Tang C, Jaillard B (2003) Origins of root-mediated pH changes in the rhizosphere and their responses to environmental constraints: a review. Plant Soil 248:43–59
Hirner A, Ladwig F, Stransky H, Okumoto S, Keinath M, Harms A, Frommer WB, Koch W (2006) Arabidopsis LHT1 is a high-affinity transporter for cellular amino acid uptake in both root epidermis and leaf mesophyll. Plant Cell 18:1931–1946
Hodge A, Stewart J, Robinson D, Griffiths BS, Fitter AH (2000) Plant N capture and microfaunal dynamics from decomposing grass and earthworm residues in soil. Soil Biol Biochem 32:1763–1772
Huang XP, Huang LM, Yue WZ (2003) The characteristics of nutrients and eutrophication in the Pearl River estuary, South China. Mar Pollut Bull 47:30–36
Jones DL, Healey JR, Willet VB, Farrar JF, Hodge A (2005) Dissolved organic nitrogen uptake by plants – an important N uptake pathway? Soil Biol Biochem 37:413–423
Kaye JP, Hart SC (1997) Competition for nitrogen between plants and soil microorganisms. Trends Ecol Evol 12:139–143
Lambers H, Juniper D, Cawthray GR, Veneklaas EJ, Martinez E (2002) The pattern of carboxylate exudation in *Banksia grandis* (Proteaceae) is affected by the form of phosphate added to the soil. Plant Soil 238:111–122
Lambers H, Shane MW, Cramer MD, Pearse SJ, Veneklaas EJ (2006) Root structure and functioning for efficient acquisition of phosphorus: matching morphological and physiological traits. Ann Bot 98:693–713
Lamont B (1972) The morphology and anatomy of proteoid roots in genus Hakea. Aust J Bot 20:155–174
Lamont B (1973) Factors affecting the distribution of proteoid roots within the root system of two *Hakea* species. Aust J Bot 21:165–187
Lamont B (1982) Mechanisms for enhancing nutrient uptake in plants with particular reference to mediterranean South Africa and Western Australia. Bot Rev 48:597–689
Lamont B (2003) Structure, ecology and physiology of root clusters – a review. Plant Soil 248:1–19
Lee Y-H, Foster J, Chen J, Voll LM, Weber APM, Tegeder M (2007) AAP1 transports uncharged amino acids into roots of *Arabidopsis thaliana*. Plant J 50:305–319
Li M, Osaki M, Rao IM, Tadano T (1997) Secretion of phytase from the roots of several plant species under phosphorus-deficient conditions. Plant Soil 195:161–169
Liang R, Li C (2003) Differences in cluster-root formation and carboxylate exudation in *Lupinus albus* L. under different nutrient deficiencies. Plant Soil 248:221–227
Lipson D, Näsholm T (2001) The unexpected versatility of plants: organic nitrogen use and availability in terrestrial ecosystems. Oecologia 128:305–316
Liu J, Uhde-Stone C, Li A, Vance C, Allan D (2001) A phosphate transporter with enhanced expression in proteoid roots of white lupin (*Lupinus albus* L.). Plant Soil 237:257–266
Miller SS, Liu J, Allan DL, Menzhuber CJ, Fedorova M, Vance CP (2001) Molecular control of acid phosphatase secretion into the rhizosphere of proteoid roots from phosphorus-stressed white lupin. Plant Physiol 127:594–606.
Nannipieri P, Kandeler E, Ruggiero P (2000) Enzyme activities and microbial and biochemical processes in soil. In: Burns RG, Dick RP (eds) Enzymes in the environment: activity, ecology, and applications. Marcel Dekker, New York, pp 1–33
Naumann G, Massonneau A, Langlade N, Dinkelaker B, Hengeler C, Romheld V, Martinoia E (2000) Physiological aspects of cluster root function and development in phosphorus-deficient white lupins (*Lupinus albus* L.) Ann Bot 85:909–919
Näsholm T, Kielland K, Ganeteg U (2009) Uptake of organic nitrogen by plants. New Phytol 182:31–48
Nduwimana J, Guenet L, Dorval I, Blayau M, Gall JYL, Le Treut A (1995) Proteases. Ann Biol Clin 53:251–264
Neumann G, Martinoia E (2002) Cluster roots – an underground adaptation for survival in extreme environments. Trends Plant Sci 7:162–167

Neumann G, Römheld V (2001) The release of root exudates as affected by the plant physiological status. In: Pinto R, Varanini Z, Nannipieri Z (eds) The rhizosphere: biochemistry and organic substances at the soil-plant interface. Marcel Dekker, New York, pp 41–93
Ortiz-Lopez A, Chang H-C, Bush DR (2000) Amino acid transporters in plants. Biochim Biophys Acta 1465:275–280
Ozawa K, Osaki M, Matsui H, Honma M, Tadano T (1995) Purification and properties of acid phosphatase secreted from lupin roots under phosphorus-deficiency conditions. Soil Sci Plant Nutr 41:461–469
Paungfoo-Lonhienne C, Lonhienne TGA, Rentsch D, Robinson N, Christie M, Webb RI, Gamage HK, Caroll BJ, Schenk PM, Schmidt S (2008) Plants can use protein as nitrogen source without assistance from other organisms. Proc Natl Acad Sci USA 105:4524–4529
Paungfoo-Lonhienne C, Schenk PM, Lonhienne TGA, Brackin R, Meier S, Rentsch D, Schmidt S (2009) Nitrogen affects cluster root formation and expression of putative peptide transporters. J Exp Bot 60:2665–2676
Playsted CWS, Johnston ME, Ramage CM, Edwards DG, Cawthray GR, Lambers H (2006) Functional significance of dauciform roots: exudation of carboxylates and acid phosphatase under phosphorus deficiency in *Caustis blakei* (Cyperaceae). New Phytol 170:491–500
Purnell HM (1960) Studies of the family Proteaceae. I. Anatomy and morphology of the roots of some Victorian species. Aust J Bot 8:38–50
Rentsh D, Schmidt S, Tegeder M (2007) Transporters for uptake and allocation of organic nitrogen compounds in plants. FEBS Lett 581:2281–2289
Roelofs RFR, Rengel Z, Cawthray GR, Dixon KW, Lambers H (2001) Exudation of carboxylates in Australian Proteaceae: chemical composition. Plant Cell Environ 24:891–903
Schmidt S, Stewart GR (1997) Waterlogging and fire impacts on nitrogen availability and utilization in a subtropical wet heathland (wallum). Plant Cell Environ 20:1231–1241
Schmidt S, Mason M, Sangtiean T, Stewart GR (2003) Do cluster roots of *Hakea actities* (Proteaceae) acquire complex organic nitrogen? Plant Soil 248:157–165
Shane MW, Lambers H (2005) Cluster roots: a curiosity in context. Plant Soil 274:101–125
Skene KR (1998) Cluster roots: some ecological considerations. J Ecol 86:1060–1064
Skene KR, James WM (2000) A comparison of the effects of auxin on cluster root initiation and development in *Gravillea robusta* Cunn. Ex R. Br. (Proteaceae) and in the genus Lupinus (Leguminosae). Plant Soil 219:221–229
Tadano T, Ozawa K, Sakai H, Osaki M, Matsui H (1993) Secretion of acid phosphatase by the roots of crop plants under phosphorus-deficient conditions and some properties of the enzyme secreted by lupin roots. Plant Soil 155(156):95–98
Tamminen P, Derome J (2005) Temporal trends in chemical parameters of upland forest soils in southern Finland. Silva Fenn 39:313–330
Tarafdar JC, Claassen N (1988) Organic phosphorus compounds as a phosphorus source for higher plants through the activity of phosphatases produced by plant roots and microorganisms. Biol Fertil Soils 5:308–312
Tarafdar JC, Claassen N (2001) Comparative efficiency of acid phosphatase originated from plant and fungal sources. J Plant Nutr Soil Sci 164:279–282
Tarafdar JC, Claassen N (2005) Preferential utilization of organic and inorganic sources of phosphorus by wheat plant. Plant Soil 27:285–293
Wasaki J, Yamamura T, Shinano T, Osaki M (2003) Secreted acid phosphatase is expressed in cluster roots of lupin in response to phosphorus deficiency. Plant Soil 248:129–136
Wasaki J, Rothe A, Kania A, Neumann G, Römheld V, Shinano T, Osaki M, Kandeler E (2005) Root exudation, phosphorus acquisition and microbial diversity in the rhizosphere of White lupine as affected by phosphorus supply and atmospheric carbon dioxide concentration. J Environ Qual 34:2157–2166
Wasaki J, Kojima S, Maruyama H, Haase S, Osaki M, Kandeler E (2008) Localization of acid phosphatase activities in the roots of white lupin plants grown under phosphorus-deficient conditions. Soil Sci Plant Nutr 54:95–102

Watt M, Evans JR (1999) Proteoid roots. Physiology roots. Physiology and development. Plant Physiol 121:317–323

Wieland G, Neumann R, Backhaus H (2001) Variation of microbial communities in soil, rhizosphere, and rhizoplane in response to crop species, soil type, and crop development. Appl Environ Microbiol 67:5849–5854

Zhou K, Yamagishi M, Osaki M, Masuda K (2008) Sugar signalling mediates cluster root formation and phosphorus starvation-induced gene expression in white lupin. J Exp Bot 59:2749–2756

Unity Is Strength: The Power of Border Cells and Border-Like Cells in Relation with Plant Defense

Azeddine Driouich, Marc-Antoine Cannesan, Flavien Dardelle, Caroline Durand, Barbara Plancot, Sophie Bernard, Marie-Laure Follet-Gueye, and Maïté Vicré-Gibouin

"If you want to go far and win the battle of your lives, stay together, stay strong, unite peacefully for the interest of your people. . . .vous savez bien que l'union fait la force!" A *highly respected man.*

Abstract Production and release of root border cells and border-like cells are fundamental processes for plant survival and development. Both types of cells are viable components of the root system that regulate its interactions with living microorganisms of the rhizosphere. Border cells are released as individual cells, whereas border-like cells remain attached to each other into small groups or as sheets after their release from the root tip. So far, border-like cells have been observed only in species belonging to the Brassicaceae family including Arabidopsis. Border cells have been largely studied in the legume species pea; in contrast, relatively little information is available on border-like cells so far due to their recent discovery. In this chapter, we present and discuss the release, organization, and the role of these cells in root protection.

1 Introduction and Definition

Plant roots invest a lot of energy in the formation and release into the rhizosphere of a population of living cells at their tips. These cells are programmed to detach from the cap during root growth either as border cells or border-like cells and represent a vital biotic boundary between the root and the rhizosphere. Border cells are experimentally defined as cells that disperse into suspension individually within seconds when the root tip is put into water (Hawes et al. 2000). The number of border cells released can vary considerably from a few hundreds to several thousands and it is generally conserved for plant species within a given family.

A. Driouich (✉) • M.-A. Cannesan • F. Dardelle • C. Durand • B. Plancot • S. Bernard • M.-L. Follet-Gueye • M. Vicré-Gibouin
Laboratoire Glycobiologie et Matrice Extracellulaire Végétale, IFRMP 23, Plate-forme de recherche en imagerie cellulaire de Haute-Normandie, Université de Rouen, 76 821 Mont Saint Aignan, Cedex, France
e-mail: azeddine.driouich@univ-rouen.fr

J.M. Vivanco and F. Baluška (eds.), *Secretions and Exudates in Biological Systems*, Signaling and Communication in Plants 12, DOI 10.1007/978-3-642-23047-9_5,

The organization of the root apical meristem (RAM) is highly correlated with the production of root border cells. In eudicotyledonous angiosperm plants, RAM can be classified into three different organizations: closed RAM (composed of highly organized tiers of cells), basic-open RAM (with cells not clearly organized into distinguishable tiers), and an intermediate-open RAM (Chapman et al. 2003; Groot et al. 2004). Hamamoto et al. (2006) demonstrated that roots with open apical organization released high numbers of individual border cells: 4,500 and 10,000 cells have been reported to be released respectively by pea and cotton roots.

Unlike border cells, border-like cells do not detach as isolated cells and are therefore defined as cells that do remain attached to each other into small groups or as sheets after their release from the root tip (Vicré et al. 2005; Driouich et al. 2007). Such cells are very clearly observed in *Arabidopsis thaliana* (Fig. 1a), but they have also been found in other Brassicaceae members including canola, radish, and cauliflower (Driouich et al. 2007; see also Fig. 1c, d). In terms of number, after a week of growth under laboratory conditions, radish root produces more border-like cells (907 ± 75) than does Arabidopsis (116 ± 10) or canola (375 ± 137) roots. So far border-like cells have not been observed in other families such as the Leguminoseae, the Solanaceae, or cereal species (see Hawes et al. 2003).

Border cells and border-like cells (Fig. 1) originate from the root-cap meristem whose cells undergo a series of divisions and differentiation giving rise successively to gravity sensing columella cells and root peripheral cells. How the release of such cells is regulated is not very well understood, but it seems to rely on developmental and environmental signals. Border cell formation and release can be switched off and on independently of root development and removal of the cells was reported to stimulate mitosis in the root-cap meristem within a few minutes (Brigham et al. 1998). In relation with this observation, it has been reported that a soluble factor (named factor B), secreted into the external medium by pea border cells, is able to control cell division in the root-cap meristem, thereby influencing border cell production and release (Brigham et al. 1998; Hawes et al. 2000). The factor is sensitive to protease hydrolysis suggesting its protein nature. Also, analysis of the Arabidopsis mutant *fez* has demonstrated that the release of border-like cells can be genetically controlled by an NAC-domain transcription factor (named FEZ) that is active in root-cap initials. The activity of FEZ in the epidermal/lateral root-cap cell initials has been shown to promote the formation of root-cap cells including the production of border-like cells (Willemsen et al. 2008). Whereas FEZ and its own negative regulator SOMBRERO are required for root-cap cell division (Willemsen et al. 2008), root-cap differentiation seems to be under the control of four transcription factors: PLETHORA 1–3 and BABYBOOM (Aida et al. 2004; Boutilier et al. 2002). These transcription factors are under the control of the hormone auxin *via* the action of auxin responsive factors.

As for environmental factors, temperature, carbon dioxide, soil-based mechanical stress, aluminum, and invasion by microorganisms have been shown to influence border cell formation, morphology, and number (Miyasaka and Hawes 2001; Curlango-Rivera et al. 2010).

It should be noted that border-like cells and border cells are not dead, unwanted "garbage" cells that are thrown away by the root into the surrounding environment.

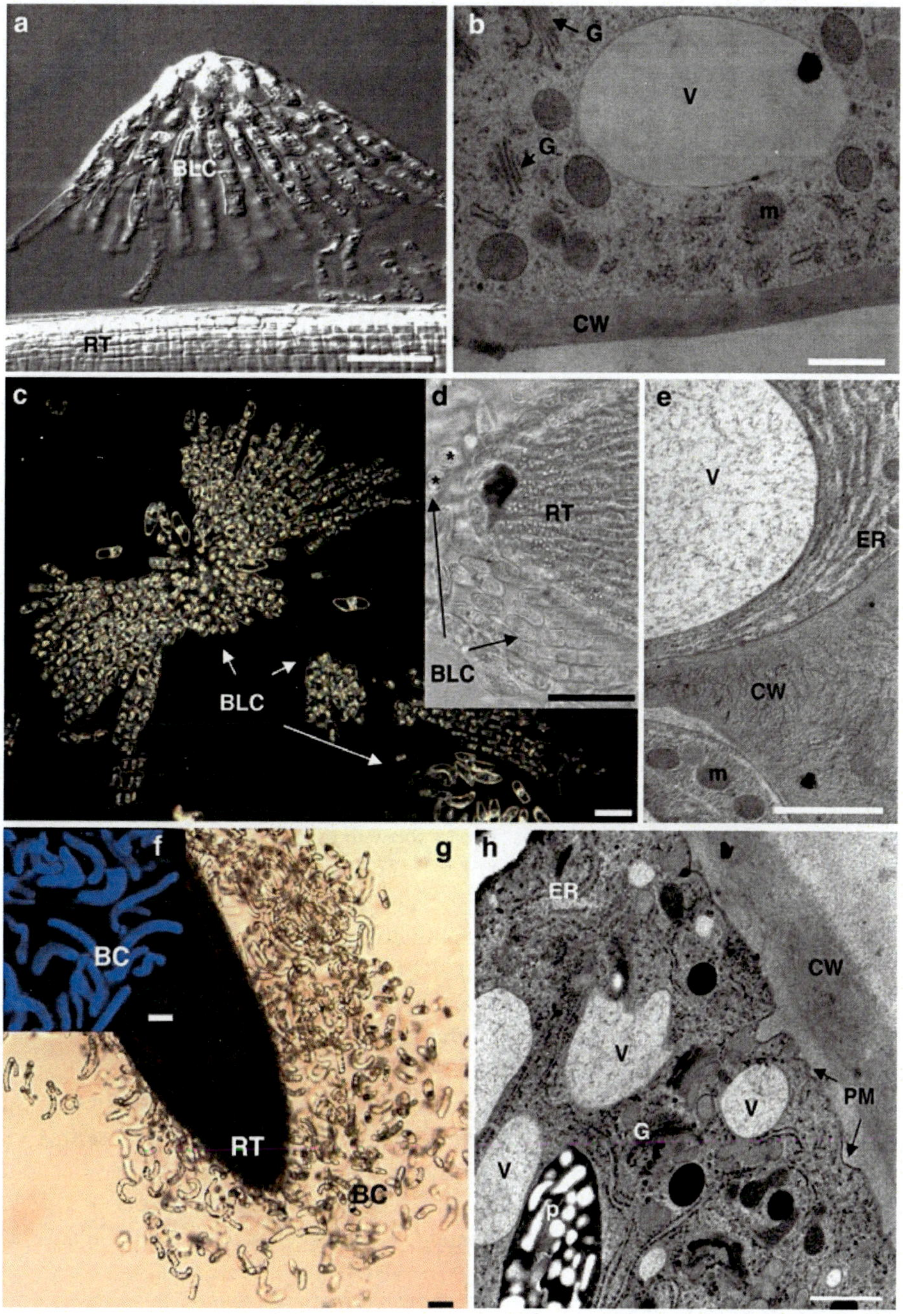

Fig. 1 Border cell and border-like cell morphology. Light and electron microscopy images of border-like cells of Arabidopsis (*Arabidopsis thaliana*) (**a**, **b**) or radish (*Raphanus sativus*) (**c**, **d**, **e**) and border cells released from pea (*Pisum sativum*) root tips (**f**, **g**, **h**). **a**, **c**, **d**, **f**, and **g**: light microscopy images of border-like cells from Arabidopsis (**a**) and radish (**c**, **d**) and border cells of pea (**f**, **g**). Unlike border cells (**f**, **g**), border-like cells (**a**, **c**, and **d**) remain attached to each other into small groups or as sheets after their release from the root tip. In **d** border-like cells released showing distinct morphologies: elongated *versus* ball-shaped (*) cells. Calcofluor staining of pea

They are released from the root tip by a controlled process as viable and metabolically active cells that survive for weeks within the soil environment and even for months under laboratory conditions (Vermeer and Mc Cully 1982; Hawes et al. 2000; Vicré et al. 2005). It is, therefore, *totally wrong* (as it is indicated in certain recent publications) to consider *A. thaliana* as a species that does not produce viable border-like cells.

2 Attachment and Organization Patterns of Border-Like Cells

The production of border-like cells from the root tip of *A. thaliana* is dependent on the stage of root development. We have observed that the release of border-like cells does occur only when seedlings are 4–5 days old and that the number of cell layers increases over time (Vicré et al. 2005). Two types of cells were observed in terms of morphology, namely ball-shaped cells, which are located at the very tip of the root and elongated cells found on the margins of the cluster (see Fig. 1c, d). Shape and size of border cells and border-like cells seem to vary depending on the species as well as on RAM organization (Table 1 and Fig. 2). Ultrastructural observations using transmission electron microscopy of high-pressure frozen/freeze-substituted cells have revealed that they notably contain a large number of mitochondria and Golgi-derived vesicles indicative of a high secretory activity at the time of their release. Similar observations were also made for radish border-like cells as well as for border cells of pea (Fig. 1b, e, h). Indeed, using immunofluorescence microscopy, it has been possible to show that border-like cells of Arabidopsis are active in secreting high amounts of polysaccharides and proteoglycans into their cell surface. Carbohydrate epitopes associated with pectins, xyloglucan (XyG), and arabinogalactan-proteins were abundant at the surface of border-like cells of Arabidopsis and radish (Figs. 3 and 4). One interesting observation made on Arabidopsis border-like cells is that XyG epitopes (recognized by CCRCM1 antibody) were not only associated with the cell wall but can also be seen detaching from the cell wall along the entire cell, forming "hairy branches" at the cell surface (Fig. 4e). Also, XyG branches can be seen bridging two neighboring cells. Such XyG bridges suggest the involvement of this polysaccharide in connecting cells together, thereby contributing to their attachment to each other (make cells adhere to each other). However, in a study aimed at investigating the role of cell wall polymers in border-like cells attachment, it has been shown that mutants with

Fig. 1 (continued) border cells (**f**). **b**, **e**, and **h**: electron micrographs of border-like cells from Arabidopsis (**b**) or radish (**e**) and border cells from pea (**h**). BLC as well as BC contain a large number of mitochondria and Golgi stack units indicative of a high secretory activity. *BC* border cells, *BLC* border-like cells, *CW* cell wall, *ER* endoplasmic reticulum, *G* Golgi stacks, *m* mitochondria, *p* plastids, *PM* plasma membrane, *RT* root tip, *V* vacuole. Bars: 50 μm (**a**, **c**, **d**, **f**, and **g**); 1 μm (**b**, **e**, and **h**)

Table 1 Variation in shape and size of border cells and border-like depending on the species as well as on RAM organization. Root tips (7–12 days old) are put into a drop of water for observation of cell release. Low adhesion means that cells are loosely attached to the root tip and to each other. Strong adhesion means cells are tightly attached to root tip and to each other. *Impatiens glandulifera* does not release border cells once the root is put into water. Under our conditions, only one layer of tightly adhered cells to root tip is seen – see also Fig. 2u

		Family	Species	RAM	BC vs BLC (adhesion)	Cell shape		Size (minimum → maximum) (μm)
						Lateral root-cap cells	BC/BLC	
Angiosperms	Eudicotyledonous	Alliaceae	*Allium schoenoprasum*	?	BLC	Elongated	Ball-shaped	20/40
		Amaranthaceae	*Amaranthus caudatus*	Closed	BLC (low adhesion)	Elongated	Ball-shaped	15/65
		Apiaceae	*Petroselinum crispum*	Intermediate	BLC (low adhesion)	Elongated	Ball-shaped	15/45
			Daucus carota	Intermediate	BLC (low adhesion)	Elongated	Ball-shaped	10/30
			Foeniculum vulgare	Intermediate	BLC (low adhesion)	Elongated	Square-shaped	10/30
		Arecaceae	*Phoenix dactylifera*	?	BLC	Elongated	Square-shaped	40/150
		Asteraceae	*Centaurea cyanus*	Closed	BLC (low adhesion)	Elongated	Square-shaped	10/40
			Senecio cineraria	Closed	BLC (low adhesion)	Elongated	Square-shaped	10/30
			Lactuca sativum	Closed	BLC (low adhesion)	Elongated	Square-shaped	10/45
		Balsaminaceae	*Impatiens glandulifera*	?	?	A layer of tightly adhered cells		
		Boraginaceae	*Myosotis sp.*	?	BLC (low adhesion)	Ball-shaped	Ball-shaped	10/25
		Brassicaceae	*Erysimum cheiri*	Closed	BLC	Elongated	Square-shaped	20/50
			Brassica rapa	Closed	BLC	Elongated	Ball-shaped	20/50
			Raphanus sativus	Closed	BLC	Elongated	Square-shaped	20/70
			Brassica napus	Closed	BLC	Elongated	Square-shaped	20/50
			Arabidopsis thaliana	Closed	BLC	Elongated	Square-shaped	15/45
			Lepidium sativum	Closed	BLC	Elongated	Ball-shaped	15/55
		Cannabaceae	*Humulus japonicus*	Intermediate	BLC (strong adhesion)	Elongated	Square-shaped	20/50
		Convolvulaceae	*Convolvulus tricolor*	Closed	BLC	Elongated	Square-shaped	15/30
			Ipomoea violacea	Closed	BLC	Elongated	Square-shaped	20/70

(continued)

Table 1 (continued)

Family	Species	RAM	BC vs BLC (adhesion)	Cell shape		Size (minimum → maximum) (μm)
				Lateral root-cap cells	BC/BLC	
Cucurbitaceae	*Cucurbita maxima*	Open	BC	Bean-shaped	Bean-shaped	20/55
Cactaceae	*Cactus sp.*	?	BLC	Elongated	Ball-shaped	10/15
Caryophylaceae	*Dianthus sp.*	Closed	BLC	Bean-shaped	Bean-shaped	15/40
Clusiaceae	*Millepertuis calycinum*	?	BLC (low adhesion)	Ball-shaped	Ball-shaped	10/25
Chenopodiaceae	*Spinacia oleracea*	?	BLC (strong adhesion)	Elongated	Ball-shaped	20/30
Fabaceae	*Pisum sativum*	Open	BC	Bean-shaped	Bean-shaped	15/50
	Vicia faba L	Open	BC	Elongated	Ball-shaped	25/80
	Phaseolus vulgaris	Open	BC	Elongated	Ball-shaped	25/80
	Medicago sativa L	Open	BC	Elongated	Ball-shaped	15/50
	Lens culinaris	Open	BC	Elongated	Ball-shaped	25/80
	Trifolium repens	Open	BLC (low adhesion)	Elongated	Ball-shaped	25/70
	Lupinus sp.	Open	BLC (low adhesion)	Elongated	Ball-shaped	20 /130
Geraniaceae	*Pelargonium zonale*	Intermediate	BLC	Bean-shaped	Bean-shaped	10/40
Lamiaceae	*Mentha spicata*	Closed	BLC	Ball-shaped	Square-shaped	10/20
	Thymus vulgaris	Closed	BLC (low adhesion)	Ball-shaped	Ball-shaped	10/20
	Salvia sp.	Closed	BLC (low adhesion)	Ball-shaped	Ball-shaped	10/20
	Lavandula angustifolia	Closed	BLC (low adhesion)	Ball-shaped	Ball-shaped	10/20
Linaceae	*Linum usitatisimum*	Intermediate	BLC	Elongated	Ball-shaped	25/250
Malvaceae	*Alcea rosea*	Intermediate	BLC (low adhesion)	Bean-shaped	Ball-shaped	15/80
	Malva dendromorpha	Intermediate	BLC (low adhesion)	Bean-shaped	Ball-shaped	20/80
Mimosaceae	*Acacia pycnantha*	?	BLC (strong adhesion)	Elongated	Ball-shaped	30/250
Papaveraceae	*Papaver rhoeas*	?	BLC	Bean-shaped	Bean-shaped	15/30
Polygonaceae	*Rumex L.*	Closed	BLC (strong adhesion)	Elongated	Ball-shaped	10/30
Polémoniaceae	*Phlox paniculata*	?	BLC (strong adhesion)	Elongated	Ball-shaped	15/30
Portulaceae	*Portulaca oleracea*	?	BLC	Bean-shaped	Bean-shaped	15/50
Primulaceae	*Primula vulgaris*	?	BLC (low adhesion)	Ball-shaped	Ball-shaped	10/20

	Ranunculaceae	*Aquilegia vulgaris*	Intermediate	BLC (low adhesion)	Elongated	Ball-shaped	10/50
	Rosaceae	*Fragaria sp.*	?	BLC (low adhesion)	Elongated	Ball-shaped	15/25
	Scrophuliaceae	*Digitalis purpurea*	?	BLC (strong adhesion)	Elongated	Ball-shaped	10/20
		Penstemon royalis	?	BLC (strong adhesion)	Elongated	Ball-shaped	10/20
		Antirrhinum majus	?	BLC (strong adhesion)	Elongated	Square-shaped	10/20
	Solanaceae	*Solanum melongena*	Closed	BLC	Elongated	Ball-shaped	10/45
		Solanum lycopersicum	Closed	BLC (low adhesion)	Elongated	Ball-shaped	10/45
	Tropaeolaceae	*Tropaeolum majus*	Intermediate	BLC (low adhesion)	Bean-shaped	Bean-shaped	10/50
	Violaceae	*Viola x wittrockiana*	?	BLC (strong adhesion)	Elongated	Ball-shaped	10/15
Monocotyledonous	Poaceae	*Zea mays*	?	BLC (low adhesion)	Elongated	Ball-shaped	15/40
		Triticum vulgaris	?	BLC (low adhesion)	Elongated	Ball-shaped	20/50
		Carex sp.	?	BLC (low adhesion)	Ball-shaped	Ball-shaped	10/20

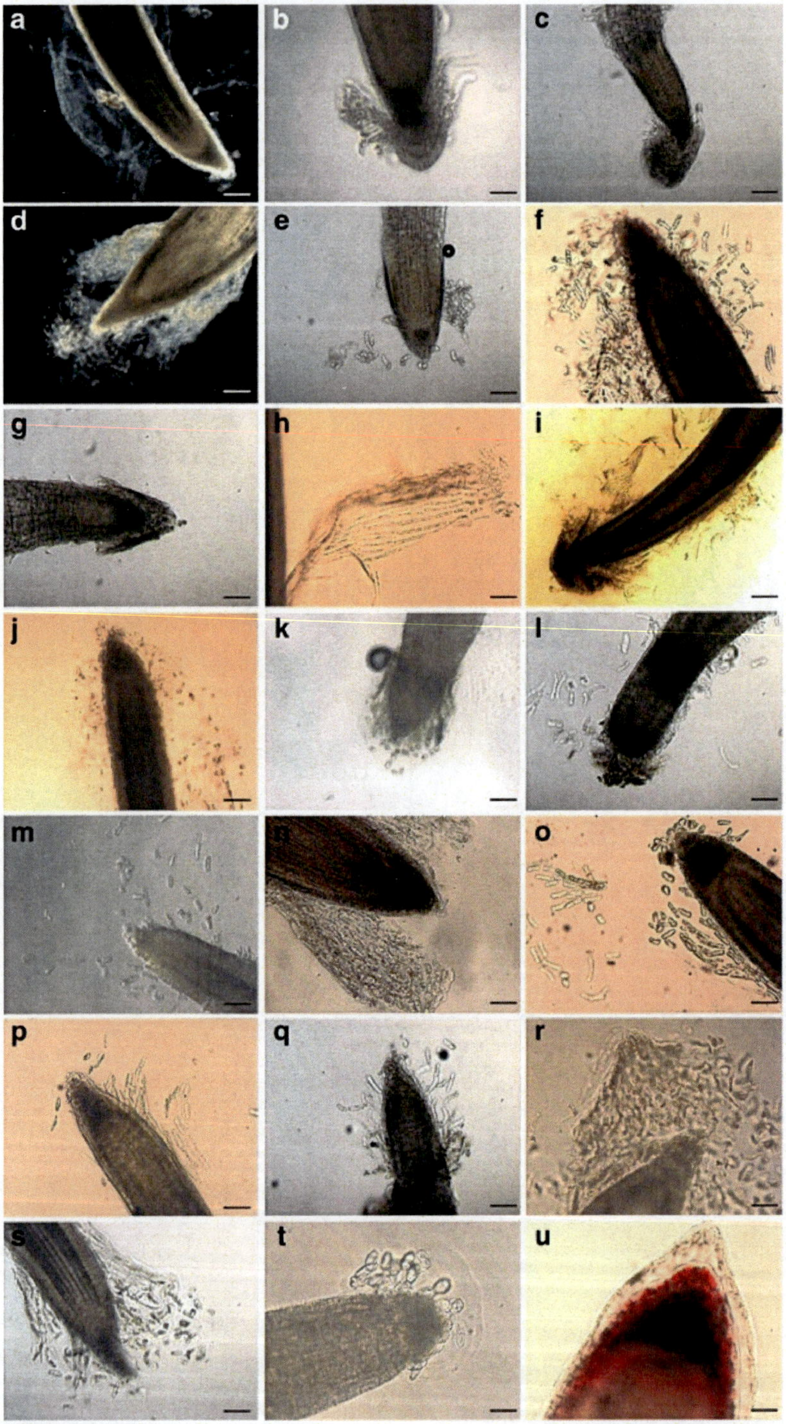

Fig. 2 Morphological phenotypes of root tips showing border cells or border-like cells of species representative of the families presented in Table 1. (**a**) *Humulus japonicus*; (**b**) *Spinacia oleracea*;

altered XyG biosynthesis (*murus 2 and murus 3*) (Vanzin et al. 2002; Madson et al. 2003) have normally attached border-like cells, suggesting that alteration of XyG structure does not cause any modification of border-like cell organization and morphology (Durand et al. 2009). However, the XyG in both mutants is altered in the side chain structure not the backbone, and therefore it would be of interest to examine the organization and morphology of border-like cells in a mutant with an altered XyG backbone structure or a mutant that has no XyG such as the *xxt1xxt2* double mutant (Cavalier et al. 2008).

A similar investigation using the *qua1* mutant has clearly demonstrated that the pectic polysaccharide homogalacturonan is responsible for border-like cell attachment. In *qua1* mutant border-like cells are converted into border cells and this conversion is accompanied by secretion of abundant mucilage released by the cells themselves while the root is growing (Fig. 4a, b and also Driouich et al. 2010). This self-produced mucilage embeds border cells allowing them to remain close to each other, a mucilage that controls unity. The composition of the mucilage is not fully known, but studies using different antibodies have shown that it is enriched in xylogalacturonan and arabinogalactan-protein epitopes (Durand et al. 2009). We have termed such an association of cells and mucilage in *qua1* mutant "a border cell biofilm" (Driouich et al. 2010) by comparison with microbial biofilms that form in response to various stress factors and that are usually composed of polysaccharides, proteins, and extracellular DNA (exDNA) (Davey and O'Toole 2000). We hypothesize that such a switch to the biofilm mode in *qua1* mutant is linked to a specialized metabolic function of border cells required for the protection of the root meristem against biotic and abiotic stress. As for bacteria, border cell mucilage might not only hold the cells together but could also protect them and facilitate cell-to-cell communication particularly during pathogen invasion of the root tip. Such a thick mucilage, which can be considered as a mucilage of unity and protection, may also have the capacity to attract and agglutinate microorganisms. It is therefore challenging to investigate whether border cells exchange specific signaling molecules and whether the mucilage has antimicrobial properties (including secreted defense proteins or peptides) as found in bacterial biofilms. Border cells and exudates from *qua1* root can be easily collected or microdissected and analyzed using proteomics, transcriptomics, and glycomics. Root mucilage isolated from many plant species has been described to be composed of up to 95% sugars and little amino acids (5%) (Bacic et al. 1986; Chaboud and Rougier 1984). Also, purified pea mucilage has been analyzed and shown to contain material similar to arabinogalactan-proteins (Knee et al. 2001). More recently, it has been shown that pea root mucilage has exDNA and antidefense proteins (Wen et al. 2007, 2009; see

---

Fig. 2 (continued) (**c**) *Penstemon royalis*; (**d**) *Acacia pycnantha*; (**e**) *Papaver rhoeas*; (**f**) *Pelargonium zonale*; (**g**) *Cactus sp.*; (**h**) *Ipomoea violacea*; (**i**) *Linum usitatisimum*; (**j**) *Amaranthus caudatus*; (**k**) *Millepertuis calycinum*; (**l**) *Myosotis sp.*; (**m**) *Primula vulgaris;* (**n**) *Solanum melongena*; (**o**) *Petroselinum crispum*; (**p**) *Lactuca sativum*; (**q**) *Fragaria sp.*; (**r**) *Thymus vulgaris*; (**s**) *Zea mays*; (**t**) *Carex sp.*; (**u**) *Impatiens glandulifera*. Bars = 100 μm (**a** and **d**) or 50 μm ((**b–g**, **i–t**) or 20 μm (**h** and **u**)

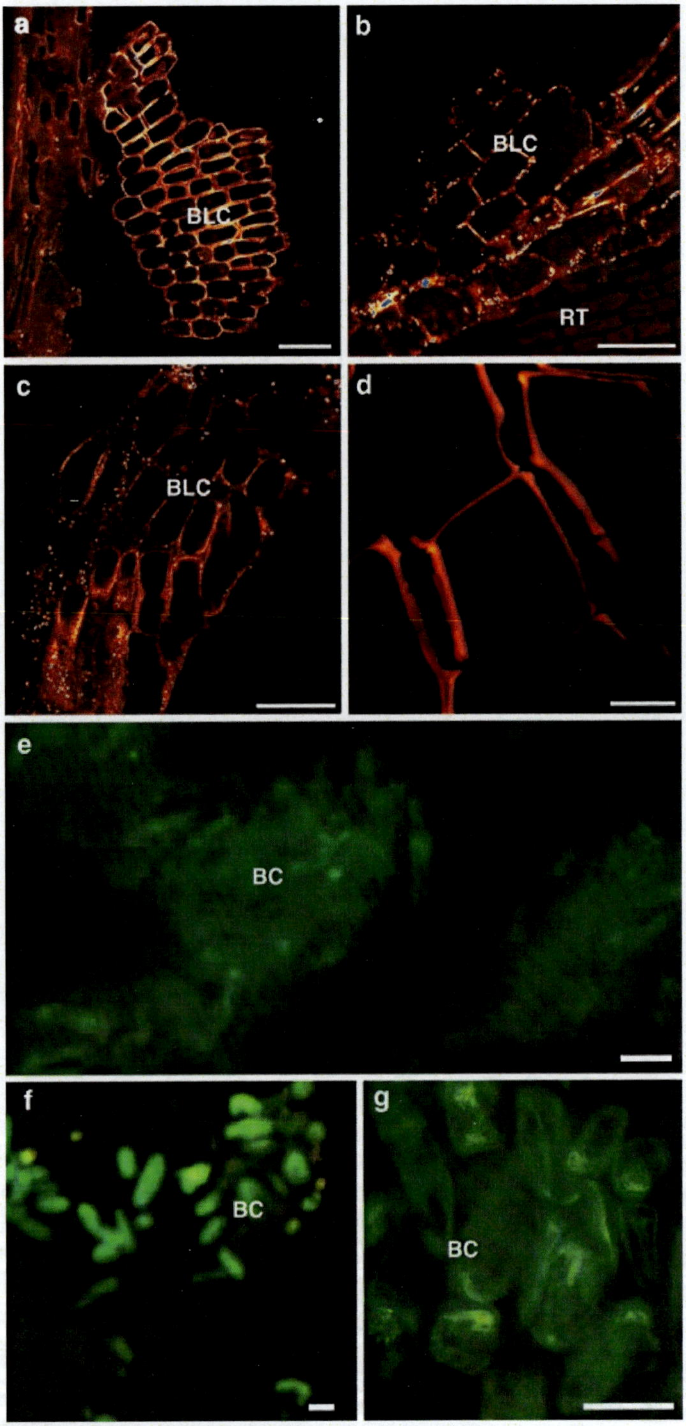

Fig. 3 Immunofluorescence labeling of border-like cells and border cells with anti-extensin (LM1), anti-arabinogalactan-protein (JIM14), and anti-pectin (LM8 and JIM5) antibodies. (**a**, **b**, **c**, and **d**)

below). Also, unlike pea border cells and unlike border cells of *qua1* mutant, border-like cells of wild-type Arabidopsis do not secrete as much mucilage as it is secreted by pea border cells.

Extending on cell wall structure of border-like cells, monosaccharide composition of the cell wall of such cells in radish and *Brassica napus* revealed the presence of a significant content in arabinose and galactose (Fig. 5; Cannesan et al. Submitted); two sugar residues that are mainly found in rhamnogalacturonan-I, arabinogalactan-proteins, and extensin. Further analysis of radish border-like cells using immunofluorescence labeling showed that extensin epitopes – recognized by LM1 or JIM11 antibodies – are strongly expressed at the cell surface of radish border-like cells (Fig. 3a, b). Thus, these cells seem to constitutively synthesize and secrete high levels of extensin that is likely to play a role as a molecular network barrier against pathogen penetration. Extensin may even be upregulated in these cells upon pathogen attack. It is well established that extensin accumulates in the cell wall of plant cells as a response to pathogen invasion (Esquerré-Tugayé and Lamport 1979; Merkouropoulos and Shirsat 2003). Also, it has been reported that both elicitor treatment and wounding lead to a rapid *in muro* insolubilization of extensin by oxidative cross-linking *via* isodityrosine motifs (Bradley et al. 1992; Brady and Fry 1997). We speculate that this might happen in the cell wall of radish border-like cells, thus creating a cross-linked protective network by the action of peroxidase and reactive oxygen species (ROS) as occurs in the cell wall of other tissues. Indeed, oxidative burst involving ROS production, established as one of the earliest response of plant cells to pathogen invasion, does occur in border-like cells of Arabidopsis treated with elicitors (Plancot et al. unpublished).

3 Border Cells Are Involved in the Protection of the Root Meristem

Border cells are fundamental to plant–microbe interactions. Within the rhizosphere, border cells are not only important in assisting the growing root to penetrate the soil, but they also provide a protecting cover surrounding the root tip against pathogens) (Gunawardena and Hawes 2002; Wen et al. 2009). Various studies have provided compelling evidence that border cells contribute significantly to the protection of the root and consequently of the entire plants. First, the number of border cell increases in response to pathogens (Cannesan et al. 2011). Second, border cells are capable of attracting, avoiding, or repelling pathogenic

Fig. 3 (continued) Radish border-like cells stained with the monoclonal antibodies LM1 (**a**) and JIM11 (**b**) specific extensin epitopes, LM8 specific for xylogalacturonan epitopes (**c**), or JIM5 specific for homogalacturonan epitopes (**d**). (**e**, **f**, and **g**) Pea border cells labeled with the monoclonal antibodies JIM14 (**e**) and JIM5 (**f** and **g**). *RT* root tip, *BLC* Border-like cells, *BC* Border cells. *Bars*: 25 μm (**d**); 50 μm (**a**, **b**, **c**, **e**, **f**, and **g**)

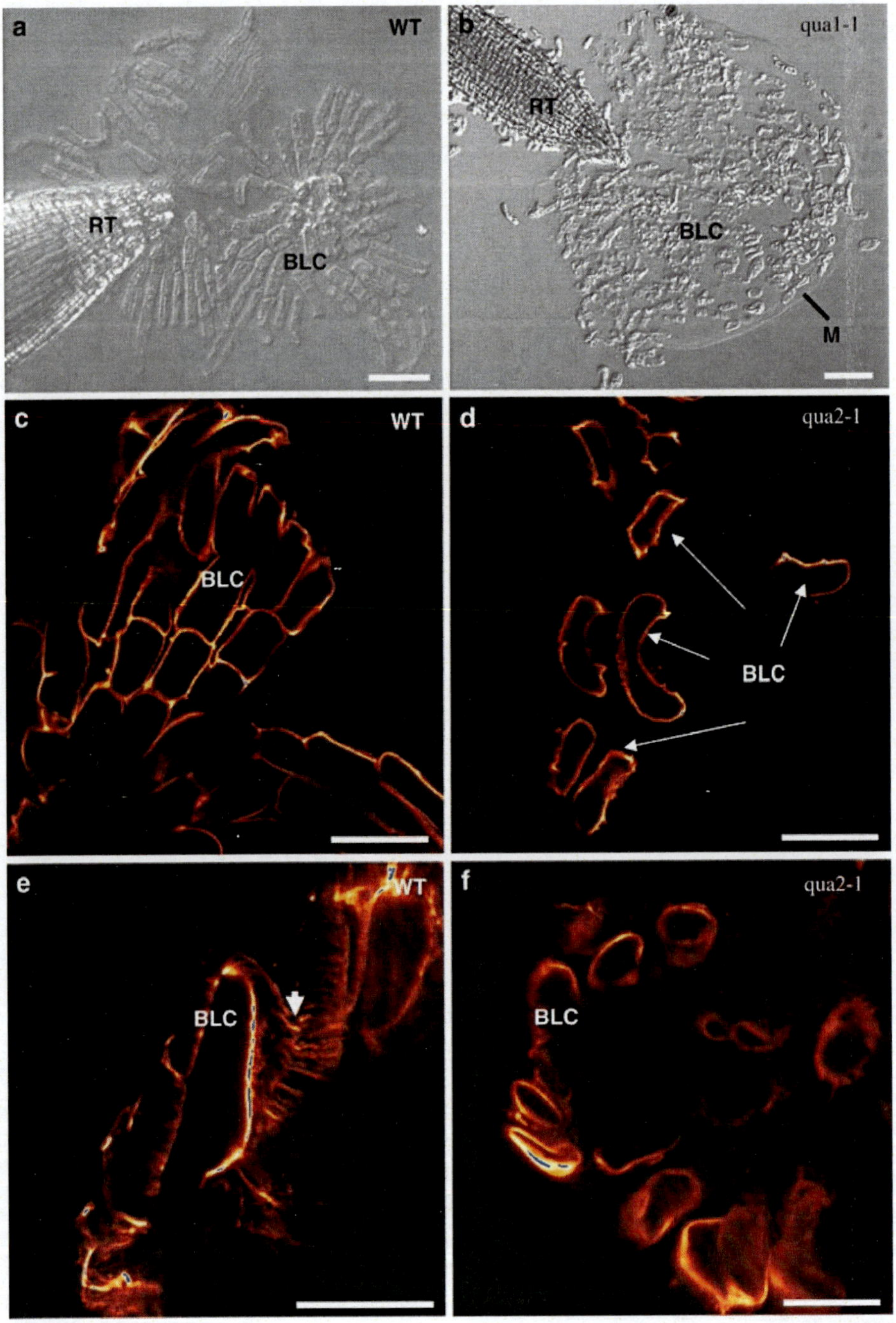

Fig. 4 Morphological phenotypes of border-like cells and polysaccharide immunostaining of cell wall epitopes in the wild-type (WT) and *quasimodo* mutants of Arabidopsis. Wild-type Columbia (**a**) and *quasimodo 1–1*, *qua1-1* (**b**) are shown. Note that in *qua 1–1*, border-like cells are converted into border cells (not attached to each other anymore) that are embedded in a thick mucilage (M). Wild-type Columbia (**c**, **e**) and *qua 2–1* mutant (**d**, **f**) border-like cells are immunostained with the

microorganisms. It has been shown that exposure of pea root to the pathogen *Nectria haematococca* results in the formation of a sort of mantle that covers the root tip (Gunawardena and Hawes 2002; Gunawardena et al. 2005). Under the microscope the mantle was shown to consist of a mixture of both border cells and fungal hyphae, and once the mantle is removed, the root tip remains free of infection indistinguishable from nontreated roots. Here, the root development and growth were unchanged from nontreated root, indicating that the apical meristems were still active and functional even with the presence of the pathogen in the surroundings. In such a case, border cells seem to fool the pathogen by acting as a decoy allowing the protection of the root tip and the apical meristems against infection. Similar observations have also been reported showing attraction and infection of cotton border cells by the fungus *Pythium dissotocum* (Goldberg et al. 1989). Border cells can also repel pathogenic bacteria by means of their secreted mucilage.

Third border cells are also capable of producing antidefense molecules. Border cells of legumes and cereals have been shown to secrete a large number of antimicrobial proteins including chitinases, peptidases, and glucanases, whose profile is modified in response to pathogenic bacteria (Wen et al. 2007; De-la-Peña et al. 2008). In addition to defensive proteins and enzymes, it has recently been shown that secretions of pea root border cells contain exDNA (Wen et al. 2009). This component is likely to exert a protective function within the secretome of border cells (Wen et al. 2009). The authors have clearly demonstrated that degradation of exDNA *via* nucleases (DNAse I or BAL31) resulted in an increased infection of root tips by *N. haematococca* supporting a role of this component in the immune response of root cells. In this regard, exDNA has long been known in many biological secretions including bacterial biofilms, snail mucigels, and white human blood cell matrices where it localizes with a number of antimicrobial peptides and proteins (Fahy et al. 1993; Allesen-Holm et al. 2006). In the case of human neutrophils, such a complex called also the NET "neutrophil extracellular trap" is capable of protecting the cells against pathogens at the sites of infection (Wartha et al. 2007; Guimarães-Costa et al. 2009). The discovery of exDNA in root tip secretions is highly exciting, but it is not easy to explain how exDNA exactly inhibits/reduces pathogen infection. This novel component of plant root secretions deserves further attention and careful investigations to unravel its mechanism of action on pathogenic microorganisms.

Another interesting role of border cells is that they have the capacity to attract and immobilize parasitic nematodes, thus limiting infection of the root. The chemical signal responsible for such an attraction is not known, but it is a heat-stable, polar fraction found in border cell secretions (Hawes et al. 2000). Also root border cells could contribute to inhibit penetration and invasion of cyst nematodes, thus limiting

Fig. 4 (continued) monoclonal antibodies, JIM5 (**c**, **d**) or CCRCM1 (antixyloglucan antibody) (**e**, **f**). In **e**, the *white arrowhead* indicates xyloglucan "fibers" at the cell surface of border-like cells. *BLC* border-like cells, *qua* quasimodo, *M* mucilage, *RT* root tip, *WT* wild type. *Bar*: 50 μm

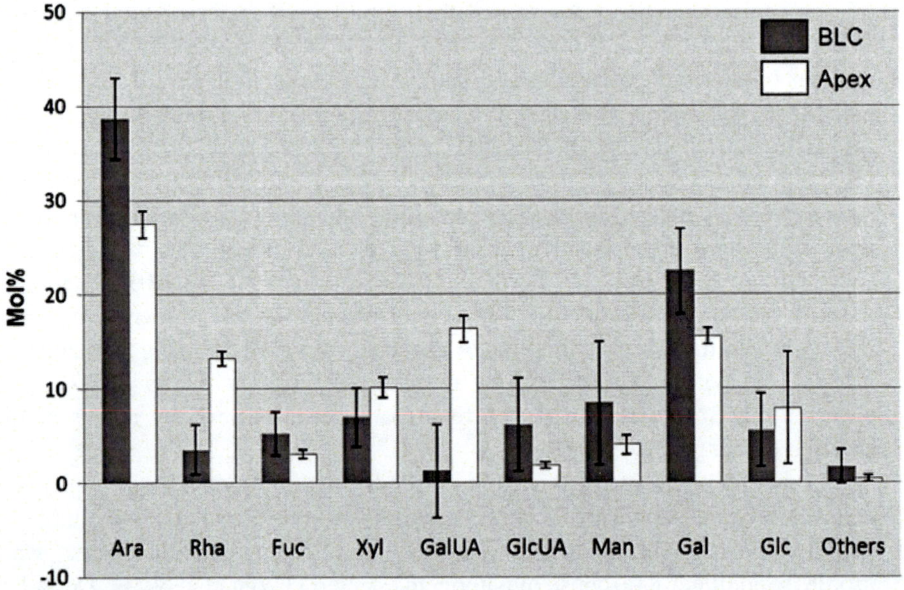

Fig. 5 Monosaccharide composition of cell walls extracted either from isolated border-like cells (BLC) or the root apex free of border cells (Apex) from radish. Note the high content of arabinosyl and galactosyl residues. *Ara* arabinose, *Fuc* Fucose, *Gal* galactose, *GalUA* galacturonic acid, *Glc* glucose, *GlcUA* glucuronic acid, *Man* mannose, *Rha* rhamnose, *Xyl* xylose

considerable loss of food crops. In this context, transgenic plants (including rice and potato) expressing protease inhibitors were shown to reduce the invasion and migration inside the root of several nematodes species (Fuller et al. 2008). For instance, transgenic potato plants that secrete a specific peptide interfering with nematode chemoreception under the control of CaMV35S promoter were able to significantly reduce the invasion by the cyst nematode *Globodera pallida* (Liu et al. 2005). Whether this peptide is expressed and secreted by root border cells in potato was not investigated. Targeting of such peptides specifically to border cells was recently reported by Lilley et al. (2010). The authors have identified a gene from Arabidopsis (MDK4-20; At5g54370) that possesses sequence homology to several genes expressed in maize root-cap cells. The promoter of this gene was then successfully used to direct expression of a nematode repellent peptide specifically to root caps, root border cells, and border-like cells of potato and Arabidopsis (Lilley et al. 2010). The MDK4-20 promoter was shown to remain active for a longer period of time than the constitutive CaMV35S promoter in detached border cells and border-like cells of the transgenic plants studied. Furthermore, the study has shown clearly that the targeted peptide reduced the establishment of the nematode *G. pallida* on root tissues. This is also an interesting discovery which paves the way to further investigations aimed at enhancing defense capabilities of border cells/border-like cells. Thus, using the same promoter and approach, it is possible to express various antimicrobial peptides or proteins specifically in border cells and border-like cells to target specific soilborne pathogens.

4 Conclusions and Future Prospects

Two centuries ago Charles Darwin (1880) highlighted in his book the importance of the root apex (including the root cap) for plant life when he wrote "... *the tip of the radicle acts like a brain*". Indeed, the root cap plays a vital role in plant root growth and health. It protects the root meristems. Both apical and root-cap meristems (stem-like cells of the root system) are fundamental in providing new cells that form new tissues for plant growth and survival. Also, it is worth noting that the root cap is the only plant organ that can be regenerated after its removal.

It is well known that the root apex absorbs all the resources (water and nutrients) required for the plant to grow from the soil. It is also well established that root tips have the capacity to develop local resistance to infection when other parts of the root are invaded by a given pathogen. Nowadays, root diseases caused by soilborne pathogens are of increasing concern as chemical pesticides are withdrawn from use in agriculture due to their high toxic effects on the environment and human health. Alternative strategies are thus required for crop protection. One of such strategies is to make use of natural molecular mechanisms of root resistance to develop novel protective compounds. Studies on border cells of the legume species pea and others have clearly shown that these are able (1) to secrete various defense molecules and (2) to confer a protective effect against pathogens. There are at least two strategies that can be used to enhance the protective capacity of root-cap cells including border cells and border-like cells. One strategy is transgenic expression of defense proteins or peptides specifically in root-cap cells. For instance, this can be done by introducing a transgene in a background of partial natural resistance to provide added protection to the plant. Studies have shown that it is possible to target genes against nematode parasitism to root tissues (Lilley et al. 2010). However, the production of transgenic crop plants is not very welcomed, especially in Europe. The other strategy is to stimulate the immune system of root cells by treatment with natural elicitors (e.g., chitin and peptidoglycans). This sort of plant "immunization" by natural stimulating agents is already used for grapevine and winter wheat protection using Iodus 40, a natural oligosaccharidic-elicitor isolated from *Laminaria digitata*, or Stifenia, an active extract from fenugreek seeds. This is probably the most preferred strategy as it is environmentally friendly and more cost-effective. Thus, crop protection along with environment protection in the context of an increasing world population is an exciting challenge for root biologists among other plant scientists.

Acknowledgment "I dedicate this chapter to Pr. I. El Hadrami, my friend and collaborator from the University of Marrakech (Morocco), with whom I have just started a collaborative program on the role of date palm root cells (including border cells) in defense against the bayoud disease caused by *Fusarium oxysporum*. He tragically passed away on November 24, 2010 in a car accident. I'll never forget his enthusiastic smile when discussing an exciting research idea." A.D Research on plant roots and crop protection in A. D laboratory is supported by le "Grand Réseau Régional de Haute Normandie 'Végétal-Agronomie et Transformation des Agro-ressources" and the University of Rouen.

References

Aida M, Beis D, Heidstra R, Willemsen V, Blilou I, Galinha C, Nussaume L, Noh YS, Amasino R, Scheres B (2004) The *PLETHORA* genes mediate patterning of the Arabidopsis root stem cell niche. Cell 119:109–120

Allesen-Holm M, Barken KB, Yang L, Klausen M, Webb JS, Kjelleberg S, Molin S, Givskov M, Tolker-Nielsen T (2006) A characterization of the DNA release in *Pseudomonas aeruginosa* cultures and biofilms. Mol Microbiol 59:1114–1128

Bacic A, Moody SF, Clarke AE (1986) Structural analysis of secreted root slime from maize (*Zea mays* L.). Plant Physiol 80:771–777

Boutilier K, Offringa R, Sharma VK, Kieft H, Ouellet T, Zhang L, Hattori J, Liu CM, Van Lammeren AA, Miki BL, Custers JB, van Lookeren Campagne MM (2002) Ectopic expression of *BABY BOOM* triggers a conversion from vegetative to embryonic growth. Plant Cell 14:1737–1749

Bradley DJ, Kjellbom P, Lamb CJ (1992) Elicitor- and wound-induced oxidative cross-linking of a proline-rich plant cell wall protein: a novel, rapid defense response. Cell 70:21–30

Brady JD, Fry SC (1997) Formation of di-isodityrosine and loss of isodityrosine in the cell walls of tomato tell-suspension cultures treated with Funga1 Elicitors or H_2O_2. Plant Physiol 115:87–92

Brigham LA, Woo HH, Wen F, Hawes MC (1998) Meristem-specific suppression of mitosis and a global switch in gene expression in the root cap of pea by endogenous signals. Plant Physiol 118:1223–1231

Cannesan MA, Gangneux C, Lanoue A, Giron D, Laval K, Hawes M, Driouich A, Vicré-Gibouin M (2011) Association between border cell responses and localized root infection by pathogenic Aphanomyces euteiches. Ann Bot 108:459–69

Cavalier DM, Lerouxel O, Neumetzler L, Yamauchi K, Reinecke A, Freshour G, Zabotina OA, Hahn MG, Burgert I, Pauly M, Raikhel NV, Keegstra K (2008) Disrupting two *Arabidopsis thaliana* xylosyltransferase genes results in plants deficient in xyloglucan, a major primary cell wall component. Plant Cell 20:1519–1537

Chaboud A, Rougier M (1984) Identification and localization of sugar components of rice (*Oryza sativa* L.) root cap mucilage. J Plant Physiol 116:323–330

Chapman K, Groot EP, Nichol S, Rost TL (2003) The pattern of root apical meristem organization and primary root determinate growth are coupled. J Plant Growth Regul 21:287–295

Curlango-Rivera R, Duclos DV, Ebolo JJ, Hawes MC (2010) Transient exposure of root tips to primary and secondary metabolites: Impact on root growth and production of border cells. Plant Soil 306:206–216

Darwin CR (1880) The power of movement in plants. John Murray, London

Davey ME, O'toole GA (2000) Microbial biofilms: from ecology to molecular genetics. Microbiol Mol Biol Rev 64:847–867

De-la-Peña C, Lei Z, Watson BS, Sumner LW, Vivanco JM (2008) Root-microbe communication through protein secretion. J Biol Chem 283:25247–25255

Driouich A, Durand C, Vicré-Gibouin M (2007) Formation and separation of root border cells. Trends Plant Sci 12:14–19

Driouich A, Durand C, Cannesan MA, Percoco G, Vicré-Gibouin M (2010) Border cells versus border-like cells: are they alike? J Exp Bot 61:3827–3831

Durand C, Vicré-Gibouin M, Follet-Gueye ML, Duponchel L, Moreau M, Lerouge P, Driouich A (2009) The organization pattern of root border-like cells of Arabidopsis is dependent on cell wall homogalacturonan. Plant Physiol 150:1411–1421

Esquerré-Tugayé MT, Lamport DTA (1979) Cell surfaces in plant-microorganism interactions: I. A structural investigation of cell wall hydroxyproline-rich glycoproteins which accumulate in fungus-infected plants. Plant Physiol 64:314–319

Fahy JV, Steiger DJ, Liu J, Basbaum CB, Finkbeiner WE, Boushey HA (1993) Markers of mucus secretion and DNA levels in induced sputum from asthmatic and from healthy subjects. Am Rev Respir Dis 147:1132–1137

Fuller VL, Lilley CJ, Urwin PE (2008) Nematode resistance. New Phytol 180:27–44

Goldberg NP, Hawes MC, Stanghellini ME (1989) Specific attraction to and infection of cotton root cap cells by zoospores of *Pythium dissotocum*. Can J Bot 67:1760–1767

Groot EP, Doyle JA, Nichol SA, Rost TL (2004) Phylogenetic distribution and evolution of root apical meristem organization in dicotyledonous angiosperms. Int J Plant Sci 165:97–105

Guimarães-Costa AB, Nascimento MT, Froment GS, Soares RP, Morgado FN, Conceição-Silva F, Saraiva EM (2009) *Leishmania amazonensis* promastigotes induce and are killed by neutrophil extracellular traps. Proc Natl Acad Sci USA 106:6748–6753

Gunawardena U, Hawes MC (2002) Tissue specific localization of root infection by fungal pathogens: role of root border cells. Mol Plant Microbe Interact 15:1128–1136

Gunawardena U, Rodriguez M, Straney D, Romeo JT, VanEtten HD, Hawes MC (2005) Tissue-specific localization of pea root infection by *Nectria haematococca*. Mechanisms and consequences. Plant Physiol 137:1363–1374

Hamamoto L, Hawes MC, Rost TL (2006) The production and release of living root cap border cells is a function of root apical meristem type in dicotyledonous angiosperm plants. Ann Bot 97:917–923

Hawes MC, Gunawardena U, Miyasaka S, Zhao X (2000) The role of root border cells in plant defense. Trends Plant Sci 5:128–133

Hawes MC, Bengough G, Cassab G, Ponce G (2003) Root caps and rhizosphere. J Plant Growth Regul 21:352–367

Knee EM, Gong FC, Gao M, Teplitski M, Jones AR, Foxworthy A, Mort AJ, Bauer WD (2001) Root mucilage from pea and its utilization by rhizosphere bacteria as a sole carbon source. Mol Plant Microbe Interact 14:775–784

Lilley CJ, Wang D, Atkinson HJ, Urwin PE (2010) Effective delivery of a nematode-repellent peptide using a root-cap-specific promoter. Plant Biotechnol J 9:151–161

Liu B, Hibbard JK, Urwin PE, Atkinson HJ (2005) The production of synthetic chemodisruptive peptides *in planta* disrupts the establishment of cyst nematode. Plant Biotechnol J 3:487–496

Madson M, Dunand C, Li X, Verma R, Vanzin GF, Caplan J, Shoue DA, Carpita NC, Reiter WD (2003) The MUR3 gene of Arabidopsis encodes a xyloglucan galactosyltransferase that is evolutionarily related to animal exostosins. Plant Cell 7:1662–1670

Merkouropoulos G, Shirsat AH (2003) The unusual Arabidopsis extensin gene *atext1* is expressed throughout plant development and is induced by a variety of biotic and abiotic stresses. Planta 217:356–366

Miyasaka SC, Hawes MC (2001) Possible role of root border cells in detection and avoidance of aluminium toxicity. Plant Physiol 125:1978–1987

Vanzin GF, Madson M, Carpita NC, Raikhel NV, Keegstra K, Reiter WD (2002) The *mur2* mutant of *Arabidopsis thaliana* lacks fucosylated xyloglucan because of a lesion in fucosyltransferase AtFUT1. Proc Natl Acad Sci USA 99:3340–3345

Vermeer J, Mc Cully ME (1982) The rhizosphere in *Zea*: new insights into its structure and development. Planta 156:45–61

Vicré M, Santaella C, Blanchet S, Gateau A, Driouich A (2005) Root border-like cells of Arabidopsis. Microscopical characterization and role in the interaction with rhizobacteria. Plant Physiol 138:998–1008

Wartha F, Beiter K, Normark S, Henriques-Normark B (2007) Neutrophil extracellular traps: casting the NET over pathogenesis. Curr Opin Microbiol 10:52–56

Wen F, Curlango-Rivera G, Hawes MC (2007) Proteins among the polysaccharides. A new perspective on root cap slime. Plant Sign Behav 2:410–412

Wen F, White GJ, VanEtten HD, Xiong Z, Hawes MC (2009) Extracellular DNA is required for root tip resistance to fungal infection. Plant Physiol 151:820–829

Willemsen V, Bauch M, Bennett T, Campilho A, Wolkenfelt H, Xu J, Haseloff J, Scheres B (2008) The NAC domain transcription factors *FEZ* and *SOMBRERO* control the orientation of cell division plane in *Arabidopsis* root stem cells. Dev Cell 15:913–922

Plant Volatiles and Other Specialized Metabolites: Synthesis, Storage, Emission, and Function

Vasiliki Falara and Eran Pichersky

Abstract Each plant species produces a set of specialized metabolites that interact with its biotic and abiotic environment to optimize plant fitness. The major classes of specialized metabolites are terpenoids, phenylpropanoids and alkaloids as well as a few classes of fatty acid and other amino acid derivatives, and they serve to protect the plants against herbivores and pathogens and attract beneficial organisms such as pollinators. These compounds are often synthesized and stored in, or emitted from, specific cells or structures that might occur in all aerial and underground parts of the plants. Here we review the limited knowledge we have about storage and emission of such compounds.

1 Introduction

Plants are primary producers – they harvest light energy, minerals from the soil and carbon dioxide from the air to synthesize carbohydrates, fats, proteins, and vitamins. Organisms higher on food chain, including humans, consume plants in order to survive. Plants have also been the source of many other compounds that humans find useful – spices, scents, and medicinal. Such compounds are usually designated as "secondary metabolites," "natural products," "botanicals," or "phytochemicals." People have known since antiquity that different plant species make, store, and emit different sets of such compounds. Work by early German ecologists in the late nineteenth century (reviewed in Hartmann 2008) showed that these compounds are made as an adaptation by a plant species to the particular challenges it confronts in its specific niche, such as attracting a specific pollinator or warding off specific pest. This knowledge was temporarily lost in the early

V. Falara • E. Pichersky (✉)
Department of Molecular, Cellular and Developmental Biology, University of Michigan, Ann Arbor, MI 48109, USA
e-mail: lelx@umich.edu

J.M. Vivanco and F. Baluška (eds.), *Secretions and Exudates in Biological Systems*, Signaling and Communication in Plants 12, DOI 10.1007/978-3-642-23047-9_6,

twentieth century, and for some time the predominant scientific opinion was that such compounds were simply "waste," by-products of biochemical reactions that had to be secreted out of the plant to cleanse the plant of such toxic compounds.

Since the 1950s, we have gained a renewed appreciation of the ecological roles of such compounds (Fraenkel 1959). We now have many examples of how specific compounds help plants survive and thrive in specific ecological situations. We also know many more such compounds – tens of thousands of such compounds have already been identified, and many more are awaiting discovery – and it has become clear than each plant species synthesizes only a small fraction of this repertoire. Because each species makes a distinct array of such compounds, and because they serve distinct roles, these compounds have now been designated as "specialized metabolites." In the last few years, much work has been devoted to understanding where and how specialized plant compounds are synthesized (Pichersky et al. 2006). Less attention has been devoted to how these compounds are stored and how they are emitted. This chapter attempts to briefly summarize our knowledge of where and how specialized compounds are synthesized, and the little knowledge we have gained on where these compounds are stored or emitted from.

Specialized metabolites occur in every part of the plants – roots, stems, leaves, flowers, fruits, and seeds. Moreover, within each organ they may be produced in just one type of cells or in multiple cells. When specialized metabolites are discussed, it is customary to divide them into chemical groups, and we will follow this approach to introduce them. However, in this chapter we are primarily concerned with the location in which they are synthesized, stored, and released, so we will devote more space to a discussion of these aspects, which will be divided along spatial boundaries.

1.1 The Major Classes of Specialized Metabolite and Their Biosynthetic Pathways

1.1.1 Terpenoids

There are tens of thousands of known terpene structures from plants. Terpenes are built from blocks of five carbons. The simplest terpene is isoprene, a 5-carbon compound that is synthesized from dimethylallyl diphosphate (DMAPP) and is emitted in large amounts from the leaves of many trees (Wilkinson et al. 2006; Singh et al. 2007; Schnitzler et al. 2010). Monoterpenes are volatile, 10-carbon compounds that are synthesized from the condensation of DMAPP and its isomer, isopentenyl diphosphate (IPP). Monoterpenes such as linalool and limonene are constituents of many floral bouquets (Dudareva et al. 1996; Deng et al. 2004; Giuliani et al. 2009). Sesquiterpenes are 15-carbon compounds made from one DMAPP molecule and two IPP molecules, are also volatile, and are found in scents as well as in leaf tissues as defense compounds (Schnee et al. 2002; Huber et al. 2005). Finally, the 20-carbon diterpenes are less volatile and often toxic, and are synthesized from one DMAPP and three IPP molecules. The synthesis of all these terpenes is catalyzed by a series of structurally and evolutionarily related enzymes called terpene synthases (Trapp and Croteau 2001; Martin et al. 2004). Both IPP and DMAPP are made in both the cytosol and the plastids, and there are terpene synthases that are active in the cytosol and some that are active in the plastids.

1.1.2 Phenylpropanoids

This is also a very large group of specialized metabolites in plants. They share a common origin from the amino acid L-phenylalanine (Phe), which itself is derived from the shikimic acid pathway that operates in the plastids. Once phenylalanine is made, however, it is transported to the cytosol and all other subcellular compartments of the cell. The phenylpropanoids are believed to be synthesized mostly in the cytosol. The first step in phenylpropanoid biosynthesis is the conversion of Phe to cinnamic acid by the enzyme PAL (phenylalanine ammonia lyase) (Yu and Jez 2008). Cinnamic acid is then converted to 4-hydroxycinnamic acid, and then linked to CoA. 4-Hydroxycinnamoyl-CoA then serves as a precursor to most of the phenylpropanoids (Yu and Jez 2008; Tanaka et al. 2008).

Extension of the 4-hydroxycinnamic acid backbone by various polyketide synthases leads to the production of various flavonoids, compounds with a basic structure of three-ring, conjugated bond system. Such compounds include the isoflavones found in legumes (e.g., genistin, daidzin) and the anthocyanin pigment molecules found throughout flowering plants. Some flavonoids have been shown to be secreted from roots and to attract mycorrhiza (Harrison 1999, 2005). Other phenylpropanoids are synthesized via a shorter polyketide extension that gives rise to a two-ring system (e.g., resveratrol) (Lanz et al. 1991). Another class of phenylpropanoid specialized compounds are synthesized via a reduction of the propenoate side chain to give rise to chavicol, eugenol, and related compounds (Gang et al. 2001; Vassao et al. 2006). Shortening of the propenoate side chain of cinnamic acid or 4-hydroxycinnamic acid gives rise to benzoic acid and 4-hydroxybenzoic acid, respectively (Orlova et al. 2006). The function of many of the phenylpropanoids is still not clear, but some are involved in defense and many of the small ones are found as components of floral scent (Bednarek et al. 2005; Oyama-Okubo et al. 2005; Tan et al. 2006; Mellway et al. 2009).

1.1.3 Alkaloids

A large group of specialized compounds in plants are called alkaloids. However, the alkaloids do not all issue from the same basic set of biochemical pathways. Rather, what they have in common is the presence of nitrogen inside at least one ring. Typically, the synthesis of an alkaloid compound begins with the decarboxylation of an amino acid such as tyrosine or tryptophan. The resulting amine is further elaborated to create additional rings with various functionalities, and moieties derived from other pathways (such as terpenoids) may be added as well. The result is often an extremely elaborated molecule such ajmaline, berberine, and morphine (Facchini et al. 1996; Stockigt and Panjikar 2007; Ziegler et al. 2009). In general, alkaloids are toxic to animals in one way or another and are believed to serve as defense compounds in plants.

1.1.4 Other Types of Specialized Metabolites

Beside these three large groups of specialized compounds, there are many others that are derived from various starting points anchored in primary metabolism. Some are derived from fatty acids, such as the methylketones (Fridman et al. 2005). Many others are derived from catabolism of amino acids, and they may contain nitrogen (but usually not as part of a ring) or sulfur (Nafisi et al. 2007). Many such compounds are volatile and impart a strong smell directly or after further decomposition. For example, onion and garlic plants contain many sulfur-containing compounds derived from cysteine (Challenger and Greenwood 1948; Jones et al. 2004) that undergo further decomposition upon damage, and Brassicaceae plants produce glucosinolates, which are derived from various amino acids which are oxidized, sulfated, and glycosylated and also decompose to give both toxic and volatile compounds upon injury to the tissue (Halkier and Gershenzon 2006).

2 Site of Synthesis, Storage and Emission

Plants have developed specific cell types and structures to support specialized metabolism at the sites where the produced metabolites are needed to play their physiological role. Since these metabolites are often toxic to the plant tissues themselves, particularly if they accumulate at high levels, plants have evolved mechanisms to sequester such compounds or otherwise protect themselves from adverse effects. Therefore, in this section discussion of examples of specialized metabolism will be organized according to the site of synthesis and the specific structures that facilitate storage, exudation, and emission.

2.1 Roots

Several other chapters in the book cover secretion from roots, so this topic will not be described in detail here. However, the roots are the sites of synthesis of many specialized metabolites that are then stored there or transported to the aerial parts of the plant. Examples of several classes of secondary metabolites secreted from roots have been described. The best-studied cases are the alkaloids produced from roots of Solanaceae species like nicotine, hyoscyamine and scopolamine (Hakkinen et al. 2005). Other alkaloids produced in the roots are emetine and cephaeline in *Psychotria ipecacuanha* (Nomura et al. 2008) and camptothecin in *Camptotheca acuminate* (Lorence et al. 2004). Most of these compounds were initially isolated and studied because of the use of root remedies in traditional medicine rather than their physiological role in the plant. Other classes of compounds are also produced in the roots of various species: fatty acid derivatives (Wu et al. 2009), naphthoquinones (Brigham et al. 1999), diterpene ginkolides (Nakanishi 2005) and labdane diterpenes (Munesada et al. 1992). An interesting example is nicotine, which is made in the roots of tobacco. Most of it stays in the roots but some is transported to the aerial parts after herbivory. There it can be further modified into the various nicotine derivatives isolated from the leaves of the plant (Siminszky et al. 2005, and see below). Since research on specialized metabolism in roots has been rather limited, so far no specific cells or structures have been found as the site of production. Nevertheless accumulation of pigments has been detected in root epidermal cells as well as their cell walls (Brigham et al. 1999).

2.2 Leaves and Stems

The most active site of synthesis and storage of specialized metabolism is often the trichomes (Fig. 1). Trichomes in general are epidermal appendages found on the aerial parts of a plant – leaves, stems, and flowers (Wagner et al. 2004; Schilmiller et al. 2008). Some trichomes are not metabolically active, and such trichomes are usually simple and unicellular (e.g., cotton fibers). Other trichomes, on the other hand, are typically multicellular. Their morphology is varied. Some include a "gland," or an enlarged subcuticular area, where specialized metabolites accumulate, and other secrete the compounds they synthesize instead of (or in addition to) storing them, often causing the plant surfaces to become "sticky" with the exuded material. The glandular trichomes may also release the material they store upon rupture, brought about by chewing insects and other herbivores, as observed for example in *Mentha piperita* (Turner et al. 2000).

It is estimated that 30% of the angiospermous plant have some type of trichomes on some parts of their aerial parts (Wagner et al. 2004). In many plants, multiple types of glandular and nonglandular trichomes are present, often on the same part of the plant. For example, plants in the Solanaceae, which include tobacco and wild and cultivated tomato species, have at least five different types of trichomes, including metabolically nonactive trichomes as well as storage and secreting ones (Antonious et al. 2005) (Fig. 1a). One type is responsible for the secretion of acyl sugars, while another type synthesizes and stores methyl-ketone and terpenoid derivatives (Fridman et al. 2005; Schilmiller et al. 2010). Mint, basil (Fig. 1d) and other Lamiaceae species also contain two types of glandular trichomes, called capitate and peltate (Gang et al. 2001). The capitates glands contain mostly hydrocarbons. The peltate glands in mints synthesize mostly terpenoids (Lange et al. 2000). In

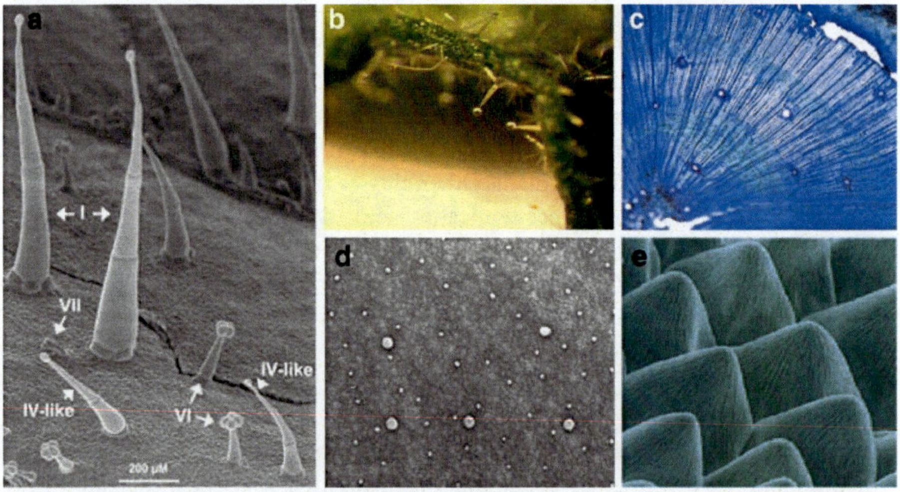

Fig. 1 Specific cells and structures for specialized metabolism synthesis. (**a**) Different types of trichomes on tomato leaf. (**b**) Resin secreting trichomes of *Cistus creticus*. (**c**) Resin-ducts on *Pinus sylvestris* stem. (**d**) Peltate and stellate trichomes on *Ocimum basillicum* leaf. (**e**) Conical cells of *Antirrhinum majus* petal epidermis. The individual frames in this figure first appeared in the following papers, and are printed by permission. (**a**) Kang J-H, Shi F, Jones AD, Marks MD, Howe GA (2010). Distortion of trichome morphology by the hairless mutation of tomato affects leaf surface chemistry. J Exp Bot 61:1053–1064; (**b, e**): Pichersky E, Noel JP, Dudareva N (2006) Biosynthesis of plant volatiles: nature's diversity and ingenuity. Science 311: 808. (**c**) http://sols.unlv.edu/Schulte/BIO426/StudentImages/Xylem/Xylem.html (**d**) Ioannidis D, Bonner L, Johnson CB (2002) UV-B is required for normal development of oil glands in *Ocimum basilicum* L. (Sweet Basil). Ann Bot 90:453–460

basil, these glands produce and store both phenylpropanoids and terpenoids, and the synthesis and storage of such compounds follow different developmental trajectories (Gang et al. 2001). In addition, differences among different basil chemotypes in the synthesis and accumulation of these two classes of compounds and of individual compounds are observed in genetically distinct cultivars (Xie et al. 2008).

Terpenoids and flavonoids are common constituents found in trichomes. Flavonoids are not typically volatile so when the trichomes are ruptured they remain on the surface of the leaves and stems. On the other hand, monoterpenes, sesquiterpenes, and some diterpenes are volatile, so when the trichomes burst they are released and quickly evaporate from the surface. Some sesquiterpene and diterpenes are further oxidized, and these compounds when released can form a stick substance on the surface that physically traps insects and protects the plants from herbivores. For example, the trichomes of the Mediterranean shrub *Cistus creticus* (Fig. 1b) and the sage *Salvia divinorum* produce labdane-type and neoclerodane diterpenes, respectively (Siebert 2004; Falara et al. 2008).

Other types of metabolites found in glandular trichomes include acylated sugars in Solanaceae species (Schilmiller et al. 2010), fatty acyl glycosides with cyclic structures in *Silene gallica*, glycosylated fatty acids from *Ibicella lutea* and *Proboscidea louisiana*, and fatty acid amides from *Medicago sativa* (Ranger et al. 2005; Asai and Fujimoto 2010; Asai et al. 2010).

Alkaloids have also been found to accumulate in glandular trichomes. As mentioned above, nicotine is produced in the roots of *Nicotiana* species (Saunders and Bush 1979). It is transported to the aerial parts of the plant, including the trichomes, particularly after the plant has been injured, and there it undergoes demethylation to form nornicotine (Siminszky et al. 2005). Neither nornicotine nor nicotine is secreted. However, in the trichomes nornicotine is *N*-acylated and then secreted on the leaf surface where it plays its defensive role against herbivores (Laue et al. 2000). Similarly, the alkaloid camptothecin accumulates in high concentrations in the glandular trichomes and secretory canals of leaves and stems of *C. acuminate*. It is thought that its biosynthesis occurs in the root and the compound is then transported to various other organs (Liu and Wang 2004).

Not all specialized compounds are synthesized and stored in glands. The "lemony" scent in lemongrass (*Cymbopogon* spp.) is due to the presence of the isomers geranial and neral (collectively called "citral"). However, these plants, in contrast to most other aromatic species, lack any type of glandular trichomes. Instead, they posses characteristic oil cells located inside the leaf blade. These cells are parenchymatic cells and are found on the adaxial side of the leaf. They are characterized by a high degree of lignification which protects the rest of the leaf tissue from the toxic citral synthesized and stored inside these cells (Lewinsohn et al. 1998).

In the Brassicaceae family, which includes the model plant species *Arabidopsis thaliana*, the sulfur- and nitrogen-rich glucosinolates are synthesized and stored in special cells in the stems and the flower stalks called S-cells (Koroleva et al. 2000) and at the periphery of the phloem parenchyma in the leaf (Shroff et al. 2008). Cells with high levels of myrosinases, the enzymes that convert glucosinolates into toxic thiocyanates, isothiocyanates and nitriles, surround these thin cell-walled S-cells. This cell-architecture helps the plant keep myrosinases and glucosinolates separate. However, when the tissue is physically crushed, or when a sucking insect such as an aphid inserts its proboscis through the myrosinase cells and the S-cells and toward the phloem, the contents of the two cells mix together and the toxic degradation products of glucosinolates are formed (Shroff et al. 2008).

A similar strategy with compartmentation of biosynthetic pathways in different cell types, ultimately leading to a dimeric compound, has been adapted by *Catharanthus roseus*, the source of vinca alkaloids, vinblastine and vincristine used in cancer chemotherapies (Noble 1990). The monomeric precursors of these compounds are catharanthine and vindoline. Catharanthine has been found in the surface wax of the leaves, while vindoline accumulates within specialized mesophyll cells, idioblasts and laticifers (Roepke et al. 2010). The biosynthesis of one of the early precursors of both compounds, the monoterpene secologanin, is initiated in internal phloem parenchyma cells, while the final biosynthetic steps, including the elaboration of secologanin and tryptamine, the other precursor, happens in the epidermal cells (Facchini and De Luca 2008). However, the exact mechanism responsible for covalently linking catharanthine and vindoline to produce the final alkaloid compound upon herbivory or wounding is still unclear.

2.3 *Stems-Specific Specialized Metabolites*

In species that bear secreting or glandular trichomes on their stems, trichome development and chemical composition follow a similar pattern with the equivalent structures on the leaves. Nevertheless, trichomes are not the only known structures recruited to specialized metabolism in the stem tissues. A well-studied example is the traumatic resin ducts in conifers (Fig. 1c). These are large canals restricted to the bark (phloem, cortex, periderm) of the main trunk and older branches that are lined up with terpene-synthesizing cells, leading to constitutive accumulation of terpenoid-based resin consisting mostly of monoterpenes and diterpenes but containing also lower levels of sesquiterpenes. Mechanical wounding, insect feeding, fungal elicitation, ethylene, and methyl-jasmonate causes additional axial traumatic resin ducts to appear within the developing xylem (Hudgins and Franceschi 2004; Byun-McKay et al. 2006; Hudgins et al. 2006; McKay et al. 2003). As in trichomes, many of the resin diterpenes are further oxidized, causing them to be less volatile and more "sticky". When trunks and branches are physically damaged, for example by boring insects, or broken, the mixture of terpenes will ooze out of the open area. The volatile monoterpenes and sesquiterpenes eventually evaporate, leaving the broken area (and sometimes the insects that bore into the bark), covered with the viscous, toxic diterpenes that harden further over time. The resin thus serves as both chemical and physical barrier. In addition, the emitted monoterpenes and sesquiterpenes could be detected by carnivorous insects that will attack the insects causing the damage to the tree in the first place (Pichersky and Gershenzon 2002).

2.3.1 Flowers

As the reproductive structures of plants, the flowers of animal-pollinated species have evolved many strategies to attract pollinators. Among the various floral characteristics, flower morphology (color and shape), scent and nectar rewards play critical roles. Specialized metabolites are the key elements in these processes: anthocyanins, carotenoids and betalains are the usual pigments responsible for the color, small usually lipophilic molecules contribute to the flower odor, while specialized metabolites in the nectar have recently been found to have repelling properties against unwanted pollinators (Grotewold 2006; Kessler and Baldwin 2007; Junker and Bluthgen 2010).

Terpenoids (monoterpenes and sesquiterpenes), phenylpropanoids and benzenoids as well as fatty acid derivatives and metabolites containing sulfur or nitrogen are among the broad range of chemicals found in floral bouquets. While the synthesis of such compounds has recently been given renewed attention and is covered briefly above, our knowledge on the site of floral volatile biosynthesis in flowers is still quite limited. Initial study of floral scent in orchids identified "osmophores", or "scent glands", as the site of synthesis and emission of volatiles (Stern et al. 1986; Vogel 1990). The term "osmophore" originally referred to a region on the epidermis that was characterized by the selective uptake and retention of the neutral red stain, specific for lipophilic material. However, whether the stained cells in such regions were the actual site of biosynthesis or just facilitated storage was not determined, nor the actual identity of the compounds that bound the dye. When genes and enzymes for scent biosynthesis were finally identified (in other species), it was shown in a few cases, for instance by in situ RNA hybridization or immunolocalization experiments, that the biosynthesis of volatile compounds is typically restricted to cells of the epidermal layers of petals and sometimes in a few other flower parts like the stigma and the style (Dudareva et al. 1996, 1998; Kolosova et al. 2001).

It was also shown that in some scented flowers, the epidermal cells of the petals have a characteristic conical shape that is thought to facilitate enhanced light absorption by pigments and to increase the surface for volatile emission (Fig. 1e) (Kolosova et al. 2001; Whitney et al. 2009). In many cases, adaxial and abaxial epidermal layers are morphologically and functionally distinct, with only the surfaces facing the potential pollinators engaged in insect attraction. Usually additional epidermal appendages (unicellular trichomes) are present that are metabolically active, providing chemical guideposts for the insect to find the nectar in order to increase pollination efficiency (Kolosova et al. 2001).

Orchids, with their sexually deceptive and food-deceptive strategies for pollination attraction, are among the plant species with the most fascinating pollinator attraction systems. Pollination by sexual deception always involves the emission of volatiles that mimic the sex pheromone of a female insect, while morphological similarity of the flower to the female insect may or may not be complete (Schiestl et al. 2000; Ayasse et al. 2003; Schiestl et al. 2003; Brodmann et al. 2009). The structures that mimic the physical appearance of the female insect, situated on the petals or the sepals, are typically the site of both scent synthesis and emission. Separating the scent-emitting parts from the rest of the flower has been shown to diminish attraction of pollinators (Peakall 2007).

2.3.2 Fruits

Fruit is an organ where plants invest a lot of energy on metabolic processes related with specialized compounds, since it is directly linked to plant fitness through seed dispersal. The presence of specialized metabolites as flavor ingredients is believed to have evolved as an aid in attracting animals that consume the fruit and deposit the seeds (sometimes after passage through the digestive system). In addition, specialized compounds with antimicrobicidal properties protect the fruit from spoilage. The most common classes of specialized metabolites found in fruit include the flavonoids, which contribute to their color and antioxidant capacity, as well as the terpenes and the phenylpropanoids, which impart to aroma and flavor (Goff and Klee 2006). The accumulation of metabolites in the fruit changes in a spatio-temporal pattern. Studies describing such patterns have so far focused on species with high commercial value (Schwab et al. 2008), while the molecular and biochemical mechanisms behind the changes in those patterns have been examined in only a few species. Even more limited is our knowledge about specific cells or structures in the fruits related to their biosynthesis.

Strawberry (*Fragaria* × *ananassa*) fruit accumulates cinnamoyl glucose esters, which are the precursors of volatile specialized metabolites such as the flavor constituents methyl cinnamate and ethyl cinnamate (Lunkenbein et al. 2006). The other major volatile of strawberry fruit is linalool, while its red color is caused by an accumulation of anthocyanidins as well as flavonols that act as copigments (Aharoni et al. 2004; Griesser et al. 2008). Another group of flavonoids accumulate to protect the fruit from pathogens that first affect the flower but are quiescent until ripening is almost completed (Halbwirth et al. 2006). Another example of fleshy fruit that serves as a model system for climacteric fruit ripening is tomato where volatiles, flavonoids and pigments all play a major role in fruit quality. Similar metabolite profiles have been found among several cultivated tomato species tested, while for a single cultivar a spatio-temporal specificity in the presence of specific metabolites was evident in different fruit tissues and the different stages of development (Moco et al. 2007).

An interesting example of what appears to be a defensive compound is seen in the fruit of the vanilla orchid. Mature green vanilla pods accumulate 4-*O*-(3-methoxy-benzaldehyde)-β-D-glucoside (glucovanillin). This intermediate is then hydrolyzed by an endogenous β-glucosidase to produce vanillin, the major aroma component of vanilla beans (Odoux and Brillouet 2009). Another profound example of aromatic seeds is the seeds of *Carum carvi* and *Anethum graveolens*, which contain large concentrations of limonene, carvone and carveols (Bouwmeester et al. 1995). In *Arabidopsis*, 3-benzoyloxypropylglucosinolate and 4-benzoyloxybutylglucosinolate are glucosinolates that preferentially accumulate in higher concentrations in developing seeds (Graser et al. 2001).

3 Conclusions

While a great deal of progress has been made identifying genes and enzymes involved in the synthesis of specialized metabolites in plants, and determining the site of synthesis of these compounds, the location and processes of storage and emission have received less attention. Progress in this area will depend in part on the development of methods to identify specific metabolites in situ. In addition, transport processes need to be investigated and the cellular machinery involved identified. Such knowledge will contribute to further understanding of plant ecological interactions as well as facilitate plant metabolic engineering.

References

Aharoni A, Giri AP, Verstappen FW, Bertea CM, Sevenier R, Sun Z, Jongsma MA, Schwab W, Bouwmeester HJ (2004) Gain and loss of fruit flavor compounds produced by wild and cultivated strawberry species. Plant Cell 16:3110–3131

Antonious GF, Kochhar TS, Simmons AM (2005) Natural products: seasonal variation in trichome counts and contents in *Lycopersicum hirsutum f. glabratum*. J Environ Sci Health B 40:619–631

Asai T, Fujimoto Y (2010) Cyclic fatty acyl glycosides in the glandular trichome exudate of *Silene gallica*. Phytochemistry 71:1410–1417

Asai T, Hara N, Fujimoto Y (2010) Fatty acid derivatives and dammarane triterpenes from the glandular trichome exudates of *Ibicella lutea* and *Proboscidea louisiana*. Phytochemistry 71:877–894

Ayasse M, Schiestl FP, Paulus HF, Ibarra F, Francke W (2003) Pollinator attraction in a sexually deceptive orchid by means of unconventional chemicals. Proc Biol Sci 270:517–522

Bednarek P, Schneider B, Svatos A, Oldham NJ, Hahlbrock K (2005) Structural complexity, differential response to infection, and tissue specificity of indolic and phenylpropanoid secondary metabolism in Arabidopsis roots. Plant Physiol 138:1058–1070

Bouwmeester H, Davies J, Toxopeus H (1995) Enantiomeric composition of carvone, limonene, and carveols in seeds of dill and annual and biennial caraway varieties. J Agric Food Chem 43:3057–3064

Brigham LA, Michaels PJ, Flores HE (1999) Cell-specific production and antimicrobial activity of naphthoquinones in roots of *lithospermum erythrorhizon*. Plant Physiol 119:417–428

Brodmann J, Twele R, Francke W, Yi-bo L, Xi-qiang S, Ayasse M (2009) Orchid mimics honey bee alarm pheromone in order to attract hornets for pollination. Curr Biol 19:1368–1372

Byun-McKay A, Godard KA, Toudefallah M, Martin DM, Alfaro R, King J, Bohlmann J, Plant AL (2006) Wound-induced terpene synthase gene expression in Sitka spruce that exhibit resistance or susceptibility to attack by the white pine weevil. Plant Physiol 140:1009–1021

Challenger F, Greenwood D (1948) Sulphur compounds of Allium; detection of n-propylthiol in the onion; the fission and methylation of diallyl disulphide in cultures of *Scopulariopsis brevicaulis*. Biochem J 43:ix

Deng C, Song G, Hu Y (2004) Application of HS-SPME and GC-MS to characterization of volatile compounds emitted from *Osmanthus* flowers. Ann Chim 94:921–927

Dudareva N, Cseke L, Blanc VM, Pichersky E (1996) Evolution of floral scent in *Clarkia*: novel patterns of S-linalool synthase gene expression in the *C. breweri* flower. Plant Cell 8:1137–1148

Dudareva N, D'Auria JC, Nam KH, Raguso RA, Pichersky E (1998) Acetyl-CoA:benzylalcohol acetyltransferase–an enzyme involved in floral scent production in Clarkia breweri. Plant J 14:297–304

Facchini PJ, De Luca V (2008) Opium poppy and Madagascar periwinkle: model non-model systems to investigate alkaloid biosynthesis in plants. Plant J 54:763–784

Facchini PJ, Penzes C, Johnson AG, Bull D (1996) Molecular characterization of berberine bridge enzyme genes from opium poppy. Plant Physiol 112:1669–1677

Falara V, Fotopoulos V, Margaritis T, Anastasaki T, Pateraki I, Bosabalidis AM, Kafetzopoulos D, Demetzos C, Pichersky E, Kanellis AK (2008) Transcriptome analysis approaches for the isolation of trichome-specific genes from the medicinal plant *Cistus creticus* subsp. *creticus*. Plant Mol Biol 68:633–651

Fraenkel GF (1959) The raison d'être if secondary plant substances. Science 129:1466–1470

Fridman E, Wang J, Iijima Y, Froehlich JE, Gang DR, Ohlrogge J, Pichersky E (2005) Metabolic, genomic, and biochemical analyses of glandular trichomes from the wild tomato species *Lycopersicon hirsutum* identify a key enzyme in the biosynthesis of methylketones. Plant Cell 17:1252–1267

Gang DR, Wang J, Dudareva N, Nam KH, Simon JE, Lewinsohn E, Pichersky E (2001) An investigation of the storage and biosynthesis of phenylpropenes in sweet basil. Plant Physiol 125:539–555

Giuliani C, Pellegrino RM, Tirillini B, Bini LM (2009) Composition of essential oils from leaves and flowers of *Stachys germanica* subsp. *salviifolia* (Ten.) gams (Labiatae) and related secretory structures. Nat Prod Commun 4:831–834

Goff SA, Klee HJ (2006) Plant volatile compounds: sensory cues for health and nutritional value? Science 311:815–819

Graser G, Oldham NJ, Brown PD, Temp U, Gershenzon J (2001) The biosynthesis of benzoic acid glucosinolate esters in *Arabidopsis thaliana*. Phytochemistry 57:23–32

Griesser M, Hoffmann T, Bellido ML, Rosati C, Fink B, Kurtzer R, Aharoni A, Munoz-Blanco J, Schwab W (2008) Redirection of flavonoid biosynthesis through the down-regulation of an anthocyanidin glucosyltransferase in ripening strawberry fruit. Plant Physiol 146:1528–1539

Grotewold E (2006) The genetics and biochemistry of floral pigments. Annu Rev Plant Biol 57:761–780

Hakkinen ST, Moyano E, Cusido RM, Palazon J, Pinol MT, Oksman-Caldentey KM (2005) Enhanced secretion of tropane alkaloids in Nicotiana tabacum hairy roots expressing heterologous hyoscyamine-6beta-hydroxylase. J Exp Bot 56:2611–2618

Halbwirth H, Puhl I, Haas U, Jezik K, Treutter D, Stich K (2006) Two-phase flavonoid formation in developing strawberry (*Fragaria x ananassa*) fruit. J Agric Food Chem 54:1479–1485

Halkier BA, Gershenzon J (2006) Biology and biochemistry of glucosinolates. Annu Rev Plant Biol 57:303–333

Harrison MJ (1999) Molecular and cellular aspects of the arbuscular mycorrhizal symbiosis. Annu Rev Plant Physiol Plant Mol Biol 50:361–389

Harrison MJ (2005) Signaling in the arbuscular mycorrhizal symbiosis. Annu Rev Microbiol 59:19–42

Hartmann T (2008) The lost origin of chemical ecology in the late 19th century. Proc Natl Acad Sci USA 105:4541–4546

Huber DP, Philippe RN, Godard KA, Sturrock RN, Bohlmann J (2005) Characterization of four terpene synthase cDNAs from methyl jasmonate-induced Douglas-fir, *Pseudotsuga menziesii*. Phytochemistry 66:1427–1439

Hudgins JW, Franceschi VR (2004) Methyl jasmonate-induced ethylene production is responsible for conifer phloem defense responses and reprogramming of stem cambial zone for traumatic resin duct formation. Plant Physiol 135:2134–2149

Hudgins JW, Ralph SG, Franceschi VR, Bohlmann J (2006) Ethylene in induced conifer defense: cDNA cloning, protein expression, and cellular and subcellular localization of 1-aminocyclopropane-1-carboxylate oxidase in resin duct and phenolic parenchyma cells. Planta 224:865–877

Jones MG, Hughes J, Tregova A, Milne J, Tomsett AB, Collin HA (2004) Biosynthesis of the flavour precursors of onion and garlic. J Exp Bot 55:1903–1918

Junker RR, Bluthgen N (2010) Floral scents repel facultative flower visitors, but attract obligate ones. Ann Bot 105:777–782

Kessler D, Baldwin IT (2007) Making sense of nectar scents: the effects of nectar secondary metabolites on floral visitors of *Nicotiana attenuata*. Plant J 49:840–854

Kolosova N, Sherman D, Karlson D, Dudareva N (2001) Cellular and subcellular localization of S-adenosyl-L-methionine:benzoic acid carboxyl methyltransferase, the enzyme responsible for biosynthesis of the volatile ester methylbenzoate in snapdragon flowers. Plant Physiol 126:956–964

Koroleva OA, Davies A, Deeken R, Thorpe MR, Tomos AD, Hedrich R (2000) Identification of a new glucosinolate-rich cell type in Arabidopsis flower stalk. Plant Physiol 124:599–608

Lange BM, Wildung MR, Stauber EJ, Sanchez C, Pouchnik D, Croteau R (2000) Probing essential oil biosynthesis and secretion by functional evaluation of expressed sequence tags from mint glandular trichomes. Proc Natl Acad Sci USA 97:2934–2939

Lanz T, Tropf S, Marner FJ, Schroder J, Schroder G (1991) The role of cysteines in polyketide synthases. Site-directed mutagenesis of resveratrol and chalcone synthases, two key enzymes in different plant-specific pathways. J Biol Chem 266:9971–9976

Laue G, Preston CA, Baldwin IT (2000) Fast track to the trichome: induction of N-acyl nornicotines precedes nicotine induction in *Nicotiana repanda*. Planta 210:510–514

Lewinsohn E, Dudai N, Tadmor Y, Katzir I, Ravid U, Putievsky E, Joel DM (1998) Histochemical localization of citral accumulation in lemongrass leaves (*Cymbopogon citratus* (DC.) Stapf., Poaceae). Ann Bot 81:35–39

Liu WZ, Wang ZF (2004) Accumulation and localization of camptothecin in young shoot of *Camptotheca acuminata*. Zhi Wu Sheng Li Yu Fen Zi Sheng Wu Xue Xue Bao 30:405–412

Lorence A, Medina-Bolivar F, Nessler CL (2004) Camptothecin and 10-hydroxycamptothecin from *Camptotheca acuminata* hairy roots. Plant Cell Rep 22:437–441

Lunkenbein S, Bellido M, Aharoni A, Salentijn EM, Kaldenhoff R, Coiner HA, Munoz-Blanco J, Schwab W (2006) Cinnamate metabolism in ripening fruit. Characterization of a UDP-glucose: cinnamate glucosyltransferase from strawberry. Plant Physiol 140:1047–1058

Martin DM, Faldt J, Bohlmann J (2004) Functional characterization of nine Norway Spruce TPS genes and evolution of gymnosperm terpene synthases of the TPS-d subfamily. Plant Physiol 135:1908–1927

McKay SA, Hunter WL, Godard KA, Wang SX, Martin DM, Bohlmann J, Plant AL (2003) Insect attack and wounding induce traumatic resin duct development and gene expression of (−)-pinene synthase in Sitka spruce. Plant Physiol 133:368–378

Mellway RD, Tran LT, Prouse MB, Campbell MM, Constabel CP (2009) The wound-, pathogen- and ultraviolet B-responsive MYB134 gene encodes an R2R3 MYB transcription factor that regulates proanthocyanidin synthesis in poplar. Plant Physiol 150:924–941

Moco S, Capanoglu E, Tikunov Y, Bino RJ, Boyacioglu D, Hall RD, Vervoort J, De Vos RC (2007) Tissue specialization at the metabolite level is perceived during the development of tomato fruit. J Exp Bot 58:4131–4146

Munesada K, Siddiqui HL, Suga T (1992) Biologically active labdane-type diterpene glycosides from the root-stalks of *Gleichenia japonica*. Phytochemistry 31:1533–1536

Nafisi M, Goregaoker S, Botanga CJ, Glawischnig E, Olsen CE, Halkier BA, Glazebrook J (2007) Arabidopsis cytochrome P450 monooxygenase 71A13 catalyzes the conversion of indole-3-acetaldoxime in camalexin synthesis. Plant Cell 19:2039–2052

Nakanishi K (2005) Terpene trilactones from *Gingko biloba*: from ancient times to the 21st century. Bioorg Med Chem 13:4987–5000

Noble RL (1990) The discovery of the vinca alkaloids–chemotherapeutic agents against cancer. Biochem Cell Biol 68:1344–1351

Nomura T, Quesada AL, Kutchan TM (2008) The new beta-D-glucosidase in terpenoid-isoquinoline alkaloid biosynthesis in *Psychotria ipecacuanha*. J Biol Chem 283:34650–34659

Odoux E, Brillouet JM (2009) Anatomy, histochemistry and biochemistry of glucovanillin, oleoresin and mucilage accumulation sites in green mature vanilla pod (*Vanilla planifolia*; Orchidaceae): a comprehensive and critical reexamination. Fruits 64:221–241

Orlova I, Marshall-Colon A, Schnepp J, Wood B, Varbanova M, Fridman E, Blakeslee JJ, Peer WA, Murphy AS, Rhodes D, Pichersky E, Dudareva N (2006) Reduction of benzenoid synthesis in petunia flowers reveals multiple pathways to benzoic acid and enhancement in auxin transport. Plant Cell 18:3458–3475

Oyama-Okubo N, Ando T, Watanabe N, Marchesi E, Uchida K, Nakayama M (2005) Emission mechanism of floral scent in *Petunia axillaris*. Biosci Biotechnol Biochem 69:773–777

Peakall R (2007) Speciation in the Orchidaceae: confronting the challenges. Mol Ecol 16:2834–2837

Pichersky E, Gershenzon J (2002) The formation and function of plant volatiles: perfumes for pollinator attraction and defense. Curr Opin Plant Biol 5:237–243

Pichersky E, Noel JP, Dudareva N (2006) Biosynthesis of plant volatiles: nature's diversity and ingenuity. Science 311:808–811

Ranger CM, Winter RE, Rottinghaus GE, Backus EA, Johnson DW (2005) Mass spectral characterization of fatty acid amides from alfalfa trichomes and their deterrence against the potato leafhopper. Phytochemistry 66:529–541

Roepke J, Salim V, Wu M, Thamm AM, Murata J, Ploss K, Boland W, De Luca V (2010) Vinca drug components accumulate exclusively in leaf exudates of Madagascar periwinkle. Proc Natl Acad Sci USA 107:15287–15292

Saunders JW, Bush LP (1979) Nicotine biosynthetic enzyme activities in *Nicotiana tabacum* L. genotypes with different alkaloid levels. Plant Physiol 64:236–240

Schiestl FP, Ayasse M, Paulus HF, Lofstedt C, Hansson BS, Ibarra F, Francke W (2000) Sex pheromone mimicry in the early spider orchid (*Ophrys sphegodes*): patterns of hydrocarbons as the key mechanism for pollination by sexual deception. J Comp Physiol A 186:567–574

Schiestl FP, Peakall R, Mant JG, Ibarra F, Schulz C, Franke S, Francke W (2003) The chemistry of sexual deception in an orchid-wasp pollination system. Science 302:437–438

Schilmiller AL, Last RL, Pichersky E (2008) Harnessing plant trichome biochemistry for the production of useful compounds. Plant J 54:702–711

Schilmiller A, Shi F, Kim J, Charbonneau AL, Holmes D, Daniel Jones A, Last RL (2010) Mass spectrometry screening reveals widespread diversity in trichome specialized metabolites of tomato chromosomal substitution lines. Plant J 62:391–403

Schnee C, Kollner TG, Gershenzon J, Degenhardt J (2002) The maize gene terpene synthase 1 encodes a sesquiterpene synthase catalyzing the formation of (E)-beta-farnesene, (E)-nerolidol, and (E, E)-farnesol after herbivore damage. Plant Physiol 130:2049–2060

Schnitzler JP, Louis S, Behnke K, Loivamaki M (2010) Poplar volatiles – biosynthesis, regulation and (eco)physiology of isoprene and stress-induced isoprenoids. Plant Biol 12:302–316

Schwab W, Davidovich-Rikanati R, Lewinsohn E (2008) Biosynthesis of plant-derived flavor compounds. Plant J 54:712–732

Shroff R, Vergara F, Muck A, Svatos A, Gershenzon J (2008) Nonuniform distribution of glucosinolates in *Arabidopsis thaliana* leaves has important consequences for plant defense. Proc Natl Acad Sci USA 105:6196–6201

Siebert DJ (2004) Localization of salvinorin A and related compounds in glandular trichomes of the psychoactive sage, *Salvia divinorum*. Ann Bot 93:763–771

Siminszky B, Gavilano L, Bowen SW, Dewey RE (2005) Conversion of nicotine to nornicotine in *Nicotiana tabacum* is mediated by CYP82E4, a cytochrome P450 monooxygenase. Proc Natl Acad Sci USA 102:14919–14924

Singh AP, Varshney CK, Singh UK (2007) Seasonal variations in isoprene emission from tropical deciduous tree species. Environ Monit Assess 131:231–235

Stern WL, Curry KJ, Whitten WM (1986) Staining fragrance glands in orchid flowers. Bull Torr Bot Club 113:288–297

Stockigt J, Panjikar S (2007) Structural biology in plant natural product biosynthesis -architecture of enzymes from monoterpenoid indole and tropane alkaloid biosynthesis. Nat Prod Rep 24:1382–1400

Tan KH, Tan LT, Nishida R (2006) Floral phenylpropanoid cocktail and architecture of *Bulbophyllum vinaceum* orchid in attracting fruit flies for pollination. J Chem Ecol 32:2429–2441

Tanaka Y, Sasaki N, Ohmiya A (2008) Biosynthesis of plant pigments: anthocyanins, betalains and carotenoids. Plant J 54:733–749

Trapp SC, Croteau RB (2001) Genomic organization of plant terpene synthases and molecular evolutionary implications. Genetics 158:811–832

Turner GW, Gershenzon J, Croteau RB (2000) Development of peltate glandular trichomes of peppermint. Plant Physiol 124:665–680

Vassao DG, Gang DR, Koeduka T, Jackson B, Pichersky E, Davin LB, Lewis NG (2006) Chavicol formation in sweet basil (*Ocimum basilicum*): cleavage of an esterified C9 hydroxyl group with NAD(P)H-dependent reduction. Org Biomol Chem 4:2733–2744

Vogel S (1990) The role of scent glands in pollination (transl. by Bhatti JS). Smithsonian Institute, Washington, DC

Wagner GJ, Wang E, Shepherd RW (2004) New approaches for studying and exploiting an old protuberance, the plant trichome. Ann Bot 93:3–11

Whitney HM, Chittka L, Bruce TJ, Glover BJ (2009) Conical epidermal cells allow bees to grip flowers and increase foraging efficiency. Curr Biol 19:948–953

Wilkinson MJ, Owen SM, Possell M, Hartwell J, Gould P, Hall A, Vickers C, Nicholas Hewitt C (2006) Circadian control of isoprene emissions from oil palm (*Elaeis guineensis*). Plant J 47:960–968

Wu L, Dixon PM, Nikolau B, Kraus GA, Widrlechner MP, Wurtele ES (2009) Metabolic profiling of Echinacea genotypes and a test of alternative taxonomic Treatments. Planta Med 75:178–183

Xie Z, Kapteyn J, Gang DR (2008) A systems biology investigation of the MEP/terpenoid and shikimate/phenylpropanoid pathways points to multiple levels of metabolic control in sweet basil glandular trichomes. Plant J 54:349–361

Yu O, Jez JM (2008) Nature's assembly line: biosynthesis of simple phenylpropanoids and polyketides. Plant J 54:750–762

Ziegler J, Facchini PJ, Geissler R, Schmidt J, Ammer C, Kramell R, Voigtlander S, Gesell A, Pienkny S, Brandt W (2009) Evolution of morphine biosynthesis in opium poppy. Phytochemistry 70:1696–1707

Lignocellulose Decomposition by Microbial Secretions

Navaneetha Santhanam, Dayakar V. Badri, Stephen R. Decker, Daniel K. Manter, Kenneth F. Reardon, and Jorge M. Vivanco

Abstract Carbon storage in terrestrial ecosystems is contingent upon the natural resistance of plant cell wall polymers to rapid biological degradation. Nevertheless, certain microorganisms have evolved remarkable means to overcome this natural resistance. Lignocellulose decomposition by microorganisms comprises an essential step in closing the loop of the global carbon cycle as it facilitates the recycling of carbon reposited in the form of structural polymers in plant cell walls. The significance of microbial decomposition of lignocellulose has recently risen to greater heights with the revisitation of the potential of lignocellulosic biomass as a valuable and abundant feedstock for the renewable energy and bioproducts industry. The scope of this chapter is to succinctly touch upon the composition of lignocellulosic biomass, the major enzymes involved in decomposing lignocellulosic biomass, and the fungi and bacteria that secrete these enzymes.

N. Santhanam
Department of Chemical and Biological Engineering, Colorado State University, Fort Collins, CO, USA

Department of Horticulture and Landscape Architecture, Colorado State University, Fort Collins, CO 80523-1370, USA

D.V. Badri • J.M. Vivanco (✉)
Department of Horticulture and Landscape Architecture, Colorado State University, Fort Collins, CO 80523-1370, USA
e-mail: j.vivanco@colostate.edu

S.R. Decker
National Renewable Energy Laboratory, Biosciences Center, Golden, CO, USA

D.K. Manter
USDA-ARS Soil-Plant-Nutrient Research Unit, Fort Collins, CO, USA

K.F. Reardon
Department of Chemical and Biological Engineering, Colorado State University, Fort Collins, CO, USA

J.M. Vivanco and F. Baluška (eds.), *Secretions and Exudates in Biological Systems*, Signaling and Communication in Plants 12, DOI 10.1007/978-3-642-23047-9_7,

1 Lignocellulose Decomposition by Microbial Secretions

Land plants sequester carbon from the atmosphere through photosynthesis and store it as organic matter, primarily in the form of lignocellulose. The term "lignocellulose" refers to plant cell wall derived biomass such as shoots, leaves and stems of living plants, wood, agricultural residues, and forest litter. Irrespective of its source, lignocellulosic biomass is composed of three major components – lignin, cellulose, and hemicelullose. Together, these polymers provide structural integrity to plants while also serving as reservoirs for carbon storage. The carbon locked up in lignocellulose is recycled back to the atmosphere by several processes – autotrophic respiration by land plants, heterotrophic respiration by microbes that oxidize soil organic carbon originating from sources such as fallen trees or plant litter, and forest and agricultural waste fires (Falkowski et al. 2000). Fires usually oxidize large amounts of organic matter in a very short time span, compared to the respirative processes. Global carbon recycling from terrestrial ecosystems is regulated by the interplay between the inherent recalcitrance of lignocellulosic biomass and the ability of certain microorganisms to overcome this barrier.

The unique degradative mechanisms developed by such microorganisms to release the carbon preserved in lignocellulosic biomass are thus extremely significant. Knowledge regarding these mechanisms is important not only in terms of understanding factors that affect the global carbon cycle, but also in terms of enhancing human utilization of plant cell wall polysaccharides for industrial applications that include timber for construction, fibers for textiles, cellulose for paper, and more recently, lignocellulosic biofuels. This chapter aims to provide an overview of some aspects of the very large and rapidly growing field of microbial lignocellulose decomposition. It begins with a brief overview of lignocellulosic biomass composition and then describes the major microbial secretions that have evolved to degrade the individual components of lignocellulosic biomass.

2 Lignocellulosic Biomass

Plant cell walls constitute about 70% of the 130 billion tons of terrestrial biomass estimated to be produced annually, worldwide, representing an abundant natural resource (Duchesne and Larson 1989). Understanding the chemistry of lignocellulosic biomass is fundamental to effective utilization of this resource. Initially, growing plant cells lay down primary cell walls comprised principally of cellulose, which allows the cells to elongate under turgor pressure as the fibrils slip past each other. Cellulose in plants occurs as bundles of unbranched individual glucan chains called elementary fibrils or microfibrils, which are its fundamental structural unit (Haigler et al. 1980). These microfibrils are intricately entwined with hemicelluloses, a heterogeneous group of branched polysaccharides. The cellulose and hemicelluloses are enveloped in a hydrated matrix made up of pectins, another class of highly branched heterogeneous acidic polysaccharides. Inter- and intrachain hydrogen bonding between these diverse polysaccharides holds the cell wall components together. The hydrophilic nature of the pectins and hemicelluloses also help retain water in the plant cell wall of living cells, a critical aspect of cell viability. In addition to the networks of pectins and hemicelluloses, primary cell walls contain 2–10% structural glycoproteins, some enzymes, and have 1–5% ionically and covalently bound minerals (O'Neill and York 2003).

After cessation of plant cell elongation, secondary walls are deposited inside primary walls, usually in a number of layers (McCann and Carpita 2008). Secondary walls contain cellulose and hemicelluloses but differ from primary cell walls as they also contain lignin, a three-dimensional phenyl propane polymer, in addition to the hydrated matrix. Despite an apparent overall uniformity in terms of components, plant cell walls are inherently extremely diverse. Different proportions of cellulose, hemicelluloses, and lignin amid different species, differences in compositions of the hemicelluloses and lignin polymer, growth stage, growth conditions, and different cell types even from different parts of a single plant contribute to the incredible diversity of plant cell walls (Pauly and Keegstra 2010). Typical proportions found in most lignocellulosic materials are in the range of 40–50% cellulose, 15–35% hemicellulose, and 12–30% lignin, with percentages expressed in terms of dry weight relative to total biomass. This diversity of structure and composition can significantly impact the usefulness of various biomass types in biofuels production.

2.1 Cellulose

Cellulose molecules consist of linear condensation polymers made up of D-anhydroglucopyranose units linked by β-1,4-glucosidic linkages. Simply stated, cellulose is a polymer made up of glucose, although, as the -1,4 linkage results in every other glucose being inverted, cellobiose is more correctly cited as the repeating unit. One end of the cellulose chain is termed the reducing end, as the hemiacetal ring is open, exposing the reducing aldehyde. The other end of the chain is called the non-reducing end because the C1 carbon in the hemiacetal is involved in the β-1,4-bond, preventing ring opening. This property provides directionality to cellulose chains. The anhydrocellobiose repeating unit for cellulose also results in the major structural difference between cellulose and starch. As each anhydroglucose unit is rotated 180° with respect to its nearest neighbor, cellulose chains are very flat, unlike helical starch chains.

A cellulose fiber contains crystalline and amorphous regions, with crystalline regions being more resistant to enzymatic degradation. X-ray diffraction data of pure crystalline cellulose from various sources have indicated that cellulose crystals are made of chains arranged in layered sheets (Gardner and Blackwel 1974). Within each sheet, the chains align parallel to each other and are linked by hydrogen bonds while the sheets are stacked by Van der Waals interactions. The hydrophobicity of cellulose sheets makes cellulose resistant to chemical hydrolysis because of the formation of a dense aqueous layer near the hydrated cellulose surface (Matthews et al. 2006) while the strong intra- and interchain hydrogen bonding networks make crystalline cellulose resistant to enzymatic hydrolysis (Nishiyama et al. 2002). The fine details of cellulose structure and the impact of these properties on the formation and deconstruction of cellulose is slowly being worked out, primarily through computational modeling techniques (Beckham et al. 2011a, b; Matthews et al. 2011).

2.2 Hemicellulose

The term "hemicellulose" is used to describe a group of branched heteropolysaccharides of the plant cell wall that is distinct from cellulose and pectin, and made up of a backbone of β-1,4-linked hexoses as well as pentoses. Classically they have been defined as the part of the plant cell wall that can be extracted with alkali but not with hot water, chelating agents or dilute acid, though this definition does not always hold (Scheller and Ulvskov 2010). Pentosans or xylans are the hemicelluloses made up of D-xylose while hexosans or mannans are the hemicelluloses made up of D-glucose and D-mannose (Kuhad et al. 1997). Polysaccharides such as galactans, arabinans, and arabinogalactans which are sometimes referred to as hemicelluloses are essentially a part of pectin, at least during initial synthesis of plant cell wall polysaccharides and, strictly speaking, may not be included among the hemicelluloses (Decker et al. 2009; Scheller and Ulvskov 2010).

Hemicelluloses can be differentiated from pectins by the fact that hemicelluloses usually have shorter side chains, are not as heavily branched, are extensively cross linked to other cell-wall components, and are, in general, not much acidic (Decker et al. 2009). Unlike cellulose and lignin, hemicelluloses are soluble and hygroscopic. They are closely associated with lignin through cinnamate acid ester linkages and with cellulose through extensive hydrogen bonding. They are also cross linked with each other through covalent and hydrogen bonding. Functionally, hemicellulose, in conjunction with cellulose and lignin, provides rigidity as well as flexibility to the plant cell wall. Because of their hydrophilic nature, they also tend to retain water in the cell wall.

In a recent review, Scheller et al. (Scheller and Ulvskov 2010) grouped hemicelluloses into xyloglucans, xylans, mannans and glucomannans, and β-($1 \rightarrow 3,1 \rightarrow 4$)-glucans. Xyloglucans are very similar to cellulose in that they have a backbone made up of β-1,4-linked anhydroglucopyranose units, but are distinguished by the highly repetitive substitutions with side chains of α-D-xylose units. The xylose residues may be further derivatized by galactose and fucose moieties. Xylans are characterized by the presence of substituents such as L-arabinose, 4-O-methyl-D-glucuronic acid, or acetyl groups on a backbone of β-1,4-linked xylopyranose units. Mannans and galactomannans comprise hemicelluloses that have backbones made up entirely of β-1,4-linked mannopyranose units while glucomannans and galactoglucomannas have a backbone made up of mannopyranose and glucopyranose units in a nonrepeating sequence.

2.3 Lignin

The name lignin has been coined from the Latin word "lignum" which literally means wood (Sarkanen and Ludwig 1971). Lignin is not a well-defined uniform chemical compound but is a complex mixture of aromatic heteropolymers made up of phenyl propane units derived primarily from three aromatic alcohols, namely, p-coumaryl, coniferyl, and sinapyl alcohol, termed monolignols. Lignin is synthesized by the enzyme-mediated dehydrogenative polymerization of these three monolignols (Boerjan et al. 2003; Buranov and Mazza 2008). The monolignols form structural elements when incorporated into the lignin polymer and are called p-hydroxyphenyl (H), guaiacyl (G), and syringyl (S) moieties, respectively. During the process of lignification, a complex three-dimensional polymer is produced through radical coupling of the moieties via β-O-4, α-O-4, β-5, β-1, 5–5, 4-O-5 and β-β linkages that do not form regular and ordered repeating units as found in cellulose (Argyropoulos et al. 2002; Froass et al. 1996; Kukkola et al. 2004). The amount and composition of lignin varies among different cell types, different taxa, and between individual cell wall layers. Lignin in plant cell walls can be broadly divided into three types: gymnosperm or softwood lignin, angiosperm or hardwood lignin, and graminaceous or grass lignin (Pearl 1967).

Hardwood lignins principally contain G and S units, with traces of H units, softwood lignins mainly consist of G units and low levels of H units, and grass lignins contain H, G, and S units in significant amounts at varying ratios (Boerjan et al. 2003). From a functional point of view, lignin provides stiffness and mechanical strength to the plant, enables transport of water and nutrients through the vascular systems, and provides resistance to the plant against biotic stresses such as plant pathogens and abiotic stresses such as changes in humidity.

3 Cellulose Decomposition by Microbial Secretions

As mentioned earlier, the crystalline nature of cellulose renders it highly resistant to enzymatic decomposition. Accessibility to cellulose in plant cell walls is further obstructed by the fact that non-cellulosic biopolymers are extensively hydrogen bonded to the surface of cellulose microfibrils and form a complex physical barrier protecting it from external biotic attack (Iiyama et al. 1994). Despite its natural intractability, certain microorganisms, insects, and animals have evolved the capability to degrade cellulose. The majority of cellulose from fallen trees and plant litter in soil is degraded by aerobic fungi and bacteria using cellulose-degrading enzymes termed cellulases. Termites and some other insects have indigenous cellulolytic symbionts in their digestive tracts while a few others have also been found to produce cellulolytic enzymes different from those produced by their gut microflora (Hess et al. 2011; Warnecke et al. 2007; Watanabe and Tokuda 2001). Mollusks such as snails, slugs, and shipworms have also been found to produce cellulose-degrading enzymes. Mastication of forage by herbivores such as cows, deer, and sheep also plays a significant role in the degradation of plant biomass in their rumen as it helps in size reduction and conditioning of the fibrous material ingested. The plant fiber is retained in their digestive tracts for a sufficiently long time and is digested by the ruminal, anaerobic bacteria and fungi (Desvaux 2006).

3.1 Cellulases

Cellulases are enzymes that catalyze the hydrolysis of the β-1-4-glycosidic linkage in cellulose. Though all cellulases share the same chemical specificity for the β-1, 4-glucoside bonds of cellulose, they have evolved different specificities toward the more macroscopic properties of cellulose such as crystallinity and degree of polymerization. Aerobic fungi and bacteria produce mixtures of extracellular individual cellulases that are capable of completely degrading crystalline cellulose. These cellulases are made up of a minimum of two structural domains, the catalytic domain (CD) and the cellulose binding module (CBM), while additional domains may also be present. The CD and the CBM are usually joined together by a short linker peptide.

The CD contains an active site where the hydrolysis of a single glucan chain takes place. Using amino acid sequence alignment of CDs, cellulases have been classified into families such that all members of a family share the same basic three-dimensional structure and have the same stereochemistry for cleavage of the β-1,4-glucosidic linkage (CAZy (Carbohydrate Active EnZymes database), (Henrissat 1991). Two stereochemically different mechanisms, namely the inverting and retaining mechanisms, are possible for cellulose hydrolysis (Beguin and Aubert 1994). By the retaining mechanism, enzymes carry out transglycosylation, producing cellobiose as the β anomer while in the inverting mechanism the first formed cellobiose is an α anomer. Despite the fact that cellulose is an unbranched homopolymer of anhydro-glucose, cellulases occur in a widely varied array of structures, mechanisms, and sequences. Currently, cellulases have been found to belong to as many as 14 families, according to the CAZy database. This considerable diversity in the types of cellulases can be attributed to the complexity and diversity of plant cell wall structure that cellulases have to overcome in order to gain access to cellulose as well as to the complexity and recalcitrance of cellulose itself (Wilson 2009).

One of the main functions of CBMs is to tether the cellulase to the insoluble cellulose surface to bring the CD close to the glucan chains (Boraston et al. 2004; Hashimoto 2006). As with the CDs, CBMs have also been classified into families based on sequence comparison, with family I CBMs being limited to fungi and family II CBMs only produced by bacteria (Tomme et al. 1996). Fungal CBMs are small, usually made up of 33–36 amino residues, whereas bacterial CBMs are larger consisting of about 100 amino acids (Wilson and Irwin 1999). Although structurally different, the CBMs of the first two families are all made up of β-sheets and have a flat face containing several aromatic and potential hydrogen-binding residues that are spaced in such a way as to stack against the glucose residues on the cellulose (Din et al. 1994; Tormo et al. 1996). These hydrophobic interactions increase binding stability. Flexible polypeptide linkers connect the CDs and CBMs, allowing them to function independently to a certain extent, while keeping them together. Linkers are found to be glycosylated, particularly in fungal cellulases while low levels of glycosylation have been detected in some aerobic bacterial cellulases (Wilson and Irwin 1999).

In order to understand the mechanism of enzymatic hydrolysis of cellulose and to simplify analyses, most biochemical studies use model substrates that are relatively free of non-cellulosic plant cell-wall-associated components (Tomme et al. 1995). Assays have been developed to characterize and differentiate cellulases based on their activity on different model substrates (Teeri 1997). On the basis of their modes of action, cellulases can be functionally classified into three different types: endocellulases, exo-cellulases or cellobiohydrolases, and the more recently discovered processive endocellulases (Wilson and Irwin 1999). Cellulose polymers that have been rendered soluble by chemical substitution, such as carboxymethyl cellulose (CMC), are model substrates that are readily degraded by endocellulases, resulting in a rapid decrease in the degree of polymerization (DP) and viscosity of the substrate (Shen et al. 1996; Wood 1968). While acting on CMC, exocellulases seem to be limited by the availability of unsubstituted chain ends on the substrate and hence do not decrease their DP or viscosity. Thus, endocellulases are thought to attack the cellulose chain randomly at internal sites on amorphous regions in the cellulose chain releasing oligosaccharides and new chain ends. A fourth group, the β-glucosidases, hydrolyze soluble cello-oligomers to glucose. They typically do not have CBMs and are thought to be membrane-associated.

Exocellulases are thought to attack chain ends on crystalline regions of cellulose, cleaving off cellobiose or glucose units, depending on which enzyme is being examined. Exocellulases can be of two different types, based on their stereospecificities, with one type attacking the cellulose chain from the reducing end and the other type attacking the chain from the non-reducing end (Barr et al. 1996). An important characteristic of some cellulases that can efficiently degrade crystalline cellulose is their "processivity" (Li et al. 2007; Teeri 1997; Varrot et al. 1999). Processivity is the ability of a cellulase molecule to adsorb to a cellulose chain, perform hydrolytic cleavage, slide along the same chain, and continue to cleave bonds without dissociating until an obstruction or the end of the chain is reached. Early evidence for this processivity was based on product formation and synergy with other endos and exos. More recently, direct visual evidence of the processivity of *Trichoderma reeseii* cel7A has been obtained using high-speed atomic force microscopy (Igarashi et al. 2009). The exo-cellulases, specifically those of the reducing end specific glycosyl hydrolase family 7, are the heavy lifters in the fungal cellulase systems. The great majority of cellulose solubilization results from cellobiose production by the family 7 exo-cellulases.

In contrast to the fungal systems, bacterial cellulose degraders do not produce GH7 or other exo-cellulases. This limitation necessitates alternative strategies for cellulose solubilization such as oligomer uptake, cellulosome-directed hydrolysis, or utilization of what have come to be referred to as processive endocellulases (Watson et al. 2009). While processivity is considered almost dogma for exo-cellulases, processive endocellulases have only recently reached the mainstream of cellulase research (Cohen et al. 2005; Parsiegla et al. 2008; Reverbel-Leroy et al. 1997; Watson et al. 2009; Zverlov et al. 2005). The functional mechanisms of processive endo-cellulases, such as those of exo-cellulases, are not yet understood.

Long before the establishment of separation, purification, and cloning methods to yield high-purity cellulases, researchers recognized the fact that efficient hydrolysis of cellulose required a complex mixture of cellulases (Reese et al. 1950; Wood 1968). The ability of two or more cellulases to work more effectively when added simultaneously rather than in succession is termed synergism (Irwin et al. 1993; Walker et al. 1993), and it is an important characteristic of the enzymatic hydrolysis of cellulose. Synergy can be found between endos and exos, as endos open up more internal chain sites for exo binding and activity, and for reducing- and non-reducing end exos, as each operates in a separate mechanism.

As cellulose is insoluble, adsorption of soluble cellulases to cellulose is a prerequisite to any hydrolytic activity, whether cleaving internal chain linkages or releasing soluble products from the chain terminus. Soluble sugars produced by enzymatic hydrolysis of cellulose range in size from monomeric glucose up to about cellohexose with size being dictated by the type and ratio of cellulases being produced by the hydrolytic microbe. The solubility of these products decreases significantly with the degree of polymerization and which oligomer (or monomer) is transported into the cell for metabolism varies by species.

3.1.1 Cellulases Secreted by Fungi

Fungi are renowned for their ability to degrade organic matter by secreting a battery of extracellular cellulases. Basidiomycota, Ascomycota, and Deuteromycota are known to have innumerable species capable of degrading wood. The most well-known genera include *Trichoderma, Aspergillus, Cladosporium, Fusarium, Geotrichum, Myrothecium, Paecilomyces*, and *Penicillium* among the Deuteromycetes, *Phanaerochaete, Poria, Schizzophyllum, Serpula*, and *Coriolus* among the Basidiomycetes, and *Chaetomium, Bulgaria*, and *Helotium* among the Ascomycetes (Lynd et al. 2002). The most primitive group of cellulolytic fungi belongs to the anaerobic chytridomycetes, which typically inhabit the digestive tracts of herbivorous animals.

Most commercial cellulase preparations are developed from the extracellular cellulases secreted by the mesophilic filamentous fungus *Hypocrea jecorina*, better known by its original name, *Trichoderma reesei* (Vinzant et al. 2001). *Trichoderma reesei* cellulases display efficiency in degrading crystalline cellulose and secrete prolific amounts of cellulases, with cellulase production yields up to as much as 50–100 g/l (Rabinovich 2006). The culture filtrate of *T. reesei* contains a multitude of cellulose-degrading enzymes because of the existence of several isoenzymes and proteolytic fragments (Kubicek 1992). To date, seven *T. reesei* genes encoding cellulases have been cloned: Cel7A, an exocellulase or cellobiohydrolase (formerly CBHI), comprising roughly 60% of the supernatant; Cel6A, another exocellulase (formerly CBHII), comprising about 20% of the supernatant; and five endocellulases Cel7B, Cel5A, Cel12A, Cel61A, Cel45A, and two β-glucosidases which make up the rest of the culture fluid (Shoemaker et al. 1983). All *T. reesei* cellulases, except for Cel12A and the glucosidases, have a two-domain structure with a catalytic domain and a cellulose-binding domain joined together by a highly glycosylated linker peptide (Kubicek 1992). Cel7A, an exocellulase with preference for the reducing end of the cellulose chain and Cel6A, an exocellulase with preference for the non-reducing end, can synergistically achieve complete, though slow, hydrolysis of crystalline cellulose, even in the absence of endocellulases (Teeri 1997).

Humicola insolens is another fungus that is used for the industrial production of cellulases (Schulein 1997). Other fungi whose cellulases have been extensively studied include *Aspergillus aculeatus*, *Talaromyces emersonii*, *Chrysosporium lucknowense*, *Agaricus bisporus*, *Aureobasidium pullulans*, *Cochlibolus carbonum*, several *Fusarium* species, several *Penicillium* species, several *Aspergillus* species, *Pleurotus ostreatus*, and *Thermoascus aurantiacus* (Wilson 2009).

3.1.2 Cellulases Secreted by Bacteria

Cellulolytic bacteria can be divided into two main categories based on their oxygen requirements. The aerobic cellulolytic bacteria are found predominantly in the order Actinomycetales and the anaerobic are found mostly in the order Clostridiales. These two categories have distinctly different strategies for degrading cellulose. While aerobic cellulolytic bacteria secrete a constellation of individual extracellular cellulases, the anaerobic *Clostridia* combine the essential glycosyl hydrolases, namely the cellulases, hemicellulases, and carbohydrate esterases onto extracellular, localized, membrane-associated, multi-enzyme complexes called cellulosomes (Lamed and Bayer 1988). The membrane-associated cellulosome is also tightly bound to the crystalline cellulose substrate, creating proximity between the bacterium and the cellulose substrate and thus minimizing diffusional losses of hydrolysis products (Lamed et al. 1987; Mayer et al. 1987). Since this chapter is about microbial secretions, cellulosomes will not be discussed in greater detail but the reader is referred to comprehensive reviews on this topic (Bayer et al. 2004; Ding et al. 2008).

The aerobic cellulolytic prokaryotes include the Gram-positive bacteria *Cellulomonas* and *Thermobifida* and the gliding bacteria *Cytophaga* and *Sporocytophaga* (Lynd et al. 2002). Most of the cellulase genes cloned so far to aid the determination of the organization and regulation of these genes have been obtained from bacterial sources as a result of the relative ease of cloning of bacterial genes. Though bacteria produce cellulases at ten times lower yields compared to fungi, a major advantage of the former is their higher growth rates and resistance to more severe growth conditions such as higher temperature and pH (Rabinovich 2006).

One of the aerobic bacterial species whose cellulases have been extensively studied is *Thermobifida fusca* (Wilson 2004). *Thermobifida fusca* YX (formerly *Thermomonospora fusca*) is a thermophilic soil bacterium belonging to the order Actinomycetales. It is a filamentous aerobe that degrades cellulose and hemicellulose. It is usually found in heated plant residues such as compost piles, decaying hay, manure heaps, and paper mill waste, and it has an optimal temperature for growth ranging from 48°C to 55°C (Wilson 2004). The extracellular crude cellulase mixture from a protease-deficient mutant has been purified and characterized and is now known to contain six different cellulases, two small cellulose-binding modules and a xylanase (Irwin et al. 1993). An intracellular β-glucosidase and an extracellular xyloglucanase have also been isolated and characterized (Irwin et al. 2003; Spiridonov and Wilson 2001). Though *T. fusca* secretes an exocellulase, Cel6B, similar to *T. reesei* Cel6A, it lacks a homolog of the potent processive Cel7A exocellulase. Other well-studied aerobic bacterial cellulases include those secreted by *Cellulomonas fimi*, *Acidothermus cellulolyticus*, *Cytophaga hutchinsonii*, *Caldibacillus cellovorans*, *Bacillus pumilus*, *Cellulomonas flavigen*, *Cellvibrio fulvus*, *Sporocytophaga myxococcoides*, *Rhodothermus marinus*, *Streptomyces reticule*, *Micromonospora chalcae*, and *Saccharophagus degradans* (Lynd et al. 2002).

3.2 Oxidative Enzymes for Cellulose Degradation

Wood-rotting Basidiomycetes and Ascomycetes have been found to possess two different strategies to degrade cellulose: one hydrolytic and one oxidative (Baldrian and Valášková 2008). Brown-rot fungi typically secrete extracellular cellulases for hydrolytic cleavage of cellulose. In addition to cellulases, white-rot fungi, so called because of the bleached appearance of the degraded biomass, secrete extracellular oxidative enzymes such as cellobiose dehydrogenase (CDH) to oxidatively enhance degradation of biomass by cellulases (Henriksson et al. 2000; Mansfield et al. 1997). Cellobiose dehydrogenases use electron acceptors such as quinones, phenoxy radicals, Fe^{3+}, Cu^{2+}, cytochrome c or tri-iodideions, but not molecular oxygen, to oxidize cellobiose, soluble cellodextrins, mannodextrins, and lactose to their corresponding lactones (Henriksson et al. 2000; Zamocky et al. 2006).

CDH displays unprecedented specificity for binding to cellulose as compared to other CBDs which usually also bind to other polysaccharides (Henriksson et al. 1997). CDH is currently hypothesized to reduce Fe^{3+} to Fe^{2+} or Cu^{2+} to Cu^{+} by oxidation of cellobiose. The reduced species, in the presence of H_2O_2 which is generated by CDH itself, are thought to produce hydroxyl radicals that modify and degrade plant cell wall polymers such as cellulose, xylan, and lignin (Baldrian and Valášková 2008; Henriksson et al. 2000). The CDH oxidative system does not require energy in the form of NADH or other compounds, but its oxidative capacity is limited by the availability of cellobiose as well as suitable electron acceptors. Many white-rot fungi and bacteria secrete CDH in conjunction with cellulases (Henriksson et al. 1997).

4 Hemicellulose Decomposition by Microbial Secretions

Hemicelluloses, with their multifarious branches and side groups, coupled with their hydrogen bonds to cellulose and covalent linkages to lignin, form a complex, heterogeneous, and robust network that provides structural integrity to the plant cell wall. Conversely, these very same properties render them difficult to degrade. Though hemicelluloses, unlike cellulose, are soluble, their branched heteropolymer nature necessitates the need for several enzymes to act in conjunction with effective degradation.

4.1 Hemicellulases

Enzymes capable of degrading the hemicellulose component of plant cell walls are collectively designated "hemicellulases." Hemicellulases are produced by fungi, terrestrial and marine bacteria, microbes inhabiting the rumen, yeast, and marine algae. Although most hemicellulases of fungal and bacterial origin are secreted extracellularly, some intracellular and cell-bound hemicellulases have also been reported (Kuhad et al. 1997). Like cellulases, hemicellulases are also modular proteins made up of catalytic domains (CDs) and cellulose-binding modules. Hemicellulases that hydrolyze glycosidic linkages are classified as glycoside hydrolases, whereas those that hydrolyze ester linkages of acetate or ferulate side groups are grouped under carbohydrate esterases. The CAZy database provides the most up-to-date information on the classification of these enzymes (Carbohydrate Active EnZyme Database, URL http://www.cazy.org/).

From a functional point of view, hemicellulose-degrading enzymes have been classified into depolymerizing and debranching enzymes (Decker et al. 2009). Depolymerizing hemicellulases hydrolyze linkages on the hemicellulose backbone and can be endo-acting if they hydrolyze the chain randomly somewhere in the middle and/or exo-acting if they attack the chain from the ends. Thus, xylanases are hemicellulases that depolymerize the β-1,4-xylopyranose backbone of xylans. The xylooligomers released by the above enzymes are further broken down into monomeric sugars by β-xylosidases. Similarly, mannanases hydrolyze the β-D-1, 4-mannopyranosyl linkages in the backbone of mannans and other hemicelluloses containing polymeric mannose such as glucomannans and galactoglucomannans. The degree of substitution, glucose to mannose ratio, and distribution of the substituents have a profound impact on the rate and extent of hydrolysis (McCleary 1991). Manno-oligomers released by the hydrolysis of these polysaccharides are further acted upon by β-mannosidases and β-glucosidases. The cellulose-like backbone of xyloglucans, with glucopyranose units having xylose attached to their C-6 position, is depolymerized by endoglucanases, xyloglucan endotransglycosylases, and exoglycosidases such as α-fucosidases and β-galactosidases. Recently, glucanases specific to xyloglucan, named xyloglucanases, have been reported to cleave the xyloglucan chain at the substituted glucose residues (Grishutin et al. 2004).

Debranching hemicellulases, which are also called accessory enzymes, cleave branches, and side chains from the backbone of hemicelluloses. Some debranching enzymes can only act on short chain oligomers released by the action of depolymerizing hemicellulases whereas others can directly break off branches and side chains from hemicelluloses with intact backbones. α-Glucuronidases hydrolyze the linkages between the xylan backbone and glucuronic acid or 4-*O*-methyl glucuronic acid. α-Arabinofurnosidases hydrolyze arabinose side chains from the xylan backbone while α-D-galactosidases cleave α-galactosyl branches from the backbone consisting of mannan units. The latter enzyme is essential for degradation of galactomannans and galactoglucomannans (Ademark et al. 1998). Esterases such as acetyl xylan esterases remove acetyl substituents from the hemicellulose backbone while those such as ferulic or coumaric acid esterases cleave hydroxycinnamic acid substituents.

Acetyl xylan esterases act on acetyl groups that occur mainly in xylan and galactoglucomannan, normally in the *O*-2 or *O*-3 position. The hydrolysis of acetyl groups results in a lowering of pH as well as decreasing the solubility of the polymer. Ferulic acid esterases release ferulate side chains that are ester-linked to xylan and found in cereal and hardwood xylans, serving as cross linkers to another substituted xylan or lignin. More recently, glucuronyl esterases that de-methylate 4-O-methyl glucuronic acids have also been described from several sources (Duranova et al. 2009; Spanikova and Biely 2006). When considered with some novel putative acetyl esterases that appear to act on sugars acetylated at positions other than O-2 or O-3, the role of acetylation in biomass may be more complex than previously thought (Biely et al. 2011; Li et al. 2008; Topakas et al. 2010). Break down of hemicelluloses requires the concerted and balanced action of appropriate depolymerizing and debranching enzymes. Rapid debranching of intact chains that have not been depolymerized causes chains to stick together, forming insoluble aggregates that precipitate, making them difficult to depolymerize (Christov and Prior 1993; Tenkanen 1998). Conversely, depolymerizing enzymes are hindered by the branches and sides chains on the backbone and effective hydrolysis requires debranching. Hemicellulose hydrolysis is found to be rapid and complete only under the simultaneous action of both types of enzymes (Christov and Prior 1993; Smith and Hartley 1983).

4.2 Hemicellulases Secreted by Fungi

Aerobic fungi such as *Aspergillus niger* and *Trichoderma reesei* secrete copious amounts of several hemicellulases, all of which act synergistically to completely decompose the hemicelluloses into mono- or disaccharides that become available to any surrounding microorganisms (de Vries et al. 2000). Other predominant fungi that secrete hemicellulases include *Fusarium oxysporum*, *Aspergillus wentii*, *Trichoderma koningii*, *Neurospora crassa*, *Corticum rolfsii*, *Penicillium wortmanni*, *Agaricus bisporus*, *Pleurotus ostreatus*, *Aspergillus awamori*, and *Polyporus sulphureus* (Kuhad et al. 1997). Secretion of these hemicellulases is usually accompanied by cellulase production, and the two systems are often co-induced. One notable exception is the thermophilic fungus *Thermomyces lanuginosus*, which produces an array of hemicellulases, but no cellulases.

4.3 Hemicellulases Secreted by Bacteria

Aerobic bacteria such as those from the genera *Bacillus* and *Cellvibrio* secrete a relatively moderate amount and a number of mainly depolymerizing hemicellulases that result in large soluble oligosaccharide products. These products are transported into the cell where their degradation is completed by membrane-bound or intracellular enzymes (Beylot et al. 2001; Shulami et al. 1999). The advantage of this degradation scheme is that the large oligosaccharide products, though available to surrounding microorganisms, can only be utilized by if they also possess the appropriate transport and hemicellulolytic capabilities (Shallom and Shoham 2003). Anaerobic bacteria produce hemicellulases along with cellulases as part of their cellulosomes. In their case, the hemicellulases are thought to serve the purpose of exposing the cellulose embedded within the hemicellulose network. Bacteria that are known to possess hemicellulase genes include *Bacillus halodurans*, *Bacillus subtilis*, *Cellulomonas fimi*, *Cellvibrio japonicus*, *Geobacillus stearothermophilus*, *Agrobacterium tumefaciens*, *Bifidobacterium longum*, *Caulobacter crescentus*, *Streptomyces coelicolor*, and *Xanthomonas campestris* (Shallom and Shoham 2003).

5 Lignin Decomposition by Microbial Secretions

Lignin in plant cell walls is always present in association with the cell wall polysaccharides, particularly hemicellulose, via covalent linkages, forming lignin–carbohydrate complexes (Buranov and Mazza 2008). The three-dimensional structure of lignin, its composition, hydrophobicity, and the formation of a dense network of lignin–carbohydrate complexes make lignin degradation extremely challenging. Notwithstanding these challenges, certain microorganisms such as the white rot fungi have figured a way out to overcome the bulky structure of lignin and are capable of mineralizing lignin into carbon dioxide while releasing the entrapped carbohydrate polymers for utilization by themselves and surrounding microorganisms. Three major enzymes – lignin peroxidases, manganese peroxidases, and laccases – have been found to be key catalytic tools used by lignin-degrading microbes. Some accessory extracellular enzymes such as glyoxal oxidases and aryl alcohol oxidases are also secreted and have been found to play a significant role in maintaining a steady supply of hydrogen peroxide, a key cofactor in peroxidase catalysis.

The discovery and characterization of lignin-degrading peroxidases by the white rot fungus *P. chrysosporium* lead to a breakthrough in the understanding of oxidative biodegradation of lignin which has been referred to as "enzymatic combustion" (Kirk and Farrell 1987). Heme peroxidases belong to a superfamily of enzymes that are responsible for a plethora of biosynthetic and degradation reactions and are divided into three classes based on their sequences and catalytic properties (Welinder 1992; Zamocky et al. 2006). Class I includes prokaryotic and organelle-localized eukaryotic heme peroxidases and catalase-peroxidases, class II includes all extracellular fungal heme peroxidases, and class III comprises all secreted plant heme peroxidases. Lignin peroxidases, manganese peroxidases, and versatile peroxidases fall under the umbrella of class II heme peroxidases.

5.1 Lignin Peroxidases (LiPs)

LiPs catalyze the hydrogen peroxide-dependent one-electron oxidation of phenolic and non-phenolic compounds and are capable of depolymerizing lignin. These enzymes are nonspecific in nature and are usually secreted as a family of isozymes whose relative composition and isoelectric points vary depending on the growth medium and nutrient conditions (Vicuña 2000). Their catalytic cycle resembles the catalytic mechanism common to all peroxidases. In each cycle, the heme cofactor is activated by hydrogen peroxide to form compound I which yields compound II through one-electron oxidation of one substrate molecule. The compound II further reacts with a second substrate molecule simultaneously returning the enzyme to its normal resting state to initiate a new catalytic cycle (Martinez et al. 2009).

A wide variety of reactions can be catalyzed by LiPs including oxidation of phenolic compounds such as ring and N-substituted anilines, cleavage of C_α–C_β and aryl C_α bonds, aromatic ring opening, and demethylation (ten Have and Teunissen 2001). Although the catalytic mechanism of LiPs is similar to that of horseradish peroxidase and other peroxidases, LiP is unique in several aspects. It has the ability to oxidize substrates of high redox potential, it has a low pH optimum and is sensitive to excess hydrogen peroxide (Wong 2009). As the structure of the plant cell wall prevents penetration by large 38–46 kDa globular enzymes such as LiPs, one of the strategies that ligninolytic peroxidases have been thought to develop to attack the intractable, complex structure of lignin is the use of redox mediators. The formation of organized arrays of lignin subunits in wood has been found to serve as pathways for electron transfer, suggesting that free radicals generated at the external surface of lignin can diffuse deeper into the polymer through a cascade of intramolecular electron-transfer reactions (ten Have and Teunissen 2001). This process is called mediation and the small aromatic compound that is oxidized for the generation of radical ions is called a mediator. Veratryl alcohol is one such mediator compound that serves as a substrate for LiPs and is produced by the ligninolytic microorganism as a secondary metabolite (de Jong et al. 1994). Besides its role as a mediator for electron-transfer reactions, it has also been found to prevent the inactivation of LiP by excess hydrogen peroxide (Collins et al. 1997).

5.2 Manganese Peroxidases

Manganese peroxidases (MnPs) are another group of heme peroxidases whose primary catalytic activity involves the hydrogen peroxide-dependent oxidation of

Mn^{2+} to Mn^{3+}, which in turn oxidizes a variety of phenolic compounds. Like LiPs, MnPs are also secreted into the culture supernatant in the form of several isozymes, and the catalytic cycle of MnP oxidation is similar to that of LiP. Native MnP is oxidized to compound I in the presence of hydrogen peroxide, compound I is reduced by Mn^{2+} and phenols to produce compound II and compound II is reduced back to its resting state by another Mn^{2+} ion (Wong 2009). The presence of certain aliphatic organic acids such as lactate, oxalate, and malonate has been found to enhance the rate of catalytic oxidation by MnPs. These organic acids, which are often synthesized de novo by white rot fungi, chelate Mn^{3+} to form stable complexes that can diffuse freely and oxidize phenolic substrates by one-electron oxidation of the substrate, yielding phenoxy radical intermediates. The complexed Mn^{3+} is only capable of oxidizing the phenolic portions of the lignin polymer and cannot independently oxidize the nonphenolic parts (Hammel et al. 1993). In the presence of mediators such as glutathione, oxidation by the Mn^{3+}–chelator complex of the mediator compounds yields highly reactive species that further undergo non-enzymatic reactions with non-phenolic parts of lignin to produce a varied mixture of end products. Mn-dependent generation of hydrogen peroxide by the oxidation of glyoxylate and oxalate has also been demonstrated (Kuan and Tien 1993). MnP differs from other peroxidases in the structure of its binding site (ten Have and Teunissen 2001).

A third type of peroxidases, designated versatile peroxidases (VPs), has been reported to consist of enzymes that have both Mn-mediated as well as Mn-independent oxidative capabilities. VPs can not only act on Mn^{2+}, such as MnPs, but can also oxidize phenolic and non-phenolic compounds such as veratryl alcohol and methoxybenzes that are typically substrates of LiPs (Mester and Field 1998; Moreira and Filho 2008). This catalytic versatility of VPs allows them to be used for oxidation of a wide variety of both high- and low redox potential substrates (Martinez et al. 2009; Wong 2009).

5.3 *Laccases*

Laccases are enzymes belonging to a class of multi-copper oxidases that catalyze the four-electron reduction of dioxygen to water with the concomitant one-electron oxidation of four substrate equivalents (Baldrian 2006; Morozova et al. 2007; Reinhammer 1984). They have the ability to catalyze reactions that run the gamut from degradation of polymers, oxidative coupling of phenolic compounds, and functionalization of polymers to ring cleavage. The active site of laccases consists of highly conserved features and is made up of four copper ions. The type 1 (T1) site has a mononuclear, paramagnetic, "blue" copper ion that imparts a characteristic light blue color to the laccase. The distinguishing property of this copper ion is its pronounced electronic absorption at a wavelength of 600 nm (Solomon et al. 1996). The type 2 (T2) site is a mononuclear paramagnetic, "non-blue" copper that is invisible in the electronic absorption spectrum but displays an ultrafine splitting in the EPR spectrum (Quintanar et al. 2005; Solomon et al. 1992, 1996). The type 3 (T3) site consists of a binuclear, diamagnetic spin-coupled pair of copper ions which make this site invisible in the EPR spectrum but visible in the UV region as a shoulder at 330 nm (Messerschmidt and Huber 1990; Solomon et al. 1996). The T2 and T3 sites together form what is referred to as a trinuclear cluster.

The mechanism of catalysis in laccases has been studied extensively (Lee et al. 2001; Yoon et al. 2007) but the exact details of the electron transfer pathway within the enzyme are not yet completely understood (Morozova et al. 2007; Skálová et al. 2009). The main steps in the catalytic cycle involve the oxidation of the substrate at the T1 site, the transfer of electrons from the T1 site to the trinuclear cluster, and the reduction of molecular oxygen to water at the trinuclear cluster. As in the case of LiPs, the oxidation of some laccase substrates yields high potential intermediates that serve as redox mediators. Mediators serve to expand the natural range of compounds that can be oxidized by laccases (Shleev et al. 2006).

5.4 Peroxide-Regenerating Enzymes

Glyoxal oxidase (GLOX) is an enzyme that is secreted into the culture supernatant of wood-rotting fungi concomitantly with other ligninolytic enzymes and is thought to perform the primary function of generating hydrogen peroxide (Cullen and Kersten 2004). Common substrates of GLOX include simple glycoaldehydes, α-hydroxycarbonyl compounds, and α-dicarbonyl compounds which may, potentially, be derived by the peroxidase-catalyzed oxidation of lignin (Hammel et al. 1994). Sequential oxidations of glycoladehyde by GLOX produce oxalate along with several equivalents of hydrogen peroxide. GLOX undergoes reversible inactivation when it is not coupled with a peroxidase system (Whittaker et al. 1996). LiP and nonphenolic substrates are capable of reactivating GLOX while phenolic compounds are found to prevent the activation by LiP. These observations indicate that GLOX secretion is regulated by a mechanism that is sensitive to peroxidase, peroxidase substrates, and peroxidase products (Cullen and Kersten 2004). Spectroscopic investigations have revealed that GLOX has an active site that contains a free-radical-coupled copper atom.

Aryl alcohol oxidase (AAO), also known as veratryl alcohol oxidase, is another extracellular enzyme in certain fungal species that is responsible for the regeneration of hydrogen peroxide in the supernatant (de Jong et al. 1994). AAO is an FAD-dependent enzyme that acts on chlorinated anisyl alcohols that are synthesized by the fungus, *de novo* from glucose.

5.5 Lignin Degradation by Fungi

Wood decay fungi have been found to be the most efficient degraders of lignin, owing to the extravagant array of oxidizing enzymes secreted by them in response to the recalcitrance of lignin (Hatakka 2005; Kirk and Cullen 1998; Kirk and Farrell 1987; Lundell et al. 2010; Orth et al. 1993; Tuor et al. 1995). On the basis the distinctive chemical and morphological features of decay resulting from their activities, wood decay fungi have been divided into three categories by Blanchette namely white-rot, soft-rot, and brown-rot (Blanchette 1995). The white-rot fungi, characterized by their ability to degrade cellulose, hemicelluloses, and lignin, are classified as Basidiomycetes. Among the white-rot fungi, two distinct patterns of decay are observed (Kirk and Cullen 1998). The first involves the simultaneous decay of cellulose, hemicellulose, and lignin. While degrading wood, the hyphae of white-rot fungi originating from spores or nearby colonies attack and invade wood cells along the lumen walls resulting in cell wall erosion. Fungal mycelia eventually fill the large voids that are created when the eroded zones in cell walls collapse and coalesce. The second pattern of decay among white-rot fungi involves selective delignification. Species such as *Phellinus pini*, *P. nigrolimitatus*, and *Inonotus dryophyllus* that are predominantly involved in this kind of decay preferentially remove lignin and hemicellulose leaving behind white zones of delignified wood (Blanchette 1995). There is considerable variation among species and even among strains within species in their ability to degrade the three major cell wall components, and the type of decay often depends on the concentration and type of lignin. Usually, the syringyl units of lignin are degraded more preferentially than the guauicyl units.

Soft-rot fungi are characterized by their occurrence in wet environments and are classified as Ascomycetes and Deuteromycetes. They are also of two distinguishable types. Type I soft-rot fungi are found within secondary walls in the form of distinct biconical or cylindrical cavities. Type II soft rot occurs when the entire secondary wall has been slowly degraded. Only the middle lamellae remain undegraded while the rest of the secondary wall decays, distinguishing soft rot from simultaneous white rot in which everything including the middle lamellae is degraded (Blanchette 1995). Brown-rot fungi, which are classified under the Basidiomycetes, differ from the white- and soft-rot fungi by the fact that they remove all cell wall polysaccharides, leaving behind brown decay that consists of mainly of lignin that has been chemically modified but essentially undegraded.

On the basis of the combinations of extracellular enzymes they secrete to attack lignin, white-rot fungi have also been grouped into three major categories by Hatakka (1994). The lignin-degrading ability of fungi grouped in this manner has been assessed by the amount of radioactive carbon dioxide released from ring-radiolabeled synthetic lignin compounds. The first group, consisting of some of the most efficient lignin degraders, includes those that secrete LiP and MnP, such as *P. chrysosporium* and *Phlebia radiata*. The second group includes the selective lignin-degrading fungi, those that secrete MnP and laccase, such as *Dichomitus squalens* and *Rigidoporus lignosus*. The third group consists of fungi such as *Phlebia ochraceofulva* and *Junghuhnia separabilima* that secrete LiP and laccase and show poor degradation of lignin.

The white-rot fungus *P. chrysosporium* is the most well-known and well-characterized lignin-degrading microorganism in terms of its physiology and biochemistry, genetics, and molecular biology with respect to the degradation of lignin (Broda et al. 1994; Kirk and Cullen 1998; Kirk and Farrell 1987). It secretes a suite of cellulases and hemicellulases as part of its primary metabolism and lignin-degrading enzymes as part of its secondary metabolism. LiPs, MnPs, and hydrogen peroxide-generating enzymes constitute the extracellular, oxidative lignin-degrading system of *P. chrysosporium*. The production of high redox potential LiPs, in the form of multiple isozymes, is triggered under conditions of carbon, nitrogen, or sulfur limitation (Kirk et al. 1986; Leisola and Fiechter 1985; Renganathan et al. 1985). MnPs are produced in the presence of Mn^{2+} in the growth medium (Bonnarme and Jeffries 1990). Readers are referred to exhaustive reviews on various aspects of this model lignin-degrading organism (Cullen 1997; Eriksson 1993; Gold and Alic 1993; Kirk and Cullen 1998; Kirk and Farrell 1987; Pease and Tien 1992; Singh and Chen 2008). Recent advances include the whole-genome sequencing (Martinez et al. 2004), proteome and secretome analyses (Abbas et al. 2005; Sato et al. 2007; Wymelenberg et al. 2005) as well as transcriptome analysis (Sato et al. 2009) of *P. chrysosporium* in relation to lignocellulose degradation. Genome, transcriptome, and secretome analyses of a brown-rot fungus, *Postia placenta*, have revealed several new, unique extracellular glycosyl hydrolases (Martinez et al. 2009). Comparison of *P. placenta* with the *P. chrysosporium* system has provided insights into a possible evolutionary transition from white- to brown-rot during which the lignin-degrading ability was lost.

5.6 Lignin Degradation by Bacterial Secretions

The intensity and success of efforts channeled into the investigation of lignin decomposition by wood decay fungi have overshadowed the study of bacterial decomposition of lignin. The results of early researchers who looked into prokaryotic degradation of lignin met with mixed reviews because of weaknesses in the methods they used for the characterization of the lignin-degrading ability of bacteria (Kirk 1971). It was not until the availability and use of sensitive ^{14}C-lignin biodegradation assays that the lignin-degrading abilities of bacteria were fully recognized.

The first study that conclusively showed bacterial modification of lignin was published by Trojanowski et al. (1977) who found several *Nocardia* strains that were capable of releasing $^{14}CO_2$ when grown on ^{14}C-lignins that were specifically labeled in their side chains or methoxyl components. The Crawford research group (Crawford 1978; Phelan et al. 1979) showed that numerous strains of *Streptomyces* released $^{14}CO_2$ by oxidizing labeled lignins. *Bacillus megaterium* was another species that had similarly been shown to have lignin-decomposing ability (Robinson and Crawford 1978). However, even the most efficient bacteria exhibited rates that were much lower than those of their fungal counterparts (Vicuña 1988). Other, non-radiolabelled, model lignin substrates have also been used to show degradation of lignin by bacteria in several genera, including

Pseudomonas, *Flavobacterium*, *Aeromonas*, and *Xanthomonas* (Crawford and Crawford 1980).

Attempts to understand the mechanism of lignin degradation by bacteria were mostly based on studies using lignin model compounds such as phenylcoumarans, cyclic lignans, biphenyls, and compounds with β-1 and β-O-4 linkages (Zimmermann 1990). Certain bacteria have also been described as having lignin-degrading abilities based on the morphological changes that they can bring about in the lignified cell walls of wood (Blanchette 1995). Tunneling, cavitation, and erosion are the three main morphological changes that can be caused by bacteria. These are mostly found in wood that is almost saturated with water, leading to micro-aerobic conditions (Blanchette 1995).

Actinomycetes have been found to possess lignin-degrading ability (Godden et al. 1992; Kirby 2005; McCarthy 1987; McCarthy and Williams 1992). They dwell most frequently in soil and in compost piles, are adapted to growth on solid substrates, with their carbon sources being predominantly insoluble, and are known to secrete an array of extracellular enzymes that attack lignocellulosic biomass. Many of them are found to have "mycelia" that are more typical of fungi than bacteria. However, despite this morphological similarity, experimental conditions that brought out the lignin-modifying capabilities of *P. chrysosporium* did not work well with Actinomycetes, indicating very different mechanisms of physiological regulation (Ball et al. 1989). Among the lignin-degrading actinomycetes, *Streptomyces viridosporus* T7A has been the most investigated species and has been shown to secrete a plethora of plant cell-wall-degrading enzymes, including peroxidases, cellulases, and esterases (Adhi et al. 1989; Borgmeyer and Crawford 1985; Giroux et al. 1988; Magnuson and Crawford 1992; Ramachandra et al. 1988; Spiker et al. 1992; Yee et al. 1996). Four extracellular lignin peroxidase isoforms have been characterized and though all the isoforms require hydrogen peroxide as an electron donor, they differ in their substrate range (Ramachandra et al. 1988). While *P. chrysosporium* secretes lignin peroxidase under nitrogen-, carbon-, or sulfur-deficient growth conditions (Kirk and Farrell 1987), *S.viridosporus* T7A was found to secrete lignin peroxidase during growth-associated conditions in nutrient-rich medium (Korus et al. 1991). Another filamentous actinomycete, *Streptomyces cyaneus* CECT 3335, has been characterized and found to possess lignin-degrading abilities (Berrocal et al. 2000). Several other streptomycete strains have been shown to degrade lignin under solid-state fermentation conditions (Arias et al. 2005; Hernandez-Coronado et al. 1997; Trigo and Ball 1994)

Recent developments in methods using mass spectrometry for detection of lignin degradation products have facilitated more convenient studies of lignin-degrading bacteria (Hernández-Coronado et al. 1998; Hernández et al. 2001; Reale et al. 2004). The relatively recent discovery that bacteria also secrete laccases (Alexandre and Zhulin 2000; Claus 2003, 2004) has led to an upsurge in the investigation of bacterial laccases, including application to the pulp and paper industry (Arias et al. 2003; Chandra et al. 2007; Raj et al. 2007).

6 Factors Affecting Lignocellulose Degradation in Soil

While plant biomass can be degraded in several different habitats, in this chapter we discuss some of the aspects of lignocellulose decomposition in soil. Current interpretations of lignocellulose-degrading enzyme activity have been achieved through the study of the growth and regulation of these enzymes in liquid or solid media under laboratory conditions. The in-situ behavior of some of the same enzymes in soil may be very different and complex because of the complex nature of soil itself and to the presence of a plethora of other microorganisms and plants that may coexist with the organism being investigated. The lignocellulose-degrading ability of enzymes in soil is affected by several factors besides the nature of the microbes that secrete these enzymes, the chemical form of nutrients readily available to the microbes and the ecosystem in which they exist (Baldrian et al. 2011). Microbial secretions, once released into soil, adsorb onto particulate matter and become stabilized through complex interactions with soil organic matter or clay minerals that are governed by soil type and size, enzyme concentrations and thermodynamics (Wallenstein and Weintraub 2008). While these soil–enzyme interactions may lead to enzyme stabilization, providing protection against proteolysis and denaturation during freeze–thaw cycles, they may also considerably lower enzyme activity by reducing accessibility to substrates.

Another important factor that regulates enzymatic activity is the soil pH as enzymes are usually sensitive and responsive to small shifts in pH. Although the presence of trace amounts of heavy metals such as Cd, Zn, or Mn is necessary for fungal growth, excess amounts of these or other heavy metals such as Hg, Cu, or Ni may inhibit microbial growth, and cause morphological and physiological changes. The resulting shifts in the proportion of fungal to bacterial populations will impact the secreted enzymatic activity (Baldrian and Snajdr 2011). Microbial secretions related to lignin and cellulose degradation are transcriptionally regulated by heavy metals and can also be adsorbed and accumulated by fungal mycelia (Baldrian 2003). However, improved litter loss has been observed in the presence of high concentrations of Mn^{2+} possibly because of the increased MnP production and in the presence of high Cu^{2+} concentrations, which has been shown to induce laccase production (Baldrian 2006).

Temperature is another significant factor that affects the production and activity of enzymes in soil (Eggen and Sveum 1999; Okeke et al. 1996). Owing to different temperature optima and differences in temperature sensitivities of different enzymes, seasonal fluctuations in soil temperature usually result in alterations of the relative rates of decomposition of the different components of lignocellulosic biomass in soil.

Interspecies interactions leading to lignocellulose degradation between fungi, bacteria, and other soil-inhabiting organisms must also be taken into consideration. The possibility and implications of enhanced lignocellulose decomposition in soil near the litter-generating parent plant should also be accounted for.

7 Final Considerations

This chapter has addressed how several microbial players could potentially degrade plant biomass using a variety of secreted enzyme mechanisms. However, it is worth noting that in nature, plant biomass degradation occurs as a cohesive process involving several organisms that in some way utilize co-evolutionary cues. Litter decomposition is very critical in determining the soil nutrient cycling and availability (Swift et al. 1979) and it has been hypothesized that some plants may encourage the development of their own soil microbial community suited to decompose their own litter rapidly (Wardle 2002). This idea has not been explicitly tested; however, there is some indirect evidence in literature that suggests it could be possible. For instance, several studies have observed that biomass degradation occurs more rapidly when litter decomposes beneath the plant species from which it is derived than beneath different plant species (Ayres et al. 2009a, b; Bocock et al. 1960; Gholz et al. 2000; Hunt et al. 1988; Vivanco and Austin 2011). This event has been referred as "home-field advantage" in the ecology realm (Gholz et al. 2000). However, there are also reports available that "home-field advantage" is not always observed under field and laboratory conditions (Chapman and Koch 2007; Prescott et al. 2000). This may be because of the differences in the litter quality, limited availability of soil microbial taxa, soil factors, and incubation time, factors that influence the magnitude of the "home-field advantage" effect (Ayres et al. 2009a). A recent report demonstrates that the litter decomposition from forest ecosystems in North America, South America, and Europe was 8% faster at home (litter decomposes beneath the plant species from which it is derived) than away (different plant species) (Ayres et al. 2009a). Thus, "home-field advantage" might be recognized as a factor that could accelerate litter decomposition and be potentially used to isolate soil microbes specific to a given source of plant biomass.

Acknowledgements This work was funded by the Colorado Center for Biorefining and Biofuels (C2B2), the DOE Office of the Biomass Program and the US Department of Energy under Contract No. DE-AC36-08-GO28308 with the National Renewable Energy Laboratory, and the National Science Foundation (MCB-0950857 to JMV).

References

Abbas A, Koc H, Liu F, Tien M (2005) Fungal degradation of wood: initial proteomic analysis of extracellular proteins of *Phanerochaete chrysosporium* grown on oak substrate. Curr Genet 47:49

Ademark P, Varga A, Medve J, Harjunpää V, Torbjörn D, Tjerneld F, Stålbrand H (1998) Softwood hemicellulose-degrading enzymes from aspergillus niger: purification and properties of a [beta]-mannanase. J Biotechnol 63:199

Adhi TP, Korus RA, Crawford DL (1989) Production of major extracellular enzymes during lignocellulose degradation by two streptomycetes in agitated submerged culture. Appl Environ Microbiol 55:1165–1168

Alexandre G, Zhulin IB (2000) Laccases are widespread in bacteria. Trends Biotechnol 18:41

Argyropoulos DS, Jurasek L, Kristofova L, Xia Z, Sun Y, Palus E (2002) Abundance and reactivity of dibenzodioxocins in softwood lignin. J Agric Food Chem 50:658–666

Arias M, Rodríguez J, Pérez M, Hernández M, Polvillo O, González-Pérez J, González-Vila F (2005) Analysis of chemical changes in picea abies wood decayed by different streptomyces strains showing evidence for biopulping procedures. Wood Sci Technol 44:179

Arias ME, Arenas M, Rodriguez J, Soliveri J, Ball AS, Hernandez M (2003) Kraft pulp biobleaching and mediated oxidation of a nonphenolic substrate by laccase from streptomyces cyaneus cect 3335. Appl Environ Microbiol 69:1953–1958

Ayres E, Steltzer H, Berg S, Wall DH (2009a) Soil biota accelerate decomposition in high-elevation forests by specializing in the breakdown of litter produced by the plant species above them. J Ecol 97:901

Ayres E, Steltzer H, Simmons BL, Simpson RT, Steinweg JM, Wallenstein MD, Mellor N, Parton WJ, Moore JC, Wall DH (2009b) Home-field advantage accelerates leaf litter decomposition in forests. Soil Biol Biochem 41:606

Baldrian P (2003) Interactions of heavy metals with white-rot fungi. Enzyme Microb Technol 32:78

Baldrian P (2006) Fungal laccases - occurrence and properties. FEMS Microbiol Rev 30:215–242

Baldrian P, Valášková V (2008) Degradation of cellulose by basidiomycetous fungi. FEMS Microbiol Rev 32:501

Baldrian P, Snajdr J (2011) Lignocellulose-degrading enzymes in soils. In: Soil enzymology, 22nd edn. Heidelberg, Springer Berlin, p 167

Baldrian P, Voříšková J, Dobiášová P, Merhautová V, Lisá L, Valášková V (2011) Production of extracellular enzymes and degradation of biopolymers by saprotrophic microfungi from the upper layers of forest soil. Plant Soil 338:111

Ball AS, Betts WB, McCarthy AJ (1989) Degradation of lignin-related compounds by actinomycetes. Appl Environ Microbiol 55:1642–1644

Barr BK, Hsieh Y-L, Ganem B, Wilson DB (1996) Identification of two functionally different classes of exocellulases. Biochemistry 35:586

Bayer EA, Belaich J-P, Shoham Y, Lamed R (2004) The cellulosomes: multienzyme machines for degradation of plant cell wall polysaccharides. Annu Rev Microbiol 58:521–554

Beckham GT, Matthews JF, Bomble YJ, Bu LT, Adney WS, Himmel ME, Nimlos MR, Crowley MF (2011a) Identification of amino acids responsible for processivity in a family 1 carbohydrate-binding module from a fungal cellulase. J Phys Chem B 114:1447–1453

Beckham GT, Matthews JF, Peters B, Bomble YJ, Himmel ME, Crowley MF (2011b) Molecular-level origins of biomass recalcitrance: decrystallization free energies for four common cellulose polymorphs. J Phys Chem B 115:4118

Beguin P, Aubert JP (1994) The biological degradation of cellulose. FEMS Microbiol Rev 13:25–58

Berrocal M, Ball AS, Huerta S, Barrasa JM, Hernández M, Pérez-Leblic MI, Arias ME (2000) Biological upgrading of wheat straw through solid-state fermentation with *streptomyces cyaneus*. Appl Microbiol Biotechnol 54:764

Beylot MH, Emami K, McKie VA, Gilbert HJ, Pell G (2001) Pseudomonas cellulosa expresses a single membrane-bound glycoside hydrolase family 51 arabinofuranosidase. Biochem J 358:599–605

Biely P, Mastihubova M, Tenkanen M, Eyzaguirre J, Li XL, Vrsanska M (2011) Action of xylan deacetylating enzymes on monoacetyl derivatives of 4-nitrophenyl glycosides of beta-d-xylopyranose and alpha-l-arabinofuranose. J Biotechnol 151:137–142

Blanchette RA (1995) Degradation of the lignocellulosic complex in wood. Can J Bot 73: S999–S1010

Bocock KL, Gilbert O, Capstick CK, Twinn DC, Waid JS, Woodman MJ (1960) Changes in leaf litter when placed on the surface of soils with contrasting humus types. J Soil Sci 11:1

Boerjan W, Ralph J, Baucher M (2003) Lignin biosynthesis. Ann Rev Plant Biol 54:519–546

Bonnarme P, Jeffries TW (1990) Mn(ii) regulation of lignin peroxidases and manganese-dependent peroxidases from lignin-degrading white rot fungi. Appl Environ Microbiol 56:210–217
Boraston AB, Bolam DN, Gilbert HJ, Davies GJ (2004) Carbohydrate-binding modules: fine-tuning polysaccharide recognition. Biochem J 382:769–781
Borgmeyer JR, Crawford DL (1985) Production and characterization of polymeric lignin degradation intermediates from two different streptomyces spp. Appl Environ Microbiol 49:273–278
Broda P, Birch P, Brooks PLJ, Copa P, Sinnott ML, Tempelaars C, Wang Q, Wyatt A, Sims P (1994) Phanerochaete chrysosporium and its natural substrate. FEMS Microbiol Rev 13:189
Buranov AU, Mazza G (2008) Lignin in straw of herbaceous crops. Ind Crops Prod 28:237–259
Chandra R, Raj A, Purohit HJ, Kapley A (2007) Characterisation and optimisation of three potential aerobic bacterial strains for kraft lignin degradation from pulp paper waste. Chemosphere 67:839
Chapman S, Koch G (2007) What type of diversity yields synergy during mixed litter decomposition in a natural forest ecosystem? Plant Soil 299:153
Christov LP, Prior BA (1993) Esterases of xylan-degrading microorganisms: production, properties, and significance. Enzyme Microb Technol 15:460
Claus H (2003) Laccases and their occurrence in prokaryotes. Arch Microbiol 179:145
Claus H (2004) Laccases: structure, reactions, distribution. Micron 35:93
Cohen R, Suzuki MR, Hammel KE (2005) Processive endoglucanase active in crystalline cellulose hydrolysis by the brown rot basidiomycete gloeophyllum trabeum. Appl Environ Microbiol 71:2412–2417
Collins PJ, Field JA, Teunissen P, Dobson AD (1997) Stabilization of lignin peroxidases in white rot fungi by tryptophan. Appl Environ Microbiol 63:2543–2548
Crawford DL (1978) Lignocellulose decomposition by selected streptomyces strains. Appl Environ Microbiol 35:1041–1045
Crawford DL, Crawford RL (1980) Microbial degradation of lignin. Enzyme Microb Technol 2:11
Cullen D (1997) Recent advances on the molecular genetics of ligninolytic fungi. J Biotechnol 53:273
Cullen D, Kersten PJ (2004) Enzymology and molecular biology of lignin degradation. The Mycota III: biochemistry and molecular biology Berlin, Heidelberg. pp 249–273
de Jong E, Field JA, de Bont JAM (1994) Aryl alcohols in the physiology of ligninolytic fungi. FEMS Microbiol Rev 13:153
de Vries RP, Kester HCM, Poulsen CH, Benen JAE, Visser J (2000) Synergy between enzymes from aspergillus involved in the degradation of plant cell wall polysaccharides. Carbohydr Res 327:401
Decker SR, Siika-Aho M, Viikari L (2009) Enzymatic depolymerization of plant cell wall hemicelluloses. Biomass Recalcitrance: Deconstructing the Plant Cell Wall for Bioenergy Blackwell Publishing Ltd. Chapter 10
Desvaux M (2006) Unravelling carbon metabolism in anaerobic cellulolytic bacteria. Biotechnol Prog 22:1229
Din N, Forsythe IJ, Burtnick LD, Gilkes NR, Miller RC, Warren RAJ, Kilburn DG (1994) The cellulose-binding domain of endoglucanase a (cena) from *Cellulomonas fimi*: evidence for the involvement of tryptophan residues in binding. Mol Microbiol 11:747
Ding S-Y, Xu Q, Crowley M, Zeng Y, Nimlos M, Lamed R, Bayer EA, Himmel ME (2008) A biophysical perspective on the cellulosome: new opportunities for biomass conversion. Curr Opin Biotechnol 19:218
Duchesne LC, Larson DW (1989) Cellulose and the evolution of plant life. Bioscience 39:238–241
Duranova M, Spanikova S, Wosten HAB, Biely P, de Vries RP (2009) Two glucuronoyl esterases of phanerochaete chrysosporium. Arch Microbiol 191:133–140
Eggen T, Sveum P (1999) Decontamination of aged creosote polluted soil: the influence of temperature, white rot fungus pleurotus ostreatus, and pre-treatment. Int Biodeter Biodegred 43:125

Eriksson K-EL (1993) Concluding remarks: where do we stand and where are we going? Lignin biodegradation and practical utilization. J Biotechnol 30:149

Falkowski P, Scholes RJ, Boyle E, Canadell J, Canfield D, Elser J, Gruber N, Hibbard K, Hogberg P, Linder S, Mackenzie FT, Moore B 3rd, Pedersen T, Rosenthal Y, Seitzinger S, Smetacek V, Steffen W (2000) The global carbon cycle: a test of our knowledge of earth as a system. Science 290:291–296

Froass PM, Ragauskas AJ, Jiang J (1996) Chemical structure of residual lignin from kraft pulp. J Wood Chem Technol 16:347–365

Gardner KH, Blackwel J (1974) Structure of native cellulose. Biopolymers 13:1975–2001

Gholz HL, Wedin DA, Smitherman SM, Harmon ME, Parton WJ (2000) Long-term dynamics of pine and hardwood litter in contrasting environments: toward a global model of decomposition. Global Change Biol 6:751

Giroux H, Vidal P, Bouchard J, Lamy F (1988) Degradation of kraft indulin lignin by streptomyces viridosporus and streptomyces badius. Appl Environ Microbiol 54:3064–3070

Godden B, Ball AS, Helvenstein P, McCarthy AJ, Penninckx MJ (1992) Towards elucidation of the lignin degradation pathway in actinomycetes. J Gen Microbiol 138:2441–2448

Gold MH, Alic M (1993) Molecular biology of the lignin-degrading basidiomycete phanerochaete chrysosporium. Microbiol Mol Biol Rev 57:605–622

Grishutin SG, Gusakov AV, Markov AV, Ustinov BB, Semenova MV, Sinitsyn AP (2004) Specific xyloglucanases as a new class of polysaccharide-degrading enzymes. Biochim Biophys Acta 1674:268–281

Haigler CH, Brown RM Jr, Benziman M (1980) Calcofluor white st alters the in vivo assembly of cellulose microfibrils. Science 210:903–906

Hammel KE, Jensen KA, Mozuch MD, Landucci LL, Tien M, Pease EA (1993) Ligninolysis by a purified lignin peroxidase. J Biol Chem 268:12274–12281

Hammel KE, Mozuch MD, Jensen KA, Kersten PJ (1994) H2o2 recycling during oxidation of the arylglycerol. Beta.-aryl ether lignin structure by lignin peroxidase and glyoxal oxidase. Biochemistry 33:13349

Hashimoto H (2006) Recent structural studies of carbohydrate-binding modules. Cell Mol Life Sci 63:2954

Hatakka A (1994) Lignin-modifying enzymes from selected white-rot fungi: production and role from in lignin degradation. FEMS Microbiol Rev 13:125

Hatakka A (2005) Biodegradation of lignin. Wiley-VCH Verlag GmbH & Co. KGaA, Weinheim

Henriksson G, Salumets A, Divne C, Pettersson G (1997) Studies o fcellulosebinding by cellobiose dehydrogenase and a comparison with cellobiohydrolase 1. Biochem J 324:833–838

Henriksson G, Johansson G, Pettersson G (2000) A critical review of cellobiose dehydrogenases. J Biotechnol 78:93–113

Henrissat B (1991) A classification of glycosyl hydrolases based on amino acid sequence similarities. Biochem J 280(Pt 2):309–316

Hernandez-Coronado MJ, Hernandez M, Centenera F, Perez-Leblic MI, Ball AS, Arias ME (1997) Chemical characterization and spectroscopic analysis of the solubilization products from wheat straw produced by streptomyces strains grown in solid-state fermentation. Microbiology 143:1359–1367

Hernández-Coronado MJ, Hernández M, Rodríguez J, Arias ME (1998) Gas chromatography/mass spectrometry as a suitable alternative technique to evaluate the ability of streptomyces to degrade lignin from lignocellulosic residues. Rapid Commun Mass Spectrom 12:1744

Hernández M, Hernández-Coronado MJ, Montiel MD, RodrIguez J, Pérez MI, Bocchini P, Galletti GC, Arias ME (2001) Pyrolysis/gas chromatography/mass spectrometry as a useful technique to evaluate the ligninolytic action of streptomycetes on wheat straw. J Anal Appl Pyrol 58–59:539

Hess M, Sczyrba A, Egan R, Kim T-W, Chokhawala H, Schroth G, Luo S, Clark DS, Chen F, Zhang T, Mackie RI, Pennacchio LA, Tringe SG, Visel A, Woyke T, Wang Z, Rubin EM

(2011) Metagenomic discovery of biomass-degrading genes and genomes from cow rumen. Science 331:463–467
Hunt HW, Ingham ER, Coleman DC, Elliott ET, Reid CPP (1988) Nitrogen limitation of production and decomposition in prairie, mountain meadow, and pine forest. Ecology 69:1009–1016
Igarashi K, Koivula A, Wada M, Kimura S, Penttila M, Samejima M (2009) High speed atomic force microscopy visualizes processive movement of trichoderma reesei cellobiohydrolase i on crystalline cellulose. J Biol Chem 284:36186–36190
Iiyama K, Lam TB-T, Stone BA (1994) Covalent cross-links in the cell wall. Plant Physiol 104:315
Irwin DC, Spezio M, Walker LP, Wilson DB (1993) Activity studies of eight purified cellulases: specificity, synergism, and binding domain effects. Biotechnol Bioeng 42:1002
Irwin DC, Cheng M, Xiang B, Rose JKC, Wilson DB (2003) Cloning, expression and characterization of a family-74 xyloglucanase from thermobifida fusca. Eur J Biochem 270:3083
Kirby R (2005) Actinomycetes and lignin degradation. Adv Appl Microbiol 58:125–168
Kirk TK (1971) Effects of microorganisms on lignin. Annu Rev Phytopathol 9:185–210
Kirk TK, Croan S, Tien M, Murtagh KE, Farrell RL (1986) Production of multiple ligninases by phanerochaete chrysosporium: effect of selected growth conditions and use of a mutant strain. Enzyme Microb Technol 8:27
Kirk TK, Farrell RL (1987) Enzymatic "combustion": The microbial degradation of lignin. Annu Rev Microbiol 41:465–501
Kirk TK, Cullen D (1998) Enzymology and molecular genetics of wood degradation by white-rot fungi. New York (USA):273–307
Korus RA, Lodha SJ, Adhi TP, Crawford DL (1991) Kinetics of peroxidase production by streptomyces viridosporus and recombinant streptomyces lividans. Biotechnol Prog 7:510
Kuan IC, Tien M (1993) Glyoxylate-supported reactions catalyzed by mn peroxidase of phanerochaete chrysosporium: activity in the absence of added hydrogen peroxide. Arch Biochem Biophys 302:447
Kubicek C (1992) The cellulase proteins of *Trichoderma reesei*: structure, multiplicity, mode of action and regulation of formation. Enzymes and products from bacteria fungi and plant cells. Springer, Berlin/Heidelberg. 45: 1
Kuhad RC, Singh A, Eriksson KE (1997) Microorganisms and enzymes involved in the degradation of plant fiber cell walls. Adv Biochem Eng Biotechnol 57:45–125
Kukkola EM, Koutaniemi S, Pollanen E, Gustafsson M, Karhunen P, Lundell TK, Saranpaa P, Kilpelainen I, Teeri TH, Fagerstedt KV (2004) The dibenzodioxocin lignin substructure is abundant in the inner part of the secondary wall in norway spruce and silver birch xylem. Planta 218:497–500
Lamed R, Naimark J, Morgenstern E, Bayer EA (1987) Specialized cell surface structures in cellulolytic bacteria. J Bacteriol 169:3792–3800
Lamed R, Bayer EA (1988) The cellulosome of clostridium-thermocellum. Adv Appl Microbiol 33:1–46
Lee SD, Antholine GS, Hedman WE, Hodgson B, Solomon EI (2001) Nature of the intermediate formed in the reduction of o2 to h2o at the trinuclear copper cluster active site in native laccase. J Am Chem Soc 124:6180–6193
Leisola MSA, Fiechter A (1985) Ligninase production in agitated conditions by phanerochaete chrysosporium. FEMS Microbiol Lett 29:33
Li XL, Skory CD, Cotta MA, Puchart V, Biely P (2008) Novel family of carbohydrate esterases, based on identification of the hypocrea jecorina acetyl esterase gene. Appl Environ Microbiol 74:7482–7489
Li Y, Irwin DC, Wilson DB (2007) Processivity, substrate binding, and mechanism of cellulose hydrolysis by thermobifida fusca cel9a. Appl Environ Microbiol 73:3165–3172
Lundell TK, Mäkelä MR, Hildén K (2010) Lignin-modifying enzymes in filamentous basidiomycetes – ecological, functional and phylogenetic review. J Basic Microbiol 50:5

Lynd LR, Weimer PJ, van Zyl WH, Pretorius IS (2002) Microbial cellulose utilization: Fundamentals and biotechnology. Microbiol Mol Biol Rev 66:506–577

Magnuson TS, Crawford DL (1992) Comparison of extracellular peroxidase- and esterase-deficient mutants of streptomyces viridosporus t7a. Appl Environ Microbiol 58:1070–1072

Mansfield SD, deJong E, Saddler JN (1997) Cellobiose dehydrogenase, an active agent in cellulose depolymerization. Appl Environ Microbiol 63:3804–3809

Martinez AT, Ruiz-Duenas FJ, Martinez MJ, Del Rio JC, Gutierrez A (2009) Enzymatic delignification of plant cell wall: from nature to mill. Curr Opin Biotechnol 20:348–357

Martinez D, Larrondo LF, Putnam N, Gelpke MDS, Huang K, Chapman J, Helfenbein KG, Ramaiya P, Detter JC, Larimer F, Coutinho PM, Henrissat B, Berka R, Cullen D, Rokhsar D (2004) Genome sequence of the lignocellulose degrading fungus phanerochaete chrysosporium strain rp78. Nat Biotech 22:695

Matthews JF, Skopec CE, Mason PE, Zuccato P, Torget RW, Sugiyama J, Himmel ME, Brady JW (2006) Computer simulation studies of microcrystalline cellulose ibeta. Carbohydr Res 341:138–152

Matthews JF, Bergenstrahle M, Beckham GT, Himmel ME, Nimlos MR, Brady JW, Crowley MF (2011) High-temperature behavior of cellulose. J Phys Chem B 115:2155–2166

Mayer F, Coughlan MP, Mori Y, Ljungdahl LG (1987) Macromolecular organization of the cellulolytic enzyme complex of clostridium thermocellum as revealed by electron microscopy. Appl Environ Microbiol 53:2785–2792

McCann MC, Carpita NC (2008) Designing the deconstruction of plant cell walls. Curr Opin Plant Biol 11:314

McCarthy AJ (1987) Lignocellulose-degrading actinomycetes. FEMS Microbiol Lett 46:145

McCarthy AJ, Williams ST (1992) Actinomycetes as agents of biodegradation in the environment – a review. Gene 115:189

McCleary BV (1991) Comparison of endolytic hydrolases that depolymerize 1,4-beta-d-mannan, 1,5-alpha-l-arabinan, and 1,4-beta-d-galactan. Enzymes in biomass conversion. Am Chem Soc 460:437–449

Messerschmidt A, Huber R (1990) The blue oxidases, ascorbate oxidase, laccase and ceruloplasmin. Modelling and structural relationships. Eur J Biochem 187:341–352

Mester TN, Field JA (1998) Characterization of a novel manganese peroxidase-lignin peroxidase hybrid isozyme produced by bjerkandera species strain bos55 in the absence of manganese. J Biol Chem 273:15412–15417

Moreira LR, Filho EX (2008) An overview of mannan structure and mannan-degrading enzyme systems. Appl Microbiol Biotechnol 79:165–178

Morozova OV, Shumakovich GP, Gorbacheva MA, Shleev SV, Yaropolov AI (2007) "blue" laccases. Biochemistry (Mosc) 72:1136–1150

Nishiyama Y, Langan P, Chanzy H (2002) Crystal structure and hydrogen-bonding system in cellulose ibeta from synchrotron x-ray and neutron fiber diffraction. J Am Chem Soc 124:9074–9082

O'Neill MA, York WS (2003) The composition and structures of primary cell walls. The plant cell wall. Rose JKC. CRC, Boca Raton, FL, pp 1–54

Okeke B, Smith J, Paterson A, Watson-Craik I (1996) Influence of environmental parameters on pentachlorophenol biotransformation in soil by *lentinula edodes* and *phanerochaete chrysosporium*. Appl Microbiol Biotechnol 45:263

Orth AB, Royse DJ, Tien M (1993) Ubiquity of lignin-degrading peroxidases among various wood-degrading fungi. Appl Environ Microbiol 59:4017–4023

Parsiegla G, Reverbel C, Tardif C, Driguez H, Haser R (2008) Structures of mutants of cellulase ce148f of clostridium cellulolyticum in complex with long hemithiocellooligosaccharides give rise to a new view of the substrate pathway during processive action. J Mol Biol 375:499–510

Pauly M, Keegstra K (2010) Plant cell wall polymers as precursors for biofuels. Curr Opin Plant Biol 13:305–312

Pearl IW (1967) The chemistry of lignin. New York:339

Pease EA, Tien M (1992) Heterogeneity and regulation of manganese peroxidases from phanerochaete chrysosporium. J Bacteriol 174:3532–3540

Phelan MB, Crawford DL, Pometto ALI (1979) Isolation of lignocellulose-decomposing actinomycetes and degradation of specifically 14c-labeled lignocelluloses by six selected streptomyces strains. Can J Microbiol 25:1270–1276

Prescott CE, Zabek LM, Staley CL, Kabzems R (2000) Decomposition of broadleaf and needle litter in forests of british columbia: Influences of litter type, forest type and litter mixtures. Can J For Res 30:1742–1750

Quintanar L, Yoon J, Aznar CP, Palmer AE, Andersson KK, Britt RD, Solomon EI (2005) Spectroscopic and electronic structure studies of the trinuclear cu cluster active site of the multicopper oxidase laccase: nature of its coordination unsaturation. J Am Chem Soc 127:13832–13845

Rabinovich ML (2006) ethanol production from materials containing cellulose: the potential of approaches developed in Russia. Prikl Biokhim Mikrobiol 42:5–32

Raj A, Reddy M, Chandra R, Purohit H, Kapley A (2007) Biodegradation of kraft-lignin by bacillus sp. Isolated from sludge of pulp and paper mill. Biodegradation 18:783

Ramachandra M, Crawford DL, Hertel G (1988) Characterization of an extracellular lignin peroxidase of the lignocellulolytic actinomycete streptomyces viridosporus. Appl Environ Microbiol 54:3057–3063

Reale S, Di Tullio A, Spreti N, De Angelis F (2004) Mass spectrometry in the biosynthetic and structural investigation of lignins. Mass Spectrom Rev 23:87

Reese ET, Siu RG, Levinson HS (1950) The biological degradation of soluble cellulose derivatives and its relationship to the mechanism of cellulose hydrolysis. J Bacteriol 59:485–497

Reinhammer B (1984) Copper proteins and copper enzymes. Lontie R. CRC Press, Boca Raton, FL, pp 1–35

Renganathan V, Miki K, Gold MH (1985) Multiple molecular forms of diarylpropane oxygenase, an h2o2-requiring, lignin-degrading enzyme from phanerochaete chrysosporium. Arch Biochem Biophys 241:304

Reverbel-Leroy C, Pages S, Belaich A, Belaich JP, Tardif C (1997) The processive endocellulase celf, a major component of the clostridium cellulolyticum cellulosome: purification and characterization of the recombinant form. J Bacteriol 179:46–52

Robinson LE, Crawford RL (1978) Degradation of 14c-labeled lignins by bacillus megaterium. FEMS Microbiol Lett 4:301

Sarkanen KV, Ludwig CH (1971) Lignin: occurrence, formation, structure and reactions. New York, p 916

Sato S, Liu F, Koc H, Tien M (2007) Expression analysis of extracellular proteins from phanerochaete chrysosporium grown on different liquid and solid substrates. Microbiology 153:3023–3033

Sato S, Feltus F, Iyer P, Tien M (2009) The first genome-level transcriptome of the wood-degrading fungus *Phanerochaete chrysosporium* grown on red oak. Curr Genet 55:273

Scheller HV, Ulvskov P (2010) Hemicelluloses. Annu Rev Plant Biol 61:263–289

Schulein M (1997) Enzymatic properties of cellulases from humicola insolens. J Biotechnol 57:71–81

Shallom D, Shoham Y (2003) Microbial hemicellulases. Curr Opin Microbiol 6:219–228

Shen H, Meinke A, Tomme P, Damude Howard G, Kwan E, Kilburn Douglas G, Miller Robert C, Warren RAJ, Gilkes Neil R (1996) Cellulomonas fimi cellobiohydrolases. Enzymatic degradation of insoluble carbohydrates. Am Chem Soc 618:174–196

Shleev S, Persson P, Shumakovich G, Mazhugo Y, Yaropolov A, Ruzgas T, Gorton L (2006) Interaction of fungal laccases and laccase-mediator systems with lignin. Enzyme Microb Tech 39:841

Shoemaker S, Watt K, Tsitovsky G, Cox R (1983) Characterization and properties of cellulases purified from trichoderma reesei strain l27. Nat Biotech 1:687

Shulami S, Gat O, Sonenshein AL, Shoham Y (1999) The glucuronic acid utilization gene cluster from bacillus stearothermophilus t-6. J Bacteriol 181:3695–3704
Singh D, Chen S (2008) The white-rot fungus phanerochaete chrysosporium conditions for the production of lignin-degrading enzymes. Appl Microbiol Biotechnol 81:399
Skálová T, Dohnálek J, Østergaard LH, Østergaard PR, Kolenko P, Dusková J, Stepánková A, Hasek J (2009) The structure of the small laccase from streptomyces coelicolor reveals a link between laccases and nitrite reductases. J Mol Biol 385:1165
Smith MM, Hartley RD (1983) Occurrence and nature of ferulic acid substitution of cell-wall polysaccharides in graminaceous plants. Carbohydr Res 118:65
Solomon EI, Baldwin MJ, Lowery MD (1992) Electronic structures of active sites in copper proteins: contributions to reactivity. Chem Rev 92:521
Solomon EI, Sundaram UM, Machonkin TE (1996) Multicopper oxidases and oxygenases. Chem Rev 96(7): PBD: Nov 1996:Medium: X; Size: pp. 2563–2605.
Spanikova S, Biely P (2006) Glucuronoyl esterase - novel carbohydrate esterase produced by schizophyllum commune. FEBS Lett 580:4597–4601
Spiker JK, Crawford DL, Thiel EC (1992) Oxidation of phenolic and non-phenolic substrates by the lignin peroxidase of *streptomyces viridosporus* t7a. Appl Microbiol Biotechnol 37:518
Spiridonov NA, Wilson DB (2001) Cloning and biochemical characterization of bglc, a β-glucosidase from the cellulolytic actinomycete *Thermobifida fusca*. Curr Microbiol 42:295
Swift MJ, Heal OW, Anderson JM (1979) Decomposition in terrestrial ecosystems. Studies in Ecology Berkeley/Los Angeles, CA 5:372
Teeri TT (1997) Crystalline cellulose degradation: new insight into the function of cellobiohydrolases. Trends Biotechnol 15:160
ten Have R, Teunissen PJM (2001) Oxidative mechanisms involved in lignin degradation by white-rot fungi. Chem Rev 101:3397
Tenkanen M (1998) Action of trichoderma reesei and aspergillus oryzae esterases in the deacetylation of hemicelluloses. Biotechnol Appl Biochem 27:19–24
Tomme P, Warren RA, Gilkes NR (1995) Cellulose hydrolysis by bacteria and fungi. Adv Microb Physiol 37:1–81
Tomme P, Warren RAJ, Miller Robert C, Kilburn Douglas G, Gilkes Neil R (1996) Cellulose-binding domains: Classification and properties. Enzymatic degradation of insoluble carbohydrates. Am Chem Soc 618:142–163
Topakas E, Kyriakopoulos S, Biely P, Hirsch J, Vafiadi C, Christakopoulos P (2010) Carbohydrate esterases of family 2 are 6-o-deacetylases. FEBS Lett 584:543–548
Tormo J, Lamed R, Chirino AJ, Morag E, Bayer EA, Shoham Y, Steitz TA (1996) Crystal structure of a bacterial family-iii cellulose-binding domain: a general mechanism for attachment to cellulose. EMBO J 15:5739–5751
Trigo C, Ball AS (1994) Is the solubilized product from the degradation of lignocellulose by actinomycetes a precursor of humic substances? Microbiology 140:3145–3152
Trojanowski J, Haider K, Sundman V (1977) Decomposition of ^{14}c-labelled lignin and phenols by a *nocardia* sp. Arch Microbiol 114:149
Tuor U, Winterhalter K, Fiechter A (1995) Enzymes of white-rot fungi involved in lignin degradation and ecological determinants for wood decay. J Biotechnol 41:1
Varrot A, Schulein M, Davies GJ (1999) Structural changes of the active site tunnel of humicola insolens cellobiohydrolase, cel6a, upon oligosaccharide binding. Biochemistry 38:8884–8891
Vicuña R (1988) Bacterial degradation of lignin. Enzyme Microb Technol 10:646
Vicuña R (2000) Ligninolysis. Mol Biotechnol 14:173
Vinzant TB, Adney WS, Decker SR, Baker JO, Kinter MT, Sherman NE, Fox JW, Himmel ME (2001) Fingerprinting trichoderma reesei hydrolases in a commercial cellulase preparation. Appl Biochem Biotechnol 91:99
Vivanco L, Austin AT (2011) Nitrogen addition stimulates forest litter decomposition and disrupts species interactions in patagonia, argentina. Global Change Biol

Walker LP, Belair CD, Wilson DB, Irwin DC (1993) Engineering cellulase mixtures by varying the mole fraction of thermomonospora fusca e5 and e3, trichoderma reesei cbhi, and caldocellum saccharolyticum beta-glucosidase. Biotechnol Bioeng 42:1019–1028

Wallenstein MD, Weintraub MN (2008) Emerging tools for measuring and modeling the in situ activity of soil extracellular enzymes. Soil Biol Biochem 40:2098

Wardle DA (2002) Communities and ecosystems: Linking the aboveground and belowground components. Princeton, NJ

Warnecke F, Luginbuhl P, Ivanova N, Ghassemian M, Richardson TH, Stege JT, Cayouette M, McHardy AC, Djordjevic G, Aboushadi N, Sorek R, Tringe SG, Podar M, Martin HG, Kunin V, Dalevi D, Madejska J, Kirton E, Platt D, Szeto E, Salamov A, Barry K, Mikhailova N, Kyrpides NC, Matson EG, Ottesen EA, Zhang X, Hernandez M, Murillo C, Acosta LG, Rigoutsos I, Tamayo G, Green BD, Chang C, Rubin EM, Mathur EJ, Robertson DE, Hugenholtz P, Leadbetter JR (2007) Metagenomic and functional analysis of hindgut microbiota of a wood-feeding higher termite. Nature 450:560

Watanabe H, Tokuda G (2001) Animal cellulases. Cell Mol Life Sci 58:1167

Watson BJ, Zhang HT, Longmire AG, Moon YH, Hutcheson SW (2009) Processive endoglucanases mediate degradation of cellulose by saccharophagus degradans. J Bacteriol 191:5697–5705

Welinder KG (1992) Superfamily of plant, fungal and bacterial peroxidases. Curr Opin Struct Biol 2:388

Whittaker MM, Kersten PJ, Nakamura N, Sanders-Loehr J, Schweizer ES, Whittaker JW (1996) Glyoxal oxidase from phanerochaete chrysosporium is a new radical-copper oxidase. J Biol Chem 271:681–687

Wilson DB, Irwin DC (1999). Genetics and properties of cellulases. Recent progress in bioconversion of lignocellulosics. Springer, Berlin, 65: 1

Wilson DB (2004) Studies of thermobifida fusca plant cell wall degrading enzymes. Chem Rec 4:72

Wilson DB (2009) Aerobic microbial cellulase systems. Blackwell Publishing Ltd.

Wong D (2009) Structure and action mechanism of ligninolytic enzymes. Appl Biochem Biotechnol 157:174

Wood TM (1968) Cellulolytic enzyme system of *trichoderma koningii*. Separation of components attacking native cotton. Biochem J 109:217–227

Wymelenberg AV, Sabat G, Martinez D, Rajangam AS, Teeri TT, Gaskell J, Kersten PJ, Cullen D (2005) The phanerochaete chrysosporium secretome: database predictions and initial mass spectrometry peptide identifications in cellulose-grown medium. J Biotechnol 118:17

Yee DC, Jahng D, Wood TK (1996) Enhanced expression and hydrogen peroxide dependence of lignin peroxidase from streptomycesviridosporus t7a. Biotechnol Prog 12:40–46

Yoon J, Liboiron BD, Sarangi R, Hodgson KO, Hedman B, Solomon EI (2007) The two oxidized forms of the trinuclear cu cluster in the multicopper oxidases and mechanism for the decay of the native intermediate. Proc Natl Acad Sci USA 104:13609–13614

Zamocky M, Ludwig R, Peterbauer C, Hallberg BM, Divne C, Nicholls P, Haltrich D (2006) Cellobiose dehydrogenase - a flavocytochrome from wood-degrading, phytopathogenic and saprotropic fungi. Curr Protein Pept Sci 7:255

Zimmermann W (1990) Degradation of lignin by bacteria. J Biotechnol 13:119

Zverlov VV, Schantz N, Schwarz WH (2005) A major new component in the cellulosome of clostridium thermocellum is a processive endo-beta-1,4-glucanase producing cellotetraose. FEMS Microbiol Lett 249:353–358

Sugary Exudates in Plant Pollination

Massimo Nepi, Patrick von Aderkas, and Ettore Pacini

Abstract Sugary secretions are present in many plants and frequently they are linked with reproductive processes. Most of the gymnosperms, both extant and extinct, possess a pollination drop, a diluted sugary secretion protruding from the micropyle, which serves for pollen capture, hydration, and transport in the ovule. It is most probable that this secretion attracted insects giving origin to a plant–insect relationship for pollination based on a sugary solution well before the raise of angiosperm. Floral nectar, a new type of sugary exudate produced by a specific secreting tissue (the nectary), evolved rapidly when the transition from naked ovule to closed carpel was completed and the pollination drops were no longer available as a food resource for insects. Floral nectar is widely distributed and very diverse in the extant angiosperms where it represents the more common reward for a large variety of pollinators. In this chapter, we highlight the evolutionary relationship between nectar and pollination drops in terms of morphology, physiology, ecology, and biochemistry.

1 Introduction

Different types of exudates are present in the Spermatophyta that are involved in reproductive processes. Several main functions can be ascribed to these exudates (Table 1). These functions are present at different moments: during pre-pollination events (nourishment of male gametophyte during development), at pollination (pollen landing site, attracting animals involved in pollination), or during

M. Nepi (✉) • E. Pacini
BIOCONNET (Biodiversity and Conservation Network), Department of Environmental Sciences "G. Sarfatti", University of Siena, Siena, Italy
e-mail: massimo.nepi@unisi.it

P. von Aderkas
Department of Biology, University of Victoria, Victoria, BC, Canada

J.M. Vivanco and F. Baluška (eds.), *Secretions and Exudates in Biological Systems*, Signaling and Communication in Plants 12, DOI 10.1007/978-3-642-23047-9_8,

Table 1 Types of exudates involved in plant reproduction, their presence in gymnosperms and angiosperms, and their functions

	Locular fluid	Ovular secretion[a]	Floral nectaries	Stigmatic exudates	Stylar fluid[b]
Presence					
Gymnosperms	X	X			
Angiosperms	X	X	X	X	X
Functions					
Developing pollen nourishment	XX				
Pollen capture		X			
Pollen rehydration		X		X	
Pollen transport inside the ovule		X			
Promoting pollen germination		X		X	
Pollen tube growth and guidance		X[c] X		X	X
Male gametophyte – female part recognition		X		X	X
Attracting animals involved in pollination		X[d]	X[e] X[f]	X	

Functions in gymnosperms are indicated in bold

[a]This class comprises the different types of ovular secretions in gymnosperms and the mycropylar secretion of angiosperms ovules

[b]Present only in the hollow styles

[c]Proteins involved in pollen tube guidance (arabinogalactan proteins) have been detected in the pollination drop of *Taxus x media* (O'Leary et al. 2004)

[d]In some Gnetales and most probably in a certain number of gymnosperm species in the Mesozoic (Ren et al. 2009)

[e]In some *Ephedra* species (Gnetales) nectaries are present on the bracts associated with the reproductive structures in male and female strobili and are involved in attracting insects

[f]In *Asclepias syriaca* (Asclepiadaceae) nectar serves also for rehydrating pollen (Kevan et al. 1989)

postpollination events (pollen rehydration, pollen transport inside the ovule, pollen–pistil recognition, pollen tube attraction toward, and interaction with the female gametophyte). In this contribution, we focused on the sugary exudates that function during pollination.

Pollination is a kind of flight characteristic of both major seed plant clades in which pollen floats away from the male part and arrives on the female one (Pacini 2009). The main differences in pollination between angiosperms and gymnosperms are the landing site and the vector of the flight. According to Lloyd and Wells (1992), a major step in angiosperm carpel evolution was a change in the landing site for pollen from a pollination drop on the naked ovule to a wet stigma on the outside of a closed carpel. Pollination drops, water-based sugary secretions elaborated by the apical part of the nucellus (Gelbart and von Aderkas 2002), appeared very early in the phylogeny of gymnosperms. According to Doyle (1945) and Rothwell (1977) they were common among pteridosperms, an extinct clade of gymnosperms also called the seed ferns. This mechanism of pollen capture was widespread during the Mesozoic (Ren et al. 2009) and has remained so to this day. As far as pollen vectors are concerned, two main types are recognized: wind (anemophily) and animals (zoophyly, especially entomophily) (Pacini 2009). Gymnosperms are considered mainly wind-pollinated, entomophily having evolved in Cycadales and Gnetales (or Chlamydospermae) and a few extinct lineages (Crepet 1974; Crepet and Friis

1987; Labandeira et al. 2007). Recently, Ren et al. (2009) pointed out the possibility of a more complex ecological scenario dominating the Middle Jurassic and Early Cretaceous in which extinct Eurasian scorpionflies (Mecoptera) with long siphonate proboscides may have fed on the gymnosperm pollination drops and most likely were involved in pollination. According to this scenario insect pollination was already present just before and at the same time of angiosperm expansion in the Early Cretaceous. This may have driven the evolution of reproductive biology in early flowering plants (Nepi et al. 2009; Ren et al. 2009). The new condition of enveloped ovules that characterizes angiosperms brought about a complication because the primary source of sugary exudates (i.e., the pollination drop) had disappeared. The presence of insects preadapted to feed on sugary liquids may cause a strong selective pressure toward the provision of sugary exudates in the form of nectar in the early angiosperms (Nepi et al. 2009). After this first step an extraordinary diversification evolved in nectar characters, insect mouthparts, and feeding preferences. These led to the actual wide and complex web of plant–animal relationships mediated by nectar that is observed in many present-day angiosperms. A micropylar exudate containing carbohydrates and proteins is still present in the enclosed ovules of some modern angiosperms (Franssen-Verheijen and Willemse 1993; Endress 1994; Fortescue and Turner 2005 and references therein), but its function is to drive the pollen tubes across the micropyle to the embryo sac. Being not involved in pollination it will not be considered in this contribution.

Independently from both the vector and the landing site, pollen arriving on the receptive surface needs to rehydrate and absorb nutrients for its further development. Dispersed pollen is generally strongly dehydrated in comparison with common plant cells. It ranges from 10 to 30% water content. It is metabolically quiescent (Nepi et al. 2001). Water uptake is essential both for restoring metabolism and for pollen germination (Shivanna 2003). On the other hand, exogenous sugars are also essential for pollen tube germination and growth (Shivanna 2003). Pollen rehydration and nourishment are perfectly accomplished by the pollination drop: it has a large quantity of water, a certain amount of sugars, and also proteins promoting pollen tube growth (Wagner et al. 2007; Nepi et al. 2009). The evolution of a novel specific receptive organ in angiosperms – the stigma – provides a way to hydrate and nourish pollen soon after landing on the stigma surface. A stigma will bind pollen and mediate its hydration, germination, and tube migration into the style. Stigmas are classified into two groups: dry stigmas, which have intact surface cells that typically protrude as papillae and are covered by a primary cell wall, a waxy cuticle, as well as a proteinaceous pellicle; and wet stigmas, which are stigmas covered with surface cells that often lyse to release a viscous surface secretion containing proteins, lipids, polysaccharides, and pigments (Edlund et al. 2004). Stigma exudates are viscous fluids and their composition is generally dominated by lipids that mediate water transfer to pollen grains, while highly diverse proteins and peptides mediate self and foreign pollen recognition (Edlund et al. 2004). An exception to this is found in *Lilium* in which an aqueous stigmatic exudate occurs (Janson et al. 1994). The disaccharide carbohydrate sucrose as well as free monosaccharides can occur in the stigmatic exudates of a few species (Shivanna 2003). It follows that stigma exudates generally cannot be considered a primary sugary secretion, sugars being very low if not absent. Wet stigmas are considered to be the primitive condition in flowering plants (Heslop-Harrison and Shivanna 1977). It is most probable that the stigmatic secretion of early angiosperms may have served as reward for pollinators. These early pollinators would have included mainly flies (Diptera), micropterigid moths, and beetles (Coleoptera) (Lloyd and Wells 1992; Endress 2010). In a few modern angiosperm species stigmatic secretions are considered attractants primarily for some insects (Dafni 1992).

From the aforementioned considerations, it is clear that only two types of sugary exudates are involved in the pollination of gymnosperms and angiosperms: pollination drops (gymnosperms) and nectar (angiosperms and a few gymnosperms). Their physiological, ecological, biochemical, and evolutionary characters are discussed in this chapter.

2 Sugar Exudates and Pollination in Gymnosperms

Secretions during reproduction are found in nearly all gymnosperms. These secretions can roughly be grouped according to pollination by either wind or insect. An intermediate type exists that can capture aerial pollen and attract insects. Most types of secretions are ovular secretions. However, nectar secretions from male buds and female reproductive axes, though far less common, are known (Kato et al. 1995). Ovular secretions are widespread, occurring in all extant families (Gelbart and von Aderkas 2002). As previously mentioned, there are some angiosperm families that also feature ovular secretions (Franssen-Verheijen and Willemse 1993; Endress 1994). Ovular secretions are ancient, having been recorded from pteridosperms, an extinct clade of gymnosperms (Rothwell 1977). These secretions represent a basal condition.

The most common type of ovular secretion is the pollination drop (Fig. 1). Liquid is secreted from the nucellus, filling the tip of the ovule and expanding to form a drop at the micropylar tip of the ovule. The drop's primary function is to capture pollen. A variation on this is found in conifers such as Douglas-fir and larch. Ovular secretions first appear many weeks after pollination. The function of this postpollination/prefertilization drop is to transport the pollen to the nucellus and to induce germination. Ovular secretions are lacking in some genera (*Tsuga*, *Abies*, and *Araucaria*). The loss of this function in evolution has occurred independently in a number of gymnosperm groups and represents a derived condition (Gelbart and von Aderkas 2002).

Pollination drops that attract insects are also an ancient condition. This is not surprising given that gymnosperm diversity was much greater in the Mesozoic than it is today (Labandeira et al. 2007). The greatest diversity of reproductive secretion types is found in the Gnetales. Nectar is known from male cones as well as from surface organs that behave as nectaries and are associated with the collars of female buds.

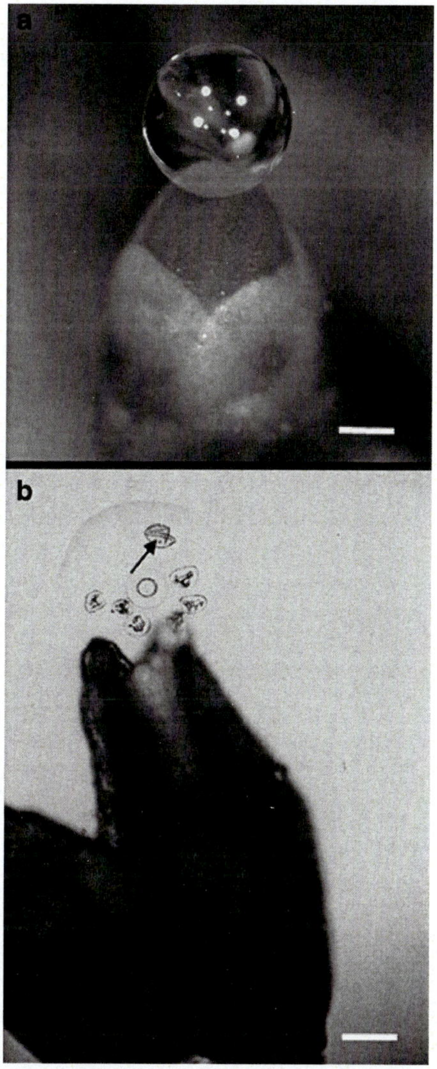

Fig. 1 Pollination drops of gymnosperms. (**a**) Pollination drop on an ovule of *Taxus* x *media*. (**b**) *Bar* = 500 μm. Ovule dissected from female cone of *Cupressus sempervirens* and observed in immersion oil. Differently hydrated pollen grains and an exine split off/(*arrow*) are visible in the pollination drop. *Bar* = 50 μm

The chemistry of pollination drops is diverse. In addition to sugars (Schumann 1902), drops have inorganic constituents such as calcium (Fujii 1903), as well as organic acids (Schumann 1902; Tison 1911), amino acids, and proteins (Ziegler 1959). Unlike angiosperms in which there are organs (i.e., stigma and style) between the ovule and the outside environment, gymnosperm ovules have no such protection. Pollination drops of gymnosperms therefore serve double duty; they not only aid reproduction, but they protect reproductive structures. The first line of defense during reproduction is the pollination drop.

The main carbohydrate components of pollination drops are glucose, fructose, and sucrose. This is to be expected, as sucrose is the main transport sugar (McWilliam 1958). Drops of insect-pollinated plants have higher concentrations of carbohydrates when compared to wind-pollinated ones. The balance between hexose sugars and sucrose also differs between these pollination types. Glucose and fructose are higher in the wind-pollinated gymnosperms, whereas sucrose concentrations tend to be much higher in insect-pollinated ones.

Compared with angiosperm nectar, gymnosperm pollination drops (and drops of gymnosperm nectar) are significantly smaller in volume. Conifers span a range from 10 nl in *Chamaecyparis lawsoniana* to 250 nl in *Taxus* x *media*. Because of difficulties of collecting sufficient sample for analysis little work has been carried out. Chemical analysis, ecological interpretation, and functional analysis have all lagged behind angiosperm nectar. The key roles ascribed to the pollination drop since its discovery by Vaucher (1841) have been pollen capture, transport, and germination (Doyle 1945). In gymnosperms such as conifers, pollination drops were of such low volume and low sugar concentration that they were thought to be of no interest to foraging insects. It was not until high concentrations of sugars were found in *Ephedra* (Ziegler 1959) that a sugar-based reward system for insect pollinators could be defined. As technologies such as genomics and proteomics expand our understanding of drop biology, we can now begin to tackle long-standing questions about the evolution of the pollination drop.

2.1 Nucellus, the Secreting Organ

Pollination drops are secreted from the nucellus (Poulis et al. 2005), which is located inside the ovule (Fig. 2). Gymnosperm ovules are morphologically very conservative. Typically, a nucellus, i.e., sporogenous tissue, is surrounded by integument (Singh 1978).

The period of pollination drop secretion is generally short taking place over a few days to a few weeks during the development of the ovule. This is in keeping with the well-noted differences among gymnosperm taxa in the period from pollination to fertilization. In some species it is only a week and in many others over a year.

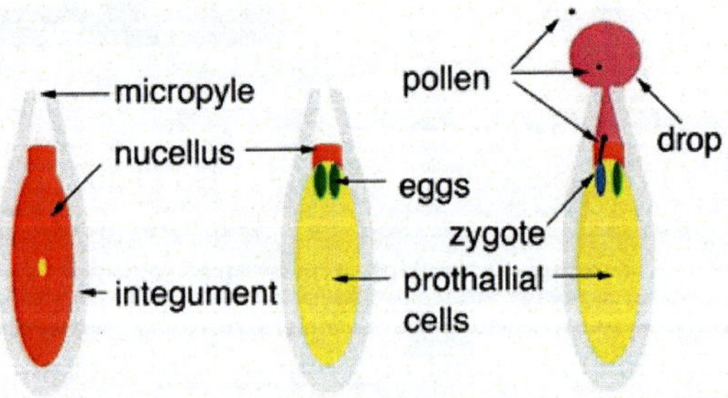

Fig. 2 Schematic diagram of pollination drop occurrence in relation to gymnosperm ovule development. The nucellus is the site of megasporogenesis and megagametophyte development. As the megagametophyte matures, it forms eggs. Drops (*pink*) are secreted by the remaining nucellus (*red*). Secretion may occur, depending on genus, prior to meiosis (e.g., *Pinus*, *Taxus*, not illustrated), or during the final stages of egg maturation as illustrated (e.g., *Larix*, *Picea*, and cycads)

Consequently, nucelluses at the time of secretion do not share similar developmental stages. In the case of *Taxus* and *Pinus*, the nucellus has not yet begun sporogenesis, i.e., meiosis (O'Leary et al. 2004), whereas in *Pseudotsuga* and *Larix*, meiosis is long past. Indeed, in these genera have fully developed megagametophytes at the time of pollination drop release (von Aderkas and Leary 1999a, b).

Nucellar structure also varies. Most commonly, the nucellus is composed of vacuolated parenchymatous cells. The nucellar cells on the surface facing the micropyle may be intact or they may have degenerated. A cuticle is usually lacking, or if present, breaks free, exposing the nucellar surface, e.g., *Pseudotsuga* (Takaso and Owens 1996). In other species, such as *Cedrus deodora*, the surface cells of the nucellus break down to form a pollen chamber (Takaso and Owens 1995). In *Picea*, a pollen chamber is present, but is not a product of breakdown but of differential growth of the nucellar cells (Owens and Molder 1979). Breakdown is, however, inevitable, as once the pollen lands on the surface of this chamber, a downward wave of cell death extends deep into the nucellus (Owens et al. 1987).

Though most do not, some nucelluses bear similarities to certain types of angiosperm nectaries. *Picea* nucellus is one of those with strong developmental similarity to a nectary. Starch builds up in elongating nucellar cells. The starch subsequently breaks down into free sugars that are then released from the surface (Owens et al. 1987). In contrast, the nucellus of most conifer species does not give the appearance of being a starch-loading tissue. Cells of *Taxus* and *Larix* are highly vacuolate. Although they do not accumulate starch, they do build up proteins. Immunolabeling with antibodies against arabinogalactan proteins (O'Leary et al. 2004) shows such buildup in cells of *Taxus* nucellus. These proteins are released into the pollination drop.

A proteomics approach has allowed us to ask whether the compounds found in the pollination drop are secreted or are merely the result of cellular breakdown. The answer is that in some they are secreted, and in others they are probably in with cellular debris associated with breakdown. Evidence of secretion is supplied by our discovery of a signal cleavage peptide that allows a protein to move across the plasmalemma into the apoplast. It was found in conifers (O'Leary et al. 2007). Evidence of breakdown is known from species such as *Picea* that exhibit extensive nucellar breakdown both before and during pollination drop formation. It is likely that these dying cells contribute a rich soup of compounds. In some extinct pteridosperms, lagenostomes were formed from cellular degeneration of the nucellar tip. These resemble a domed pollen chamber. Some also had an extensive protrusion called a salpinx (Greek trumpet). Liquid would fill the chamber and the salpinx, extending as a drop. In *Welwitschia* and *Ephedra*, erect lignified micropylar extensions of the ovule play an analogous role (Rydin et al. 2010).

Although most pollination mechanisms employ ovular secretions that are water-based, there is one other type of secretion, the microdrop. It is not found in the nucellus, but on integumentary flaps at the ovule's tip in certain conifers (Owens et al. 1981, 1987). These drops are composed of lipids. Microdrops trap pollen. Sometime later when the pollination drop is secreted, micropylar extensions release the pollen to the pollination drop.

2.2 Pollination Drop Production

Physiological studies are few and mainly concentrated in *Pseudotsuga*, *Ephedra*, and *Taxus*. Molecular biological studies of the cellular origins of constituents are restricted to a handful of conifers. These confirm the nucellar origins of the constituents.

Analysis of pollination drops typically shows dominance by sucrose, glucose, and fructose. However, nucellar sugar chemistry has not been studied. A testable model of sugar secretion might be as follows. Phloem sap is unloaded into the ovuliferous scale, where it is transformed and transported as pre-pollination drop sucrose across the nucellus, both through apoplast and symplast. In those species in which sucrose is only a minor sugar constituent of pollination drops, there may be modification of sucrose by symplastic and/or apoplastic invertases. Sugars are discharged from the nucellus. Mass flow of water from symplasm likely occurs into high local apoplastic concentrations of sugar, resulting in liquid filling the pollen chamber or the micropyle end, and then exuding from the ovule tip to form a drop.

The question of regulation remains unresolved. In part, this is due to the wide range of secretory behaviors. No one explanation is expected. Some species secrete repeatedly, but once pollination has occurred, they cease production. Studies of Pinaceae show a quick retraction of the drop following pollen capture. Secretion then stops (Doyle and O'Leary 1935; McWilliam 1958; Ho 1985; Owens et al. 1981, 1987). This is not without controversy, as repeated pollination drop secretion has been recorded for *Pinus radiata* (Lill and Sweet 1977). There are conifers that produce drops once. In *Phyllocladus*, a genus in the Podocarpaceae, drops retract after pollination. Secretion stops. Yet other conifers continue to secrete following pollination. For example, *Podocarpus* pollination drops are repeatedly secreted. Since the surrounding surfaces are wax-free, this repeated secretion and retraction amounts to pollen scavenging (Tomlinson et al. 1997).

The only physiological study is that of Ziegler (1959) in which he showed that secretion was a passive event. When he used metabolic poisons to kill the cells of the nucellus, drop formation continued unabated.

To date, nucellar tissue has not been analyzed for gene expression, protein expression, or hormone levels. Sweet and Lewis (1969) predicted that the nucellus may be able to detect auxin secreted from pollen. The authors ascribed a potential role for elevated levels of auxin in the prevention of ovule abortion. Analysis of nucellus for auxin levels before and after pollination has yet to be carried out.

2.2.1 Pollination Drop Secretion: Daily Patterns

Secretion follows a diurnal pattern, with some exceptions. Pollination drops occur during the night and early morning. In *Chamaecyparis nootkatensis* (Owens et al. 1980) and in *Pinus* (Doyle and O'Leary 1935) secretion is at night, with the drop disappearing early in the day. However, daytime secretion has been noted in some species of *Gnetum* (Pearson 1909) and *Ephedra* (Bino et al. 1984). Trees have a diurnal cycle of hydraulic pressure. At dawn trees usually have the most water, but by noon, pressure drops are significant. Another reason that formation is during the late night–early morning period is that this period has the highest relative humidity. As night temperatures are lower, relative humidity is higher (McWilliam 1958). Extremes in humidity, such as drought, affect water potential of the tree and cones, resulting in failure to produce drops. Rainfall is considered detrimental to some species, as it washes away both pollination drops and pollen (Tison 1911). Although mornings are generally better for pollination drops, postpollination/prefertilization drops are produced most of the day. An hour-by-hour simultaneous analysis of tree hydraulic pressure and ovular secretion was carried out in *Pseudotsuga menziesii*. As expected, the diurnal hydraulic pressure pattern for all trees tested was high in the morning and low in the afternoon. The profile of drop secretion did not match this diurnal pattern. Instead, it showed strong tree-by-tree variation, with some trees producing mostly during the day and others mostly at night (O'Leary and von Aderkas 2006). This implies that drop secretion physiology is not regulated by whole-tree physiology, but at the cone or even at the ovule level. Another example of cone level control is seen in *Picea engelmannii*. Drops are secreted in a wave from the proximal to distal portions of cones (Owens et al. 1987).

Daytime secretion can also be brought about by controlling the conditions during pollen reception. By potting small trees, growing them in greenhouses and inducing cones with hormone treatment, daytime secretion occurred readily in *Chamaecyparis lawsoniana* (Wagner et al. 2007). Removal of *Juniperus communis* branches to a room with 50% relative humidity allowed drops to be collected during the daytime (Mugnaini et al. 2007a). Similarly, interior spruce trees potted in a greenhouse and kept at 85% relative humidity produced a greater number of pollination drops than in a control treatment (Runions and Owens 1996). Such drops were also larger.

2.3 Ovular Secretion Characteristics

Ovular secretion biology integrates morphology with biochemistry and physiology. In addition there are spatial and temporal components. The morphology of the ovule and the surrounding tissues has been of interest because the presence or absence of waxy hydrophilic surfaces literally shapes the secretion into either a drop or into a more amorphous secretion. This is particularly notable in species in the Podocarpaceae (Tomlinson et al. 1997). From a biological standpoint, a very useful separation of secretion types is into drops from wind-pollinated or insect-pollinated systems.

Wind-pollinated species have evolved the greatest diversity of pollination mechanisms, and consequently have a wide range of drop types. These range from internal or *in ovulo* secretion of *Pseudotsuga* and *Larix* (postpollinatinon/prefertilization drops), to pollination drops that form neat little spheres on the tips of ovules (most gymnosperms), to pollination drops that scavenge pollen across larger surfaces of the ovuliferous scale, such as those of *Podocarpus* spp.

Insect-pollinated species in the orders Cycadales and Gnetales only have one type of pollination drop. However, some gnetophytes additionally possess nectaries on male buds and elsewhere.

Insect- and wind-pollinated species typically differ in the volumes of these drops. Insects seek a sweet liquid reward. Increased volume of sweet liquid tends to attract larger insects. The drop volumes of insect-pollinated species span three orders of magnitude. They are low in cycads, such as *Zamia pumila* (15–50 nl – Tang 1987) and high in the gnetophyte *Ephedra trifurca* (500–4200 nl – Niklas et al. 1986).

Drop volumes of aerially pollinated gymnosperms span only two orders of magnitude, e.g. 10 nl in *Chamaecyparis lawsoniana* to 250 nl in *Taxus* x *media* (von Aderkas unpublished). As has been shown from mathematical simulations and field studies, differences in drop volume contribute only a minor amount to the efficiency of pollen capture in *Ephedra trifurca*, a wind-pollinated species (Niklas et al. 1986; Niklas and Kerchner 1986; Niklas and Buchmann 1987). A much larger role is played by ovule form, leading to a statistically significant preferential selection of conspecific pollen in comparison to foreign pollen (Niklas and Buchmann 1987). The drop plays a passive role in pollen capture, a role that is relatively independent of its volume (Buchmann et al. 1989). In wind-pollinated conifers with postpollination/prefertilization drops such as *Larix* and *Pseudotsuga*, the volumes are also relatively low (25–50 nl – von Aderkas and Leary 1999a,b).

2.3.1 Sugar Concentration

The predominant sugars (sucrose, glucose, and fructose) range from a low of 12.5 mg/ml in *Pinus* (McWilliam 1958) to a high of 250 mg/ml in *Ephedra* (Ziegler 1959). In a study of sugar composition of six species of gymnosperms the average sugar concentration was approximately 80 mg/ml (Nepi et al. unpublished). It is often stated that drops compensate for evaporation by balanced secretion, but there are no experimental data supporting such claims. Concentration may not always be stable. A single pollination drop may last up to 12 days (Mugnaini et al. 2007a) during which it is influenced by local relative humidity. The drops of *Podocarpus* are secreted repeatedly. Initially, the drop pours out onto surrounding surfaces of ovuliferous scales to scavenge for pollen. Then it is absorbed back into the ovule. This cycle is repeated (Tomlinson et al. 1997). It is hard to imagine that the concentration can remain constant. In cases where pollination drops are coupled with rainwater scavenging (Greenwood 1986; Runions and Owens 1996) dilution must occur.

Sugar concentrations are also altered by enzymes. Secretion of invertase into pollination drops directly alters composition and concentration. Extracellular invertases have been shown to actively catalyze sucrose into fructose and glucose in postpollination/prefertilization drops of *Pseudotsuga menziesii*. This doubles the osmolality of the solution.

Concentration is also probably altered during absorption of the drop. Absorption can be passive or active. Passive absorption by evaporation is the more common of the two. As drops evaporate the concentration levels must rise. As drops age, they also appear to become more viscous. Drops of *Larix* and *Pseudotsuga* occasionally become more viscous as the season advanced. Although the volume of the drop is steady, its concentration may be slowly rising. Tison (1911) was the first of many to make similar anecdotal but unquantified comments. What the effect of this few-fold increase in concentration may have on conspecific and foreign pollen, not to mention opportunistic pathogens, is unknown.

In some conifer families, pollen causes the retreat of the drop within a few minutes, e.g., Pinaceae (McWilliam 1958; Owens et al. 1987), Cupressaceae (Owens et al. 1980, Mugnaini et al. 2007a, b), and Podocarpaceae (Tomlinson et al. 1997). The drop does not form again. Concentration changes may occur as solutes are reabsorbed in the nucellus. Such a quick response is faster than the nucellus could generate a response mediated by signaling and gene expression.

Sugar Concentrations and Insect Pollination

Since pollinating insects seek rewards, insect-pollinated gymnosperms can be split into those offering pollen rewards (cycads) and those offering sweet pollination drops or nectar (gnetophytes). Among gnetophytes, *Gnetum* (Kato et al. 1995) and *Welwitschia* (Wetschnig and Depisch 1999) are insect-pollinated, but *Ephedra* can be insect-pollinated (*E. foemina* – Porsch 1910), wind-pollinated (*E. trifurca* – Buchmann et al. 1989), or both (*E. aphylla* – Meeuse et al. 1990). Sucrose concentrations range from 150 to 250 mg/ml (Meeuse et al. 1990; Ziegler 1959). Sugar concentrations in *Gnetum cuspidatum* are similar to *Ephedra* (Kato et al. 1995). Changes in relative humidity are thought to account for variation in total sugar concentrations (30–130 mg/ml) on female strobili of *G. gnemon* (Kato et al. 1995). However much relative humidity may affect sugar concentrations, high concentrations represent an adaptation in themselves. It takes more work to remove the water from more highly concentrated drops. Such sugary drops last longer. As a result they are available longer for insects.

Cycads are weevil-pollinated. Weevils that visit male strobili and feed on pollen become covered in pollen. They then eat their way into the female strobili, which they partially consume. Pollen haphazardly falls off their bodies and lands near ovules. A pollination drop captures pollen near the edge of the micropyle and then recedes inside the ovule leading to seed set. There is no evidence to suggest that weevils consume the pollination drops. In the case of *Zamia pumila*, one of the better studied of the cycads when it comes to pollination drops, weevil feeding behavior does not even overlap with secretion. Weevils feed during late morning and afternoon, whereas pollination drop secretion is a phenomenon restricted to late night and early morning (Tang 1987). Sugar concentration is around 100 mg/ml.

Although this concentration is intermediate between conifers and gnetophytes and therefore potentially interesting to reward-seeking insects, the volumes are far from enticing, e.g., less than 50 nl/drop in *Zamia pumila* (Tang 1987). The insect is merely a vector of pollen, rewarded by food from male strobili (pollen) and female strobili (tissues): any pollination drop feeding is incidental.

A number of conifers are thought to be insect-pollinated (Ren et al. 2009). One type of evidence is the presence of insects found in fossilized strobili. Another is phylogenetic analysis that shows insects with sucking mouthparts evolved well before angiosperms. This provides good grounds to believe that extinct conifers, which were at their most diverse during the Mesozoic, may have evolved insect pollination.

Male strobili of *Gnetum* produce nectar that also attracts insects. Though the nectar drop volume was less (98 nl) than that of pollination drops (152–190 nl), the range of sugar concentration was similar, e.g., 60–100 mg/ml (Kato et al. 1995).

Sugar Concentrations and Wind Pollination

The major sugars in pollination drops of wind-pollinated species are identical to those found in insect-pollinated ones. Instead of sucrose being predominant, often it is glucose and fructose that have the highest concentrations. The total amount of sugar in wind-pollinated species is, on average, less than in insect-pollinated species. No wind-pollinated species for which sugar measurements have been published exceeds 100 mg/ml (*Taxus* – Ziegler 1959). Most have averages nearer 12.5 mg/ml recorded for *Pinus nigra* (McWilliam 1958).

Other carbohydrates have been detected, including mannitol (*Juniperus communis* – Mugnaini et al. 2007a), rhamnose, arabinose, mannose (*Cephalotaxus drupacea* – Seridi-Benkaddour and Chesnoy 1988), galacturonic acid (*Taxus baccata* and *Thuja orientalis* – Seridi-Benkaddour and Chesnoy 1988), and galactose (*Cephalotaxus drupacea* – Seridi-Benkaddour and Chesnoy 1988; *Welwitschia mirabilis* – Carafa et al. 1992).

2.3.2 Proteins

Of all the constituents, proteins were the last to be discovered. Protein is abundant in these drops, ranging from 1 to 2 mg/ml (von Aderkas unpublished). Acid phosphatase, first discovered by a histochemical analysis of nucellus of *Taxus baccata* (Ziegler 1959), was also detected histochemically in *Welwitschia mirabilis* (Carafa et al. 1992). With the advent of more widely available tandem mass spectrometers, it became possible to sequence and identify proteins. To date, 27 proteins have been identified. Some of these proteins are found in multiple species. There must be many more, as we have only sampled four families (Cupressaceae, Pinaceae, Taxaceae, and Welwitschiaceae) and seven species (*Chamaecyparis lawsoniana*, *Juniperus communis*, *J. oxycedrus*, *Larix* x *marschlinsii*, *Pseudotsuga menziesii*, *Taxus baccata*, and *Welwitschia mirabilis*). These enzymes can be grouped according to putative function as follows:

1. Carbohydrate modification (Poulis et al. 2005)
2. Defense against fungi and bacteria (O'Leary et al. 2007; Wagner et al. 2007)
3. Prevention of ice-damage, e.g., antifreeze proteins (Wagner et al. 2007)
4. Protein cleavage (Poulis et al. 2005; Wagner et al. 2007)
5. Pollen tube guidance, e.g., arabinogalactan proteins (O'Leary et al. 2004).

It is one thing to establish an identity from sequence, and quite another to show that the enzymes do what their name suggests. To that end, we are currently adding depth to this analysis by establishing in situ enzyme function. Already a picture is emerging of a dynamic apoplastic secretion that actively defends the ovule during pollination. The numerous protein classes and the relatively high concentration of proteins found to date indicate a substantial investment by the ovule in its pollination drop.

Pollination drop composition is consistent across the period of secretion. In *Larix* × *marschlinsii*, volume varied, but protein composition did not. Volumes varied by tree and in some trees volumes varied from year to year (O'Leary et al. 2007). However, HPLC-generated protein profiles remained constant across hours, days, weeks, years, or genotypes (O'Leary 2004). In *Chamaecyparis lawsoniana*, no differences were found between drops from pollinated and unpollinated ovules (Wagner et al. 2007).

2.3.3 Amino Acids and Secondary Compounds

Amino acids are present though not abundant. To date, only 16 of 21 amino acids have been identified in the analysis of free amino acids in drops (Seridi-Benkaddour and Chesnoy 1988). Most of these studies are restricted to conifers. Whether amino acid composition of insect-pollinated gymnosperms is different from that of wind-pollinated ones has not been established.

Pollination drop analysis to date has revealed few secondary compounds. Closer attention will probably reveal many more compounds; as labs get access to more sensitive chemistry this gap in our knowledge will be filled. The only class of compounds to be fairly widespread is organic acids. The presence of dicarboxylic acids was reported by Schumann (1902), Fujii (1903), and Tison (1911). At the time, it was fashionable to ascribe a chemotactic role to these acids. Pollen was thought to orient itself on organic acid gradients within the nucellus. However, this has never been conclusively shown. We have detected succinic acid (von Aderkas unpublished). Rather than testing for chemotaxis, it may be worth establishing whether this acid plays a role in defense.

2.4 Are Pollination Drops the Basal Condition in gymnosperms?

Pollination drops are a basal reproductive characteristic in gymnosperms. Primitive gymnosperms have specialized modifications for pollination drop secretion, such as lagenostome and salpinx (Singh 1978). Drop biochemistry shows common composition: all drops have carbohydrates, inorganic compounds, amino acids, and proteins. Proteomics has provided evidence that pollination drops represent a substantial investment on the part of the ovule. A number of proteins are conservatively represented in most secretions (Wagner et al. 2007). Taken together, these facts appear to make a compelling case for the conservative role of pollination drops in pollen capture in all gymnosperms.

It has been suggested that the ancestral pollination mechanism of conifers is a combination of ovular secretion of pollination drops and ovular scavenging of rainwater (Runions and Owens 1996). This is meant to be an adaptation to drier conditions of ancestral climates. Support for the view that exogenous water plays a role in the pollination of a number of pinaceous and cupressaceous genera comes from observations of the effects of humidity (Runions et al. 1995; Runions and Owens 1996) and rainwater (Greenwood 1986; Colangeli and Owens 1990; Chandler and Owens 2003) on pollen capture. Much of this argument relies on saccate pollen behavior in water. Recently, Leslie (2010) showed that flotation-based pollination is effective in preferential filtering of conspecific pollen in both *Pinus nigra* and *Pinus mugo*. It was his opinion that the secretion of a pollination drop alone is sufficient for the operation of the mechanism. Rainwater was not essential or even contributory.

3 Angiosperm Nectaries and Their Involvement in Pollination

Although nectaries are restricted to very few gymnosperm species, they are widespread in angiosperm flowers. Angiosperm floral nectaries show an impressive morphological variety (Bernardello 2007; Pacini et al. 2003). They may be located at surface level in the organ bearing them, form an outgrowth on the organ, or be concealed deep within the organ. Unlike other floral structures, the relative positions of which are conserved throughout angiosperm evolution, the nectary is not located in the same position in all plants (Fahn 1979). From the ecological point of view, the diversity in nectary location is strictly linked to the large diversity of pollinators and their foraging behavior. Fahn (1979) formulated a topographical classification of floral nectaries, which was successively revised by Schmid (1988). A septal nectary, so named because it is concealed in the septal region between adjacent carpels, is the most common type of floral nectary in monocotyledons. It is largely absent in dicotyledons (Smets et al. 2000). It is also known as a gynopleural nectary (Smets and Cresens 1988). According to Rudall (2002), septal nectaries have been lost several times in monocot evolution, probably in association with the emergence of different pollination syndromes.

A nectary is generally composed of the following parts: (1) nectary epidermis; (2) nectary parenchyma, which are the layer(s) of small cells with densely staining cytoplasm present beneath the epidermis; and (3) subnectary parenchyma that is made up of larger more loosely packed cells compared with to nectary parenchyma (Nepi 2007). Vascular bundles may be found in the nectary or subnectary parenchyma. These bundles may be composed of both xylem and phloem or solely of phloem. Much of the anatomical variability of the nectary is related to epidermis and nectary parenchyma characters. The epidermis is the site of nectar release via stomata, secretory hairs, specialized cuticle, or transfer cell-like cells (Nepi 2007). Nectary parenchyma is the tissue most involved in nectar production. Much of the more evident ultrastructural variability is related to plastid development (Nepi 2007). The more common types of plastids found in the nectary parenchyma are chloro-amyloplasts, in which accumulated starch is the direct product of photosynthesis, and amyloplasts, in which large reserves of starch are stored that are not directly derived from photosynthesis. Species with high rates of nectar production have large reserves of starch in their nectary parenchyma cells (Nepi and Pacini 2007).

Both the floral nectary and its nectar are involved in pollination because the nectary is strategically positioned in the flower. When pollinators visit the flower in search of nectar they unintentionally load pollen grains and/or discharge them on the stigma. In some species, the plant may receive other additional services from nectar-feeding pollinators. For example, *Croton suberosus* is pollinated mainly by the wasp *Polistes instabilis* that visits its flowers in search of nectar. The wasp also defends the plant from foliar herbivores (Narbona and Dirzo 2010).

It is evident that the site of nectar presentation, i.e., where nectar accumulates, plays the key role in directing pollinator movement within the flower and thus pollen loading and unloading. Nectar accumulates on the surface of the nectary (primary presentation; Fig. 3), but sometimes it flows far away (secondary presentation; Fig. 4) (Nepi and Pacini 2007). Secondary presentation is carried out by means of nectar ducts where nectar flow away from the nectary is driven by capillary forces, secretion pressure, or gravity (Vogel 1998). In *Aloe* species the different types of secondary presentation are associated with different types of floral morphology and pollinators (Nepi et al. 2006).

In a curious exception, *Asclepias syriaca* (Asclepiadeaceae) nectar is not only the reward for pollinators, but it is also the natural germination medium for pollen (Kevan et al. 1989). The nectary of this plant is located in the stigmatic chamber; the nectar flows into nectar reservoirs called cucullui via a capillary duct. It is from these reservoirs that the pollinators suck the sugary secretion. Nectar that wets the stigmatic chamber aids germination of pollen grains packed in the pollinia inserted by pollinators.

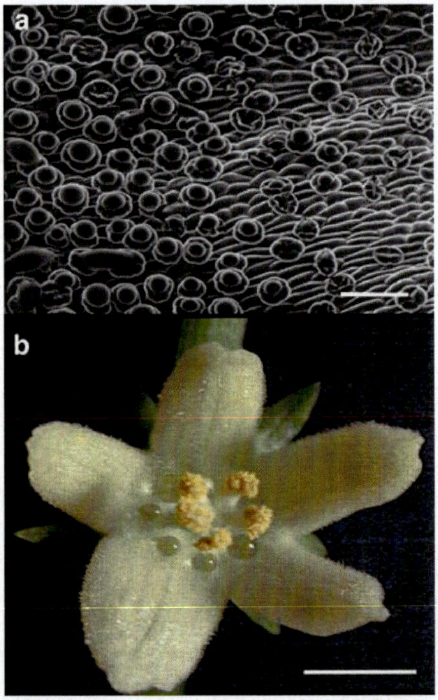

Fig. 3 Floral nectar primary presentation. (**a**) Nectar exudation by multicellular capitate trichomes in *Cyclanthera pedata* (scanning electron micrograph). The density of trichomes increases centripetally. *Bar* = 500 μm. (**b**) Nectar drops presentation in a male flower of *Sechium edule*. Nectar is produced by multicellular trichomes located in small cavity close to the base of the partially fused filaments. *Bar* = 0.5 cm

3.1 Nectar Production

Several models of nectary functioning have been proposed (for a review, see Nepi and Pacini 2007). Unfortunately, these models of nectar secretion are restricted to few species and just two approaches, namely ultrastructural or physiological ones that are never combined to greater effect. According to these models, nectar production involves three main steps:

1. Unloading of phloem sap into the subnectary and/or nectary parenchyma. Although nectaries may have active chloroplasts, carbohydrates for nectar production are mostly imported from phloem sap by the involvement of companion cells and parenchyma cells associated with the vascular bundles.
2. Transformation and transport of the prenectar across the nectar-secreting cells involving both the apoplastic and the symplastic spaces.
3. Nectar discharge outside the nectary in most cases through stomata or sometimes through modified cuticle or cuticle microchannels (Nepi 2007 and references therein).

The more recent model of the mechanism of nectary functioning was proposed by Vassilyev (2010). The model is based on a pressure-driven mass flow in the nectary apoplasm, while prenectar sugars diffuse from the sieve tubes through the symplasm to the nectary cells, where nectar is formed and sugars cross the plasma membrane by active transport. The pressure driving the mass flow originates from

Fig. 4 Floral nectary secondary presentation in *Aloe*. Nectar is produced by septal nectary inside the ovary and is released outside in different floral regions by means of capillary ducts. (**a**) Densely packed, erect flowers in the inflorescence of *A. castanea*. Note the *dark colored* nectar inside the corolla tube. *Bar* = 0.5 cm. (**b**) A flower of *A. castanea* with the corolla and filaments partially removed. Nectar accumulates at the top of the ovary in the space between the filaments. *Bar* = 0.5 cm. (**c**) The loose inflorescence of *A. greatheadii* var. *davyana* with bending mature flowers. *Bar* = 1.5 cm. (**d**) A flower of *A. greatheadii* var. *davyana* with the corolla and filaments partially removed. Nectar accumulates around the base of the ovary (*arrow*) in a kind of nectar reservoir formed by an enlargement of the basal part of the corolla tube. *Bar* = 0.5 cm

the water influx in the apoplasm from the symplasm along a sugar concentration gradient. Parenchyma cells associated with the vascular bundle absorb mainly sucrose from the sieve elements and secrete it into the apoplasm causing a water influx into the free space of the nectary. Sucrose hydrolysis by invertase – that was found recently associated with the nectary cell walls of *Arabidopsis thaliana* and essential for nectar production (Ruhlmann et al. 2010) – increases the osmotic concentration and draws more water into the nectary, thereby maintaining a favorable concentration gradient for the diffusion of sucrose through the nectary tissue.

A considerable water influx into the nectary can also be caused by the hydrolysis of starch reserves in the nectary parenchyma cells. This frequently happens just prior to the initiation of nectar production (Nepi et al. 1996; Nepi 2007 and references therein; Ren et al. 2007). The release of a substantial amount of soluble sugars increases cellular osmolarity, causing a rapid influx of water into the nectary. The result is an increased hydrostatic pressure of nectar that subsequently drives nectar flow through the stomata (Ren et al. 2007). The rapid influx of water also results in vacuole enlargement in nectary parenchyma cells at the onset of secretion (Nepi et al. 1996; Nepi 2007). Starch reserves may have another important function beyond simple provision of sugars for nectar. The rate of starch hydrolysis may regulate the hydrostatic pressure within the nectary and thus the nectar production rate. The expression of genes involved in starch synthesis and breakdown is tightly linked to the developmental stages of the nectary. In particular, starch catabolism is correlated with nectar release prior to anthesis (Ren et al. 2007).

Nectary parenchyma cells are, in most cases, the site of transformation of prenectar into nectar. Prenectar permeates the apoplasm. It also diffuses through the symplast, from parenchyma cell to parenchyma cell via plasmodesmata. Prenectar can be absorbed from the apoplasm by parenchyma cells, transformed, and secreted again in the apoplasm by eccrine or granulocrine secretion (Nepi 2007). Parenchyma cells are responsible for the selective reabsorption of sugars and their cycling through several biochemical pathways as was recently demonstrated by a study that combined magnetic resonance imaging and spectroscopy (Wenzler et al. 2008).

Nectar discharge outside the nectary can be a result of the increased hydrostatic pressure in the nectary as reported above. A common feature of parenchymatous nectaries is the presence of stomata on the nectary surface. Stomata are the site of nectar discharge. Nectary stomata are considered "modified" with respect to leaf stomata because they are have little or no ability to control stomatal closure. In species that lack such stomata, nectar release occurs through the nectary's cuticle. The cuticle is either thin allowing easy passage or it possesses pores or microchannels. These are in either the form of fibrillar outgrowths of epidermal cell wall or narrow tubular interruptions in the continuity of the cuticle (Nepi 2007 and references therein). In other instances, the cuticle initially appears completely impermeable and the nectar accumulates in a subcuticular space formed by separation of the cuticle from the epidermis. As secretion continues, the cuticle stretches, becoming very thin. It has not been determined whether the nectar is released by rupture of the cuticle or whether thin areas of the stretched cuticle become permeable to nectar (Nepi 2007).

How all of these processes are regulated and coordinated is almost completely unknown. The involvement of plant hormones in nectar production is not very clear. It was demonstrated that auxin inhibited nectar production (Shuel 1978; Aloni et al. 2005). Recently, Radhika et al. (2010) demonstrated hormonal stimulation of floral nectar production by jasmonates. This result integrates floral nectar secretion into the complex network of oxylipid-mediated developmental processes.

3.2 *Nectar Characters: Volume, Concentration, and Composition*

Nectar characteristics, including volume, solute concentration, and composition coevolved with pollinators. This is because nectar is the main food source for a large number of animals, in particular, flying animals. Large volumes of dilute floral nectar typify bird-pollinated plants, while bee- and fly-pollinated flowers produce low volumes of nectar that are more highly concentrated (Nicolson 2007). Flowers pollinated by long-tongued bees, butterflies, moths, hummingbirds, and Old World bats tend to produce sucrose-rich nectars, whereas those pollinated by flies, short-tongued bees, perching birds, and New World bats tend to produce hexose-rich nectars (Baker and Baker 1983a).

In drier environments, pollinators may seek nectar more for its water rather than for its sugar content. This appears to be the case in bird-pollinated South African aloes that flower during the dry winter season, producing large amounts of diluted nectar (Nicolson and Nepi 2005).

Recently, Herrera et al. (2009) provided a quantitative survey of yeast densities in floral nectar and found that densities of 10^3–10^4 cells per mm^3 are common. The presence of yeast in floral nectar alters important characteristics of nectar, including total sugar concentration and relative proportions of constituent sugars (sucrose, glucose, and fructose) (Herrera et al. 2008). They recommend testing for the presence of yeast when studying the composition of floral nectar.

3.2.1 Nectar Volume and Concentration

Generally speaking, nectar is considered a water-sugary solution, the volume and concentration of which vary widely both at inter- and intraspecies level (Nepi and Pacini 2007). The range of variability for nectar volume is from about 50 nL, as in single florets of Asteraceae (Wist and Davis 2006) to 9.4 mL in *Ochroma lagopus* (Bombacaceae), a large white bat-pollinated flower (Opler 1983). There is a positive correlation between nectar volume per flower and the size of the flower, just as there is a correlation between nectar volume and the size of pollinators (Nicolson 2007). The variability of floral nectar volume at the intraspecific level can be correlated with the age of the flower, the position of the flower on the plant or within the inflorescence, or with certain environmental parameters (T and RH) (Opler 1983; Pacini and Nepi 2007). Nectar volume is strongly influenced by evaporation. Old flowers frequently contain low volumes of concentrated nectar because of water evaporation (Nicolson and Nepi 2005; Nepi and Pacini 2007). Sugar concentration ranges from less than 10% (w/w) in *Aloe castanea* (*Asphodelaceae*, Nicolson and Nepi 2005) to almost 70% in *Carum carvi* (Apiaceae, Langenberger and Davis 2002).

Concentration is a very important nectar character from an ecological point of view because it has a positive and direct correlation with viscosity. As viscosity increases, more time and energy are required by nectar consumers to suck nectar. This is especially true of insects with long, extensible mouthparts, and birds with long, narrow beaks such as hummingbirds. Viscosity greatly affects their foraging behavior (Borrell and Krenn 2006, and references therein). Depending on the amount of shelter from evaporation that floral structures provide to nectar, there can be postsecretory changes in nectar concentration due to equilibration to ambient humidity (Corbet et al. 1979). Coevolution of nectar and pollinator has resulted in specific pollinators favoring certain nectar sugar concentrations. Concentration can affect insect physiology. Dilute nectar may fail to attract pollinators because the energy reward is too low, but when nectar concentration is high it can affect the rate of both ingestion and sugar intake. High concentrations can even damage their digestive systems, and disrupt osmoregulation (Heyneman 1983; Harder 1986; Nicolson 1998). Although the nectar volumes may change greatly over the course of the day (Gilbert et al. 1991; Nocentini 2010), nectar sugar concentrations are relatively stable in most flowering species (Borrell and Krenn 2006; Nocentini 2010).

3.2.2 Nectar Composition

According to the more recent studies, nectar is much more than a water-sugary solution provided by plants as reward for pollinators. Nectar is a complex milieu containing mainly sugars but also a wide range of other substances that are present in lower or trace amounts. These include amino acids, organic acids, lipids, inorganic ions, vitamins, volatiles, and alkaloids. They function as either attractants or repellents (Baker and Baker 1983b; Sangaravelan et al. 2005; Kessler and Baldwin 2007; Nicolson and Thorburg 2007; Kessler et al. 2008).

While the past literature on nectar composition was mainly focused on sugars, recent papers have underlined the ecological importance of secondary compounds involved in modulating pollinator behavior during flower visits (Kessler and Baldwin 2007; Kessler et al. 2008).

Sugars

Since sugars are so abundant as to be the main food resource of pollinators, sugars are also the most thoroughly studied substances of nectar. Sucrose and its monomers glucose and fructose are the most common sugars. Sucrose is the preferred compound for carbon transfer in plants (Akazawa and Okamoto 1980). This photosynthate flows through the phloem and enters both developing nectaries and ovules, which act as sinks. Sucrose is very frequently found in angiosperm floral nectar, although it is not ubiquitous. In an extensive study (765 species) of nectar sugar composition Baker and Baker (1983a) found that sucrose was present in the 89% of the species. In most cases it occurs in combination with glucose and fructose. Only 10% of the examined species have glucose and fructose without any sucrose. Glucose and fructose found in nectar are derived generally by the hydrolysis of the disaccharide sucrose by means of an invertase. As a result, these monosaccharides are often present in almost equal proportion. Deviation from 1:1 ratio may indicate that more than simple hydrolysis in the nectary is involved. In such cases, hexose sugars may be selectively reabsorbed from the prenectar after hydrolysis of sucrose (Nepi and Stpiczynska 2008). Another way in which unbalanced hexose ratios may occur is that either glucose or fructose may enter various biochemical pathways before being secreted (Wenzler et al. 2008).

Baker and Baker (1983a) categorized nectar according to sugar ratios, defined as the ratio by weight of sucrose to the combined hexose sugars, S/(G + F). Four classes of nectar were recognized: sucrose-dominant, sucrose-rich, hexose-rich, and hexose-dominant. This terminology was largely adopted and it is widespread. More recent authors have criticized the use of sugar ratios as misleading, because undue emphasis is placed on the sucrose content of nectar (Nicolson and Thorburg 2007). They recommend describing the sugar profile according to the relative percentages of the three main sugars.

An exception to the dominance of the "big three" sugars is reported in species of *Protea* and *Fourea* (Proteaceae), both of which are pollinated by rodents. Pentose xylose comprises up to 39% of total sugar (Jackson and Nicolson 2002). Presence of xylose in angiosperm floral nectars is rare; it is considered to be a deterrent to insects and birds (Nicolson and Thorburg 2007).

Minor sugars, i.e., present in low amounts, are found on occasion. Arabinose, mannose, melezitose, raffinose, and stachyose were reported in several species (Nicolson 2007).

Amino Acids

Although present in nectar in much lower amount than sugars, amino acids nevertheless serve as food while contributing to nectar taste. All 20 of the normal amino acids found in protein have been identified in various plant nectars. Essential amino acids may be an important nitrogen source for nectarivorous pollinators (Nicolson and Thorburg 2007). Amino acid-rich nectar improves butterfly fecundity (Mevi-Schütz and Erhardt 1997). Proline seems to have a special importance for insect. It not only contributes a taste preferred by insects (Alm et al. 1990), but it stimulates the insect's salt cell, a labellar chemosensory receptor of insects, resulting in the increased feeding behavior (Hansen et al. 1998). Proline is metabolized very rapidly. In honeybees, proline is the most abundant amino acid in the hemolymph and is required for egg laying (Nicolson and Thorburg 2007 and references therein). Oxidative proline degradation provides an efficient, short burst of energy that is utilized during the initial lift phase of insect flight. In comparison, sugars such as glucose are used for extended flight (Carter et al. 2005). Phenylalanine, an amino acid that also stimulates the sugar cell of insects, has been commonly detected in nectar. In species found in Mediterranean scrub communities, e.g., phrygana, phenylanine is relatively abundant, particularly in plants of the Lamiaceae (Petanidou 2007).

Proteins

Proteins have been detected in floral nectar since the 1930s (Beutler 1935). Advances in molecular biology have allowed better characterization of nectar protein profiles (Carter and Thornburg 2004; Carter et al. 2007; Kram et al. 2008; Hillwig et al. 2010). Proteins are not involved in attracting or repelling animals. Two general functional classes of proteins are found in nectar: carbohydrate-metabolizing enzymes (invertase, transfructosidase, and transglucosidase) and proteins that defend against microorganisms.

Metabolic enzymes are evidence that nectar is not a static product. However, the regulation of enzyme activity remains unstudied. Invertase, the hydrolyzing enzyme of sucrose into glucose and fructose that is responsible for the basic sugar composition, is of particular importance because it allows postsecretory transformation of the nectar sugar profile. The presence of invertase in nectar has been known since an early report in *Tilia* sp. (Beutler 1935). There is a surprising lack of characterization of invertase activity in more recent studies of floral nectar.

Nectar is an excellent media in which fungi and bacteria will grow, especially when it is openly exposed. Approximately 30–40% of samples contained yeasts (Herrera et al. 2009). It appears that plants defend their sugary secretions, protecting them for their reproductive purposes, while preventing their use as a base for microbes to further attack the plant's reproductive systems. A diversity of defense protein types is found in floral nectar (Carter and Thornburg 2000, 2004; Naqvi et al. 2005; Carter et al. 2007; Kram et al. 2008; Hillwig et al. 2010). A novel biochemical pathway was discovered – the nectar redox cycle – that protects nectar by maintaining high levels of hydrogen peroxide (Nicolson and Thorburg 2007 and references therein). This cycle was initially found in ornamental tobacco and in some other unrelated species (Carter and Thornburg 2000). *Petunia* also belonging to the Solanaceae does not have this redox cycle (Hillwig et al. 2010). Instead, it has ribonuclease activity that is responsible for antimicrobial activity in its nectar.

As Baker and Baker (1983b) wrote nearly 30 years ago, "It is certain that many enzymes must occur in floral nectar and their identification and distribution must precede any conclusion as to their function". At present the study of nectar proteins by means of the sequencing and biochemical characterization is at its beginning. Given the limited number of studies to date, the surprising diversity of proteins already found provides great optimism that there are other proteins and other mechanisms to be discovered.

Secondary Compounds

Secondary metabolites, including tannins, phenols, alkaloids, and terpenes, have been found in floral nectar in more than 21 angiosperm families (Adler 2000). Although these compounds have been known since the 1970s, they were considered to be toxic deterrents against predators (Baker and Baker 1983b). Recently, researchers have discovered that these compounds may play an important role in managing visitors' behavior. Flowers face a multidimensional challenge: they need to attract visitors, to compel them to vector pollen with the least investment in rewards, and to repel nectar robbers at the same time. All of this is in the service of maximizing fitness. The bouquet of secondary compounds may serve a number of these objectives. Kessler et al. (2008) discovered that both repellent and attractant were required to maximize capsule production and flower visitation by native pollinators, whereas nicotine reduced nectar robbing by non-pollinating animals.

Nicotine decreased visiting time of pollinators as well as the volume of nectar that they collected per flower. However, nicotine increased the number of visits. It was hypothesized that nectar repellents maximize the number of flower visitors per volume of nectar. Plants are not only able to keep nectar volume small but maximize outcrossing (Kessler and Baldwin 2007). According to Sangaravelan et al. (2005) naturally occurring concentrations of secondary compounds such as caffeine, nicotine, anabasine, and amygadaline did not deter insects. Secondary compounds can be regarded as post-ingestion stimulants to pollinators. Low concentrations of psychoactive alkaloids nicotine and caffeine increased feeding significantly. These compounds may have been part of the reward.

Secondary compounds found in nectar are frequently volatile organic compounds (VOCs). In excess of 1,700 VOCs are known. They contribute to the enormous variability of floral scents (Raguso 2009) that allow plant–animal interactions. Nectar scents may also include compounds emitted by surrounding floral tissues, such as a hydrophilic subset of compounds including geraniol, linalool, or jasmone. This suggests passive absorption by nectar. It may also suggest that the floral tissues other than nectaries are able to produce molecules targeted to nectar (Raguso 2004).

Secondary compounds also include pigments that color nectar, creating a visual signal to attract or repel animals (Fig. 4). Worldwide, pigments are found in the nectar of 67 divergent taxa that are geographically widely distributed (Hansen et al. 2007). The most common nectar colors range spectrally from yellow to red, but colors such as brown, black, green, and blue are also found (Hansen et al. 2007). Colored nectar is often correlated with one or more of the following three parameters: (1) vertebrate pollination, (2) insularity – many species are from islands or isolated mainland habitats, and (3) altitude – many such species are found at relatively high altitudes (Hansen et al. 2007).

Secondary compounds are involved in managing animal visits by both visual and olfactory signals, for example, the bitter-tasting dark floral nectar of *Aloe vryheidensis* because of a high content of phenolic compounds. The phenolic component appears to function as a floral filter by using visual cues to attract some animals, e.g., birds, while using taste to deter others, e.g., bees (Johnson et al. 2007).

Secondary compounds in nectar are known to have antimicrobial properties (Adler 2000 and references therein). Compounds may not only provide a direct defense from microbial invasion, but can indirectly protect the consuming animal. Manson et al. (2010) demonstrated that the consumption of a nectar alkaloid gelsemine found in *Gelsemium sempervirens* reduces pathogen loads in bumblebees. It also protects bees from infection, which, in the long run, improves their foraging efficiency.

3.3 Evolutionary Plasticity of Nectaries and Nectars

Most of the angiosperms orders have species with floral nectar. Floral nectar is present in a large number of unrelated angiosperm taxa (see Fig. 7 in Bernardello 2007). Floral nectaries have arisen independently a number of times (Brown 1938; Meeuse 1978; Friis and Endress 1990). Nectaries have also been lost in some lineages, and in some instances reacquired (Bernardello 2007). Unlike other floral structures whose relative positions are conserved in angiosperm evolution, the nectary is not located in the same position in all plants (Bernardello 2007). The molecular basis for such great nectary "plasticity" was recently discovered by Baum et al. (2001). Since the nectary location is independent of the ABC floral homeotic genes responsible for floral organ positioning, the nectary is "free" to move about the flower in response to natural selection. This positional flexibility is further increased by the way in which nectar accumulates outside the nectary; in other words, it is secondary presentation (Pacini et al. 2003). In addition to the morphological and structural plasticity, nectar shows substantial plasticity in chemical composition, particularly of sugars. Dupont et al. (2004) found a clear dichotomy between hexose-rich and sucrose-rich nectar in Macaronesian *Echium* spp. The former are pollinated by birds and the latter by insects. They concluded that sugar composition is more labile than floral morphology. In these island plant species, chemical composition changes in response to opportunistic flower-visiting birds. In contrast, plants on mainland areas are not thought to have their evolution driven by these same types of opportunistic pollinators. Nevertheless in other groups of plants, for example in the Aloiideae (van Wyk et al. 1993), sugar profile is relatively conserved within genera, subgenera, or sections even when the species are pollinated by different animals. As stated by Krömer et al. (2008), phylogenetic conservatism is more or less relaxed within different taxonomical groups and, probably, also in different ecological contexts.

4 Conclusions and Future Perspectives

Although gymnosperm pollination drops and angiosperm floral nectars are ontogenetically, anatomically, and physiologically distinct, they may share evolutionary links. Insect–plant relationships for pollination that are based on sugary exudates, i.e., pollination drops, were established well before the rise of angiosperms (Labandeira et al. 2007; Ren et al. 2009), even though wind has remained the main gymnosperm pollen vector. Insects preadapted to feed on sugary secretions were then "co-opted" by the more specialized reproductive structures (flowers) of angiosperms that readily exploited entomophily by secreting a sugary exudate in the form of nectar. Judging from the dominance of glucose and fructose in most pollination drops of modern gymnosperms, hexose-rich composition is likely to be a plesiomorphic characteristic and dominated probably also in the period of major radiation of gymnosperms. The establishment of a true entomophily in gymnosperms such as gnetophytes and in primitive angiosperms resulted in further specialization of nectars in terms of food reward. With the rise of more specialized pollinators, sugar concentrations increased, as did sucrose concentrations. Sucrose, in contrast to the more easily digestible monosaccharides favored by an extensive array of nonspecialized pollinators (flies, wasps, and beetles), is linked to more specialized pollinators (long-tongued bees and butterflies) (Nicolson 2007; Petanidou 2007). The first appearance in the fossil record of advanced pollinator groups, including many Hymenoptera and glossate Lepidoptera, occurred approximately 140 million years ago, which is very close to the first appearance of angiosperms (Willis and McElwain 2002). The food preferences of these two groups of insects – especially Lepidoptera – for sucrose-rich nectars (Baker and Baker 1983b) may have driven the selection of individual plants that produced increased carbohydrate rewards.

Liquid stigmatic exudate on the stigma of the primitive plant *Pseudowintera colorata* (Winteraceae) is the main reward for pollinators such as flies, beetles, thrips, and primitive moths (Lloyd and Wells 1992). This suggests that the primitive pollination mechanism in the ancestors of the angiosperms may have been had an unspecialized syndrome involving pollination by a fluctuating spectrum of opportunistic pollinators. Stigmatic exudates, which are known to occur in a number of other primitive angiosperms, may have been the first type of reward in angiosperms. Nectaries are rare in early branching lineages of modern angiosperms, which would appear to indicate that this condition is plesiomorphic. Curiously, the stigmatic exudate of *P. colorata* is reabsorbed following pollen deposition and, if artificially removed, is then produced again the next day (Lloyd and Wells 1992). This situation is analogous to that found in many pollination drops (Mugnaini et al. 2007a, b). These similarities may be a consequence of the evolution of pollen receptive surfaces, from the pollination drop of gymnosperms to the wet stigma of early angiosperms (Lloyd and Wells 1992). If stigmatic secretions were the first reward in an early angiosperm, nectaries may have evolved sooner or later in order to decrease damage to the stigma surface due to insect foraging activity. At least hypothetically, an evolutionary continuum linking pollination drop, stigma secretion, and nectar is possible to envision.

In the absence of genomic studies of gymnosperms, proteomics provide evidence of shared functions of pollination drops and nectar (Wagner et al. 2007; Park and Thornburg 2009; Gonzalez-Teuber et al. 2009). In both secretions the dominant proteins are enzymes involved in either carbohydrate metabolism or defense against microorganisms. Invertases and chitinases have been found in both pollination drops (Wagner et al. 2007) and nectar both floral and extrafloral (Gonzalez-Teuber et al. 2009), indicating a certain grade of conservatism of the two functional classes of proteins that are needed to maintain their osmolarity and composition.

Although proteins are longer known in nectar than in pollination drops, they have been identified in pollination drops of greater number of gymnosperm species than it have been done for angiosperm nectar. In both cases the analyzed species are too low to draw any kind of phylogenetic perspective. Much remains to be discovered about sugary secretions involved in pollination by interested biochemists and biologists. For gymnosperms it is to be considered that genomic investigations are nonexistent. This should improve in the very near future as increasing gymnosperm sequence information becomes available. The most urgent requirement is to prove the effective biological function of constituents identified in pollination drops and nectar. Integrating the physiology and biochemistry with gymnosperm and angiosperm pollination ecology, most notably in insect pollination, will provide valuable insights into the evolution of seed plants.

References

Adler LS (2000) The ecological significance of toxic nectar. Oikos 91:409–420

Akazawa T, Okamoto K (1980) Biosynthesis and metabolism of sucrose. In: Stumpf PK, Conn EE (eds) The biochemistry of plants, a comprehensive treatise. Academic, New York, pp 199–220

Alm J, Ohnmeiss TE, Lanza J, Vriesenga L (1990) Preference of cabbage white butterflies and honey bees for nectar that contains amino acids. Oecologia 84:53–57

Aloni R, Aloni E, Langhans M, Ullrich CI (2005) Role of auxin in regulating *Arabidopsis* flower development. Planta 223:315–328

Baker HG, Baker I (1983a) Floral nectar sugar constituents in relation to pollinator type. In: Little RJ, Jones CE (eds) Handbook of pollination biology. Scientific and Academic Editions, New York, pp 117–141

Baker HG, Baker I (1983b) A brief historical review of the chemistry of floral nectar. In: Bentley B, Elias T (eds) The biology of nectaries. Columbia University Press, New York, pp 126–151

Baum SF, Eshed Y, Bowman JL (2001) The *Arabidopsis* nectary is an ABC-independent floral structure. Development 128:4657–4667

Bernardello G (2007) A systematic survey of floral nectaries. In: Nicolson SW, Nepi M, Pacini E (eds) Nectaries and nectar. Springer, Dordrecht, pp 19–128

Beutler R (1935) Nectar. Bee World 24:106–162

Bino RJ, Dafni A, Meeuse ADJ (1984) Entomophily in the dioecious gymnosperm *Ephedra aphylla* Fork. (= *E. alte* A. Mey.), with some notes on *E. campylopoda* C. A. Mey. I. Verh Koning Ned Akad Wet Amst Ser C 87:1–13

Borrell BJ, Krenn HW (2006) Nectar feeding in long-proboscid insects. In: Harrel A, Speck T, Rowe NP (eds) Ecology and biomechanics. Taylor and Francis, Boca Raton, FL, pp 185–212

Brown WH (1938) The bearing of nectaries on the phylogeny of flowering plants. Proc Am Philosoph Soc 79:549–595

Buchmann SL, O'Rourke MK, Niklas KJ (1989) Aerodynamics of *Ephedra trifurca*. III. Selective pollen capture by pollination droplets. Bot Gaz 150:122–131

Carafa AM, Carratù G, Pizzolongo P (1992) Anatomical observations on the nucellar apex of *Welwitschia mirabilis* and the chemical composition of the micropylar drop. Sex Plant Reprod 5:275–279

Carter C, Healy R, O'Tool NM, Naqvi SMS, Ren G, Park S, Beattie GA, Horner HT, Thornburg RW (2007) Tobacco nectaries express a novel NADPH oxidase implicated in the defence of floral reproductive tissues against microorganisms. Plant Physiol 143:389–399

Carter C, Thornburg RW (2000) Tobacco nectarin I: Purification and characterization as a germin-like, manganese superoxide dismutase implicated in the defence of floral reproductive tissues. J Biol Chem 275:36726–36733

Carter C, Thornburg RW (2004) Is the nectar redox cycle a floral defence against microbial attack? Trends Plant Sci 9:320–324

Carter C, Sharoni S, Yehonatan L, Palmer RG, Thornburg R (2005) A novel role for proline in plant floral nectars. Naturwissenschaften 93:72–79

Chandler LM, Owens JN (2003) The pollination mechanism of *Abies amabilis*. Can J For Res 34:1071–1080

Colangeli AM, Owens JN (1990) The relationship between time of pollination, pollination efficiency and cone size in western redcedar (*Thuja plicata*). Can J For Res 25:439–443

Corbet SA, Willmer PG, Beament JWL, Unwin DM, Prys-Jones OE (1979) Post-secretory determinants of sugar concentration in nectar. Plant Cell Environ 2:293–308

Crepet WL (1974) Investigations of North American cycadeoids: the reproductive biology of Cycadeoidea. Paleontographica 148B:144–169

Crepet WL, Friis EM (1987) The evolution of insect pollination mechanisms in angiosperms. In: Friis EM, Chaloner WG, Crane PR (eds) The origin of angiosperms and their biological consequences. Cambridge University Press, Cambridge, pp 181–201

Dafni A (1992) Pollination ecology – a practical approach. Oxford University Press, Oxford

Doyle J (1945) Developmental lines in pollination mechanisms in the Coniferales. Scientific Proc Royal Dublin Soc 24:43–62

Doyle J, O'Leary M (1935) Pollination in *Pinus*. Sci Proc Roy Dublin Soc 21:181–190

Dupont YL, Hansen DM, Rasmussen JT, Olesen JM (2004) Evolutionary changes in nectar sugar composition associated with switches between bird and insect pollination: the Canarian bird-flower element revisited. Funct Ecol 18:670–676

Edlund AF, Swanson R, Preuss D (2004) Pollen and stigma structure and function: the role of diversity in pollination. Plant Cell 16:S84–S97

Endress PK (1994) Floral structure and evolution of primitive angiosperms: recent advances. Pl Syst Evol 192:79–97

Endress PK (2010) The evolution of floral biology in basal angiosperms. Philos Trans R Soc B 365:411–421

Fahn A (1979) Secretory tissues in plants. Academic, London

Fortescue JA, Turner DW (2005) The occurrence of a micropylar exudate in *Musa* and *Ensete* (Musaceae). Sci Hort 104:445–461

Franssen-Verheijen MAW, Willemse MTM (1993) Micropylar exudate in *Gasteria* (Aloaceae) and its possible function in pollen tube growth. Am J Bot 80:253–262

Friis EM, Endress PK (1990) Origin and evolution of angiosperm flowers. Adv Bot Res 17:99–162

Fujii K (1903) Über die Besäubungstropfen der Gymnospermen. Ber Deut Bot Ges 21:211–217

Gelbart G, von Aderkas P (2002) Ovular secretions as part of pollination mechanisms in conifers. Ann For Sci 59:345–357

Gilbert FS, Haines N, Dickson K (1991) Empty flowers. Funct Ecol 5:29–39

Gonzalez-Teuber M, Eilmus S, Muck A, Svatos A, Heil M (2009) Pathogenesis-related proteins protect extrafloral nectar from microbial infestation. Plant J 58:464–473

Greenwood MS (1986) Gene exchange in loblolly pine: the relation between pollination mechanism, female receptivity and pollen availability. Am J Bot 73:1443–1451

Hansen DM, Olesen JM, Mione T, Johnson SD, Müller CB (2007) Coloured nectar: distribution, ecology, and evolution of an enigmatic floral trait. Biol Rev 82:83–111

Hansen K, Wacht S, Seebauer H, Schnuch M (1998) New aspects of chemoreception in flies. Ann NY Acad Sci 855:143–147

Harder L (1986) Effects of nectar concentration and flower depth on flower handling efficiency of bumble bees. Oecologia 69:309–315

Herrera CM, de Vega C, Canto A, Pozo MI (2009) Yeasts in floral nectar: a quantitative survey. Ann Bot 103:1415–1423

Herrera CM, García IM, Pérez R (2008) Invisible floral larcenies: microbial communities degrade floral nectar of bumble bee-pollinated plants. Ecology 89:2369–2376

Heslop-Harrison J, Shivanna KR (1977) The receptive surface of the angiosperm stigma. Ann Bot 50:831–842

Heyneman A (1983) Optimal sugar concentrations of floral nectars – dependence on sugar intake efficiency and foraging costs. Oecologia 60:198–213

Hillwig MS, Liu X, Liu G, Thornburg RW, MacIntosh GC (2010) *Petunia* nectar proteins have ribonuclease activity. J Exp Bot 61:2951–2965

Ho RJ (1985) Effect of repeated pollination upon filled seed in white spruce. Can J For Res 15:1195–1197

Jackson S, Nicolson SW (2002) Xylose as nectar sugar: from biochemistry to ecology. Comp Biochem Physiol 131:613–620

Janson J, Reinders MC, Valkering AGM, Van Tuyl JM, Keijzer CJ (1994) Pistil exudate production and pollen tube growth in *Lilium longiflorum* Thunb. Ann Bot 73:437–446

Johnson SD, Hargreaves AL, Brown M (2007) Dark, bitter-tasting nectar functions as a filter of flower visitors in a bird-pollinated plant. Ecology 87:2709–2716

Kato M, Inoue T, Nagamitsu T (1995) Pollination biology of *Gnetum* (Gnetaceae) in a lowland mixed dipterocarp forest in Sarawak. Am J Bot 82:862–868

Kessler D, Baldwin T (2007) Making sense of nectar scents: the effects of nectar secondary metabolites on floral visitors to *Nicotiana attenuata*. Plant J 49:840–854

Kessler D, Gasse K, Baldwin IT (2008) Field experiments with transformed plants reveal the sense of floral scents. Science 321:1200–1202

Kevan PG, Eisikowitch D, Rathwell B (1989) The role of nectar in the germination of pollen in *Asclepias syriaca* L. Bot Gaz 150:266–270

Kram B, Bainbridge E, Perera M, Carter C (2008) Identification, cloning and characterization of a GDSL lipase secreted into the nectar of *Jacaranda mimosifolia*. Plant Mol Biol 68:173–183

Krömer T, Kessler M, Lohaus G, Schmidt-Lebuhn AN (2008) Nectar sugar composition and concentration in relation to pollination syndromes in Bromeliaceae. Plant Biol 10:502–511

Labandeira CC, Kvaçek J, Mostovski MB (2007) Pollination drops, pollen, and insect pollination in Mesozoic gymnosperms. Taxon 56:663–695

Langenberger MW, Davis AR (2002) Temporal changes in floral nectar production, reabsorption and composition associated with dichogamy in annual caraway (*Carum carvi*; Apiaceae). Am J Bot 89:1588–1598

Leslie AB (2010) Flotation preferentially selects saccate pollen during conifer pollination. New Phytol 188:273–279

Lill BS, Sweet GB (1977) Pollination in *Pinus radiata*. N Z J For Sci 7:21–34

McWilliam JR (1958) The role of the micropyle in the pollination of *Pinus*. Bot Gaz 120:109–117

Lloyd DG, Wells MS (1992) Reproductive biology of a primitive angiosperm, *Pseudowintera colorata* (Winteraceae), and the evolution of pollination systems in the Anthophyta. Plant Syst Evol 181:77–95

Manson JS, Otterstatter MC, Thomson JD (2010) Consumption of a nectar alkaloid reduces pathogen load in bumble bees. Oecologia 162:81–89

Meeuse ADJ (1978) Nectarial secretion, floral evolution, and the pollination syndrome in early angiosperm. Proc Kon Ned Akad Wetensch Amst, Series C 81:300–326

Meeuse ADJ, de Meijer AH, Mohr OWP, Wellinga SM (1990) Entomophily in the dioecious gymnosperm *Ephedra aphylla* Forsk. Israel J Bot 39:113–123

Mevi-Schütz J, Erhardt A (1997) Amino acids in nectar enhance butterfly fecundity: a long-awaited link. Am Nat 105:411–419

Mugnaini S, Nepi M, Guarnieri M, Piotto B, Pacini E (2007a) Pollination drop in *Juniperus communis*: response to deposited material. Ann Bot 100:1475–1481

Mugnaini S, Nepi M, Guarnieri M, Piotto B, Pacini E (2007b) Pollination drop withdrawal in *Juniperus communis*: response to biotic and abiotic particles. Caryologia 60:182–184

Naqvi S, Harper A, Carter C, Ren G, Guirgis A, York WS, Thornburg RW (2005) Tobacco Nectarin IV is a specific inhibitor of fungal xylosidases secreted into the nectar of ornamental tobacco plants. Plant Physiol 139:1389–1400

Narbona E, Dirzo R (2010) A reassesment of the function of floral nectar in *Croton suberosus* (Euphorbiaceae): a renard for plant defenders and pollinators. Am J Bot 97:672–679

Nepi M (2007) Nectary structure and ultrastructure. In: Nicolson SW, Nepi M, Pacini E (eds) Nectaries and nectar. Springer, Dordrecht, pp 129–166
Nepi M, Ciampolini F, Pacini E (1996) Development and ultrastructure of *Cucurbita pepo* nectaries of male flowers. Ann Bot 78:95–104
Nepi M, Franchi GG, Pacini E (2001) Pollen hydration status at dispersal: cytophysiological features and strategies. Protoplasma 216:171–180
Nepi M, Human H, Nicolson SW, Cresti L, Pacini E (2006) Nectary structure and nectar presentation in *Aloe castanea* and *A. greatheadii* var. davyana (Asphodelaceae). Plant Syst Evol 257:45–55
Nepi M, Pacini E (2007) Nectar production and presentation. In: Nicolson SW, Nepi M, Pacini E (eds) Nectaries and nectar. Springer, Dordrecht, pp 215–264
Nepi M, Stpiczynska M (2008) The complexity of nectar: secretion and resorption dynamically regulate nectar features. Naturwissenshaften 95:177–184
Nepi M, von Aderkas P, Wagner R, Mugnaini S, Coulter A, Pacini E (2009) Nectar and pollination drops: how different are they? Ann Bot 104:205–219
Nicolson SW (2007) Nectar consumers. In: Nicolson S, Nepi M, Pacini E (eds) Nectaries and nectar. Springer, Dordrecht, pp 289–342
Nicolson SW (1998) The importance of osmosis in nectar secretion and its consumption by insects. Am Zool 38:418–425
Nicolson SW, Nepi M (2005) Dilute nectar in dry atmospheres: nectar secretion patterns in *Aloe castanea* (Asphodelaceae). Int J Plant Sci 166:227–233
Nicolson SW, Thorburg RW (2007) Nectar chemistry. In: Nicolson SW, Nepi M, Pacini E (eds) Nectaries and nectar. Springer, Dordrecht, pp 215–264
Niklas KJ, Buchmann SL (1987) The aerodynamics of pollen capture in two sympatric species of *Ephedra trifurca*. Evolution 42:104–123
Niklas KJ, Kerchner V (1986) Aerodynamics of *Ephedra trifurca*. II. Computer modelling of pollination efficiencies. J Math Biol 24:1–24
Niklas KJ, Buchmann SL, Kerchner V (1986) Aerodynamics of *Ephedra trifurca*. I. Pollen grain velocity fields around stems bearing ovules. Am J Bot 73:966–979
Nocentini D (2010) Produzione di nettare e impollinazione in *Cerinthe major* L. (Boraginaceae). MSc Thesis, University of Siena
O'Leary SJB (2004) Proteins in the ovular secretion of conifers. PhD Thesis, University of Victoria, Canada
O'Leary SJB, von Aderkas P (2006) Postpollination drop production in hybrid larch is not related to the diurnal pattern of xylem water potential. Trees 20:61–66
O'Leary SJB, Joseph C, von Aderkas P (2004) Origin of arabinogalactan proteins in the pollination drop of *Taxus* x *media*. Aust J For Sci 121:35–46
O'Leary SJB, Poulis BAD, von Aderkas P (2007) The identification of two thaumatin-like proteins (TLPs) in the pollination drop of hybrid yew that may play a role in pathogen defence during pollen collection. Tree Physiol 27:1649–1659
Opler PA (1983) Nectar production in tropical ecosystem. In: Bentley B, Elias T (eds) The biology of nectaries. Columbia University Press, New York, pp 30–79
Owens JN, Molder M (1979) Sexual reproduction of white spruce (*Picea glauca*). Can J Bot 57:152–169
Owens JN, Simpson SJ, Molder M (1980) The pollination mechanism in yellow cypress (*Chamaecyparis nootkatensis*). Can J For Res 10:564–572
Owens JN, Simpson SJ, Molder M (1981) Sexual reproduction of *Pinus contorta*. I. Pollen development, the pollination mechanism and early ovule development. Can J Bot 59.1828–1843
Owens JN, Simpson SJ, Caron GE (1987) The pollination mechanism of Engelmann spruce (*Picea engelmannii*). Can J Bot 65:1439–1450
Pacini E (2009) Pollination. In: Jørgensen SE, Fath BD (eds) Encyclopedia of ecology, vol 5. Elsevier, Oxford, pp 2857–2861

Pacini E, Nepi M, Vesprini JL (2003) Nectary biodiversity: a short review. Plant Syst Evol 238:7–21

Park S, Thornburg RW (2009) Biochemistry of nectar proteins. J Plant Biol 52:27–34

Pearson HHW (1909) Further observations on *Welwitschia*. Philos Trans Roy Soc Lond Ser B 200:331–402

Petanidou T (2007) Floral nectars in Mediterranean habitats. In: Nicolson SW, Nepi M, Pacini E (eds) Nectaries and nectar. Springer, Dordrecht, pp 343–375

Porsch O (1910) *Ephedra campylopoda* CA Mey., eine entomophile Gymnosperme. Ber Dtsch Bot Ges 28:404–412

Poulis BAD, O'Leary SJB, Haddow JD, von Aderkas P (2005) Identification of proteins present in the Douglas fir ovular secretion: an insight into conifer pollen selection and development. Int J Plant Sci 166:733–739

Radhika V, Kost C, Boland W, Heil M (2010) The role of Jasmonates in floral nectar secretion. PlosOne 5:e9265

Raguso RA (2004) Why are some floral nectars scented? Ecology 85:1486–1494

Raguso RA (2009) Floral scent in a whole-plant context: moving beyond pollinator attraction. Funct Ecol 23:837–840

Ren D, Labandeira CC, Santiago-Blay JA, Rasnitsyn A, Shih CK, Bashkuev A, Logan MAV, Hotton CL, Dilcher D (2009) A probable pollination mode before angiosperms: Eurasian, long-proboscid scorpionflies. Science 326:840–847

Ren G, Healy RA, Klyne AM, Horner HT, James MG, Thornburg RW (2007) Transient starch metabolism in ornamental tobacco floral nectaries regulates nectar composition and release. Plant Sci 173:277–290

Rothwell G (1977) Evidence for a pollination-drop mechanism in Paleozoic pteridosperms. Science 198:1251–1252

Rudall PJ (2002) Homologies of inferior ovaries and septal nectaries in monocotyledons. Int J Plant Sci 163:261–276

Ruhlmann JM, Kram BW, Carter CJ (2010) Cell wall invertase 4 is required for nectar production in *Arabidopsis*. J Exp Bot 61:395–404

Runions CJ, Owens JN (1996) Pollen scavenging and rain involvement in the pollination mechanism of interior spruce. Can J Bot 74:115–124

Runions CJ, Catalano GL, Owens JN (1995) Pollination mechanisms of seed orchard interior spruce. Can J For Res 25:1434–1444

Rydin C, Khodabandeh A, Endress PK (2010) The female reproductive unit of *Ephedra* (Gnetales): comparative morphology and evolutionary perspectives. Bot J Linn Soc 163:387–430

Sangaravelan N, Nee'man G, Inbar M, Izhaki I (2005) Feeding responses of free-flying honeybees, to secondary compounds mimicking floral nectars. J Chem Ecol 31:2791–2804

Schmid R (1988) Reproductive versus extra-reproductive nectaries - historical perspective and terminological recommendations. Bot Rev 54:179–232

Schumann K (1902) Über dei weibliche Blüte der Coniferen. Verh Bot Ver Prov Brand 44:23–42

Seridi-Benkaddour R, Chesnoy L (1988) Secretion and composition of the pollination drop of *Cephalotaxus drupacea* (Gymnosperm, Cephalotaxaceae). In: Cresti M, Gore P, Pacini E (eds) Sexual reproduction in higher plants. Springer, Berlin, pp 345–350

Shuel R (1978) Nectar secretion in excised flowers. V. Effects of indoleacetic acid and sugar supply on distribution of [14C]-sucrose in flower tissues and nectar. Can J Bot 56:564–571

Shivanna KR (2003) Pollen biology and biotechnology. Science Publishers Inc., Enfield (USA)

Singh H (1978) Embryology of Gymnosperms. Gebrüder Borntraegerm, Stuttgart

Smets EF, Cresens EM (1988) Types of floral nectaries and the concept of 'character' and 'character state' – a reconsideration. Acta Bot Neerl 37:121–128

Smets EF, Ronse Decraene LP, Caris P, Rudall PJ (2000) Floral nectaries in monocotyledons: distribution and evolution. In: Wilson KL, Morrison DA (eds) Monocots: systematics and evolution. CSIRO, Melbourne, pp 230–240

Sweet GB, Lewis PN (1969) A diffusible auxin from *Pinus radiata* pollen and its possible role in stimulating ovule development. Planta 89:380–384
Takaso T, Owens JN (1995) Pollination drop and microdrop secretions in *Cedrus*. Int J Plant Sci 156:640–649
Takaso T, Owens JN (1996) Postpollination-prezygotic ovular secretions into the micropylar canal in *Pseudotsuga menziesii*. J Plant Res 109:147–160
Tang W (1987) Insect pollination in the cycad *Zamia pumila* (Zamiaceae). Am J Bot 74:90–99
Tison A (1911) Remarques sur les gouttelettes collectrices des ovules des conifères. Mem Soc Linn Norm 23:51–64
Tomlinson PB, Braggins JE, Rattenbury JA (1997) Contrasted pollen capture mechanisms in Phyllocladaceae and certain Podocarpaceae (Coniferales). Am J Bot 84:214–223
van Wyk B-E, Whitehead CS, Glen HF, Hardy DS, van Jaarsveld EJ, Smiths GF (1993) Nectar sugar composition in the subfamily Alooideae (Asphodelaceae). Biochem Syst Ecol 21:249–253
Vassilyev AE (2010) On the mechanism of nectar secretion: revisited. Ann Bot 105:349–354
Vaucher JP (1841) Histoire physiologique des plantes d'Europe, vol 4. Marc Aurel Frères, Paris
Vogel S (1998) Remarkable nectaries: structure, ecology, organophyletic perspectives III. Nectar ducts. Flora 193:113–131
von Aderkas P, Leary C (1999a) Micropylar exudates in Douglas fir – timing and volume of production. Sex Plant Reprod 11:354–536
von Aderkas P, Leary C (1999b) Ovular secretions in the micropylar canal of larches (*Larix kaempferi*, *L.* x *eurolepis*). Can J Bot 77:531–536
Wagner R, Mugnaini S, Sniezko R, Hardie D, Poulis B, Nepi M, Pacini E, von Aderkas P (2007) Proteomic evaluation of gymnosperm pollination drop proteins indicates highly conserved and complex biological functions. Sex Plant Reprod 20:181–189
Wenzler M, Holscher D, Oerther T, Schneide B (2008) Nectar formation and floral nectary anatomy of *Anigozanthos flavidus*: a combined magnetic resonance imaging and spectroscopy study. J Exp Bot 59:3425–3434
Wetschnig W, Depisch B (1999) Pollination biology of *Welwitschia mirabilis* Hook. Phyton 39:167–183
Willis KJ, McElwain JC (2002) The evolution of plants. Oxford University Press, Oxford
Wist TJ, Davis AR (2006) Floral nectar production and nectary anatomy and ultrastructure of *Echinacea purpurea* (Asteraceae). Ann Bot 97:177–193
Ziegler H (1959) Über die Zusammensetzung des "Bestäubungstropfens" und Mechanismus seiner Secretion. Planta 52:587–599

Nectar Secretion: Its Ecological Context and Physiological Regulation

María Escalante-Pérez and Martin Heil

Abstract Plants secrete nectar to attract pollinators and indirect defenders. The chemical contents of both floral and extrafloral nectar appear adapted to attract and nourish these two classes of animal mutualists. Being rich in sugars and amino acids, however, nectar also requires protection from nectar robbers and infecting micro-organisms. This role is mainly fulfilled by nectar proteins (nectarins) and by secondary compounds such as alkaloids. Although much on the chemical ecology of nectar and the phenotypic patterns of its secretion is known, we have only limited knowledge on the molecular control of the synthesis of nectar components and of nectar secretion. Likewise, carbohydrates are uploaded from the phloem into the nectariferous tissue, where they might move via an apoplastic or a vesicle-bound, symplastic pathway. Cell wall invertases play a central role in creating the required source–sink relations and controlling the sucrose/hexose ratio of nectars. No information exits on the sites of synthesis of non-carbohydrate nectar components such as proteins and alkaloids, although it appears likely that at least the bulk of nectarins is synthesized in the nectariferous tissue itself. Most of the common model species do not depend on nectar secretion, and it might be this fact that has hindered nectar research over the last 50 years. We recommend the use of contemporary "omics" techniques in comparative approaches to understand how plants synthesize and secrete nectar.

1 Introduction

Nectar serves for the attraction and nutrition of animal partners that are engaged in two major types of plant–animal mutualisms, in which plants make use of the mobility of animals: pollination and indirect defence (Heil 2008, 2011;

M. Escalante-Pérez • M. Heil (✉)
Departamento de Ingeniería Genética, CINVESTAV – Irapuato, Km. 9.6 Libramiento Norte, CP 36821 Irapuato, Guanajuato, Mexico
e-mail: mheil@ira.cinvestav.mx

J.M. Vivanco and F. Baluška (eds.), *Secretions and Exudates in Biological Systems*, Signaling and Communication in Plants 12, DOI 10.1007/978-3-642-23047-9_9,

Brandenburg et al. 2009). Both types of mutualisms can also be enabled by other types of resources, but floral and extrafloral nectar appear by far the most important plant rewards in this context. By definition, floral nectar (FN) is secreted within the flower (Fig. 1) and serves pollination, whereas extrafloral nectar (EFN) is secreted commonly, but not necessarily, outside the inflorescences (Fig. 2) and serves the indirect defence of the plant by intensifying its tritrophic interactions with carnivores (Bentley 1977; Elias 1983).

Although its general importance is being widely appreciated, nectar remains a surprisingly understudied discipline within plant science (Sect. 6). For example, little is known about nectar components other than the sugars and amino acids (Nicolson et al. 2007; González-Teuber and Heil 2009a), and new classes of substances continue to be detected in nectar. Even less is known about the synthesis

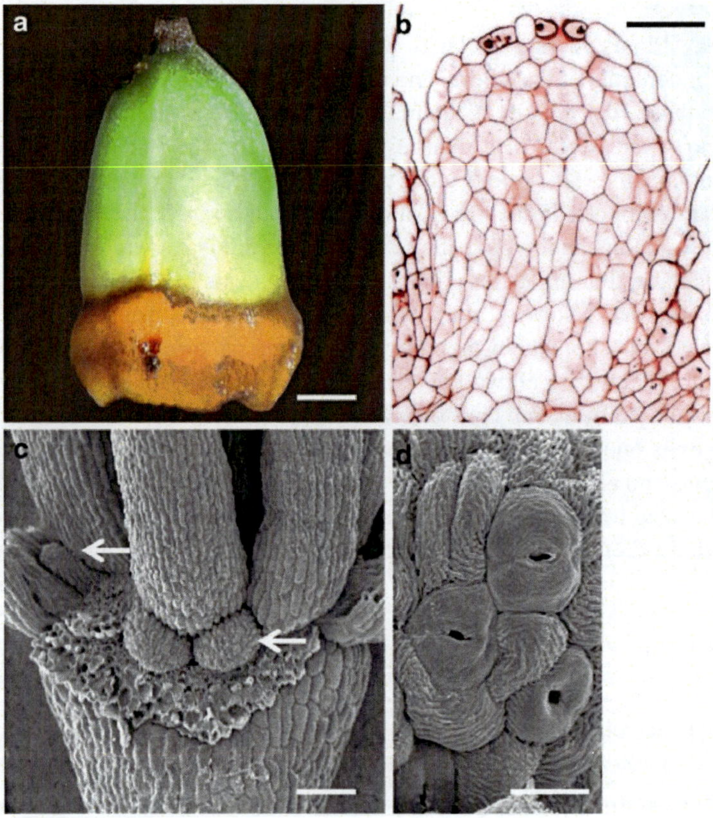

Fig. 1 Floral nectaries of *Arabidopsis thaliana* and ornamental tobacco. (**a**) Light microscopy of fresh, isolated S12 tobacco gynoecium showing the *light green* ovary and the *orange* nectary at its base, *bar* = 1 mm. (**b**) Light microscopy of 1-μm-thick sections of *Arabidopsis thaliana* floral nectaries stained showing two stomata, *bar* = 25 μm. (**c**) Scanning electron micrograph of dissected *Arabidopsis* flower showing two types of nectaries (*arrows*) located between petals (*removed*) and stamens, *bar* = 100 μm. (**d**) Enlargement of medial nectary (*right arrow* of **c**), *bar* = 25 μm. Reproduced from Ren et al. (2007b). Photographs courtesy of Harry T. Horner and Robert Thornburg (**a**) and Rosanne Healy (**b–d**)

Fig. 2 Extrafloral nectary size depends on function. The size and anatomical complexity of extrafloral nectaries are often related to their function. Obligate myrmecophytes house and nourish

of nectar components and the regulation of their secretion. In this chapter, we review the general knowledge that is available on FN as well as EFN and focus particularly on those fields in which several exciting discoveries have been made over the last decade.

For example, jasmonic acid (JA) has been identified as a hormone that affects the secretion of both FN (Radhika et al. 2010) and EFN (Heil et al. 2001, 2004; Heil 2004), and a recently discovered gene that encodes an apoplastic invertase in *Arabidopsis* (*Arabidopsis thaliana*) represents the first gene whose function is required for FN secretion (Ruhlmann et al. 2010). Nectarins (nectar proteins) were identified in tobacco (*Nicotiana* spp.) FN, *Acacia* EFN and pollination droplets of gymnosperms. The function of these nectarins is likely to be the protection against microbial infection (Carter and Thornburg 2004; Carter et al. 2007; Wagner et al. 2007; Kram et al. 2008; González-Teuber et al. 2009, 2010). Other enzymes were found to play central roles in the post-secretory tailoring of nectar chemistry (Heil et al. 2005; Kram et al. 2008).

It is still not known, however, how nectar secretion takes place and how it is controlled at the physiological, cellular and genetic levels, how plants can adapt nectar secretion to consumption rates and/or consumer identity, and how nectar that has not been consumed can be re-absorbed, although all these phenomena are well described at the phenotypic level (Pacini et al. 2003; Pacini and Nepi 2007; Heil 2011). We therefore conclude our chapter with a short discussion on how contemporary methods could be used to "wake up the sleeping beauty" of nectar research. Ongoing research has produced many expressed sequence tags from nectaries or used gene chips to investigate large-scale transcriptional changes during nectar secretion (Kram et al. 2009; Hampton et al. 2010), and the first proteomes have been obtained from nectars of various species (Peumans et al. 1997; González-Teuber et al. 2009; Park and Thornburg 2009; Hillwig et al. 2010). The field is ready for breakthroughs with regard to the (bio)chemistry of nectar and the physiological and genetic regulations of its synthesis and secretion.

2 The Role of Nectar in Plant Reproduction and Defence

How important are nectars in the above-mentioned mutualisms? Pollinators of higher plants are most commonly insects and birds. They are attracted to flowers by pollination syndromes combining flower shape, colour and odour with nectar and pollen at different quantities and qualities. Pollination by insects is generally

Fig. 2 (continued) specific ant colonies and are characterized by large nectaries that secrete a chemically highly complex EFN. (**a**) Extrafloral nectary of *Acacia cornigera*; (**b**) *Pseudomyrmex peperi* ants feeding on extrafloral nectaries of *A. collinsii*. By contrast, the extrafloral nectaries of lima bean, *Phaseolus lunatus* (**c**: *white arrows*), serve only in the facultative attraction of defenders, are localized on stipules, much smaller, and secrete a chemically more simple EFN. *Bars*: 1 cm. Photographs taken by Antonio Cisneros (**a**), Martin Heil (**b**) and Christian Kost (**c**) assumed to increase the specificity of the pollen transport among conspecific flowers and thereby guarantee adequate fertilization and outcrossing (de la Barrera and Nobel 2004; Brandenburg et al. 2009). In spite of its apparent frequency and presumed importance, however, little experimental evidence exists for the positive effect of plant nectar on pollination success and, ultimately, plant fitness. Among the Orchidaceae, for example, nectariferous species are more successful at setting fruits than nectariless species (Neiland and Wilcock 1998). However, we are not aware of similar correlational evidence from other study systems.

Although we can assume that the mere presence of nectar clearly benefits plants by increasing pollination rates (Neiland and Wilcock 1998), we must ask: does more nectar always mean more service, and what can plants do to minimize the resulting costs? Apparently, only two studies have investigated the correlation among FN consumption rates and pollinator efficiency, and both have found contrasting patterns. In wild tobacco (*Nicotiana attenuata*), pollinators are attracted from distance via the nectar-released odour, in which particularly benzyl acetone played a crucial role in increasing the number of pollinator visits. Surprisingly, the nicotine that is present in tobacco FN reduced the nectar uptake during single visits. The combination of these compounds ultimately increased the number of floral visits, particularly by hummingbirds (Kessler and Baldwin 2007; Kessler et al. 2008). By combining a long-distance attractant with a presumably bitter taste of the nectar, wild tobacco appears to have developed a strategy to maximize outcrossing with minimum investment in FN amounts. Can all plants minimize their energy investment in nectar and still maintain a high number of pollinator visits? Flowers of *Petunia* plants that were bred to produce reduced FN amounts with unchanged morphological traits of the flower paid for this attempt to "cheat" their pollinators with reduced visitation frequency by *Manduca sexta* moths and a concomitant reduction in seed production (A. Brandenburg, personal communication, and Brandenburg 2009). Do plants gain or lose fitness when reducing their FN secretion? Generalizations are impossible based on these two studies. However, it seems likely that the *Petunia* system represents the more usual scenario because *Nicotiana attenuata*, owing to its special ecology, does usually not compete for pollinators with other species.

The importance of FN in pollination has apparently never been questioned and it might be this fact that prevented quantitative experimental studies and detailed cost–benefit analyses from being planned, conducted, and published. By contrast, the ecological role of EFN has been controversially discussed (Sect. 6). Has this situation led to a better empirical investigation? In fact, the answer is "yes". During the sixties and seventies of the last century, literally hundreds of ant exclusion experiments demonstrated that the presence of ants can benefit plants by reducing overall herbivore pressure (Bentley 1977; Davidson and McKey 1993; Heil and McKey 2003). Inducing plants to produce more EFN has positive effects on the number of ant workers foraging on plants (Heil et al. 2001; Kost and Heil 2005). Several studies in different ecosystem have demonstrated that higher EFN availability can increase the survival rates of ant workers (Lach et al. 2009) and other predators (Limburg and Rosenheim 2001) and also ant activity and aggressiveness

(Sobrinho et al. 2002; Ness 2006; Heil et al. 2009). Studies using wild cotton (*Gossypium thurberi*) demonstrated that fewer ants visited plants with experimentally reduced numbers of extrafloral nectaries; leaf damage on these plants was higher and seed number was lower compared with plants with natural levels of EFN, indicating that EFN mediates the benefits of ants (Rudgers 2004). Indeed, indirect defence via ants represents one of the few anti-herbivore defence strategies for which a clear effect on net herbivory rates and plant fitness has been shown for different species (Chamberlain and Holland 2009). In summary, a positive correlation of investment with benefit for the plant has been shown for EFN, but the generality of this assumption has yet to be proved for FN.

3 Nectar Chemistry and Function

Plants secrete nectar to attract pollinators and defenders. Therefore, nectar needs to be chemically attractive to these insects. Nectars in general contain sugars and amino acids in an aqueous solution, which can be supplemented by proteins, lipids, and even attractive volatiles (Heil 2011). The content of essential amino acids is often enriched, and consumers are known to respond positively to this enrichment (see below). In short, nectar is chemically designed to present an appetizing meal to a large group of mutualistic animals. However, every reward that partners exchange in a mutualistic interaction is prone to being abused by exploiters: non-mutualistic species that do not render the respective benefit to the reward-producing partner (Bronstein 2001). How can plants protect their nectar from these "nectar thieves"? In the following chapter we discuss how the chemical properties of nectar create its "Janus-like" face: nectar presents a nice, attractive face to its legitimate consumers but might appear very different from the perspective of putative exploiters (Fig. 3).

3.1 The Attractive Components of Nectar

Carbohydrates and free amino acids in the nectar are most important for the function of attraction. Because this aspect of the chemical composition of nectar has been reviewed repeatedly (Nicolson et al. 2007; González-Teuber and Heil 2009a), we provide only a short overview here. Most authors assume nectar to be adapted in its composition and concentration to the nutritional preferences of the consumers (Baker and Baker 1982, 1983; Pacini et al. 2003; Johnson and Nicolson 2008; Kromer et al. 2008). Even nectar viscosity has been reported to determine the spectrum of pollinators, because certain animals might not be able to consume too viscous nectars (Kohler et al. 2010). Besides this physical trait, the composition of the nectar determines, at least in part, the spectrum of nectar consumers, because animals differ in their nutritive requirements. For example, hummingbirds, butterflies, moths and long-tongued bees usually prefer sucrose-rich floral nectars, as do most ant species that feed on EFN, whereas short-tongued bees and flies prefer FN rich in hexoses (Blüthgen and Fiedler 2004; Nepi and Stpiczyńska 2008; González-Teuber and Heil 2009a; Nepi et al. 2009). However, some nectarivorous birds and ants lack the sucrose-cleaving enzyme invertase and are thus not able to assimilate sucrose, and hence prefer sucrose-free nectars (Martínez del Rio 1990; Heil et al. 2005).

Sugars represent the dominant compound class in nectar and are usually about 100–1,000 times more concentrated than amino acids. However, amino acids can significantly affect the attractiveness of nectar. Birds and bats can also gain nitrogen from other sources, whereas many adult insects feed only on liquids. We can therefore assume that insect-pollinated flowers should possess more amino acids

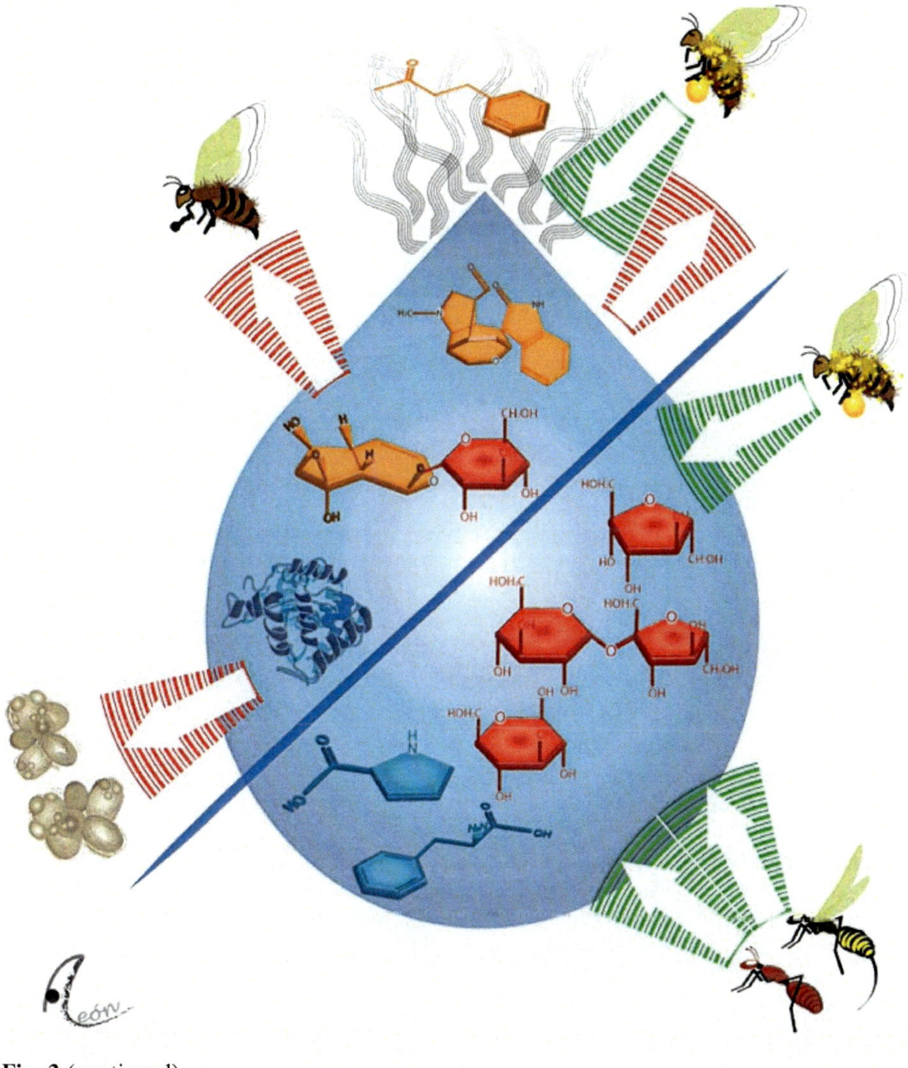

Fig. 3 (continued)

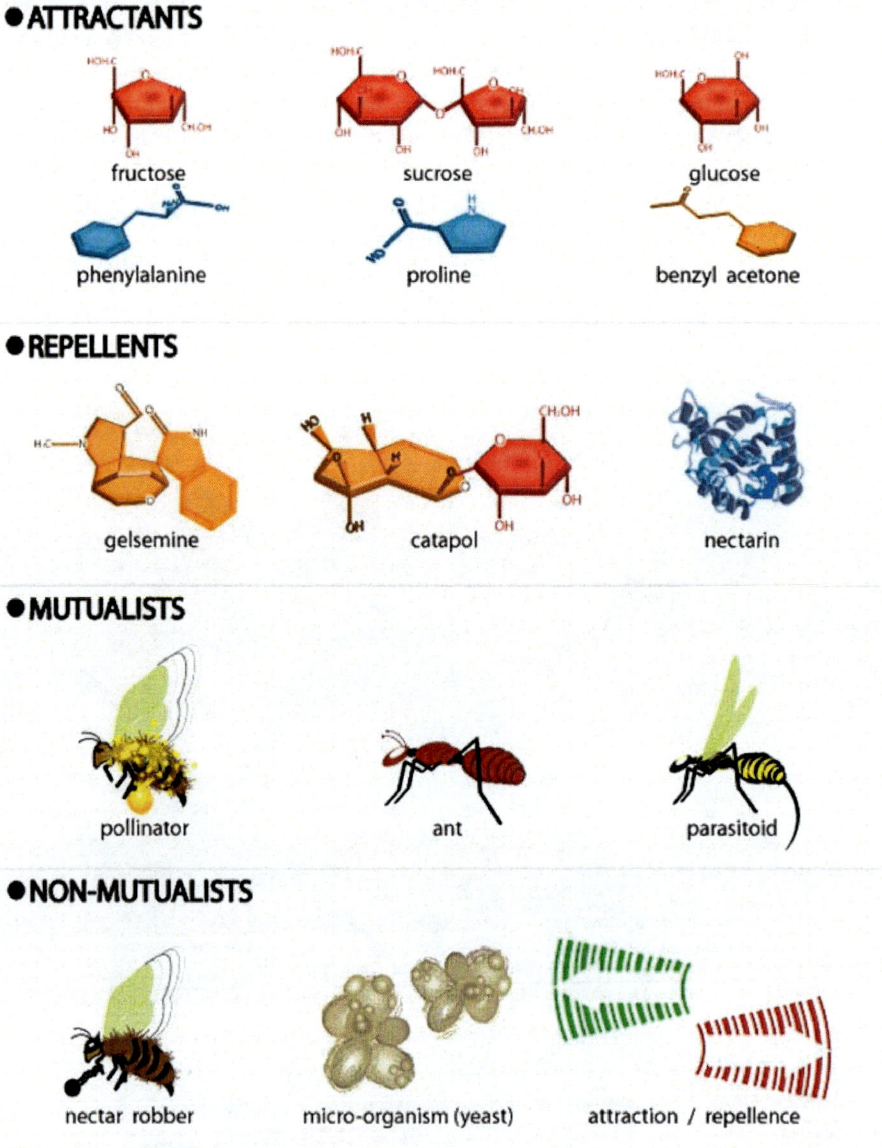

Fig. 3 The Janus-faced nature of nectar. Nectar chemistry serves both the attraction of mutualists such as pollinators and defenders and the protection from nectar robbers and nectar-infecting microorganisms. The most important attractive classes of compounds are mono- and disaccharides, amino acids and volatile components such as benzyl acetone. Repellent effects are exhibited by secondary compounds such as gelsemine and iridoid glycosides. Interestingly, gelsemine has also repels legitimate pollinators. Nectar proteins (nectarins) mainly serve its protection from microbial infections

in their nectar than vertebrate-pollinated flowers, although high amino acid concentrations in FNs of bird-pollinated flowers have been reported in specific cases (Nicolson 2007). High amino acid concentrations have indeed been reported for FNs from flowers that are adapted to butterflies (Baker and Baker 1982), flies (Potter and Bertin 1988), or bees (Petanidou et al. 2006). Similarly, ants prefer nectars rich in amino acids, and ants as well as many insect pollinators can show strong preferences for specific, usually essential amino acids (Blüthgen and Fiedler 2004; Carter et al. 2006; González-Teuber and Heil 2009b).

Which other compound classes are involved in the attractive function of nectar? That flowers attract pollinators via attractive odours is known since centuries and readily used by the perfume industry. In fact, flower petals are still the most important source of natural odours. By contrast, nectar odours were only recently considered as a relevant signal for pollinators (Raguso 2004). Butterflies and moths preferred artificial flowers containing scented nectar over those that contained pure sugar solutions (Weiss 2001), and parasitoid wasps localized cotton (*Gossypium hirsutum*) EFN using only its odours (Röse et al. 2006). Even spiders can use odours to localize nectar (Patt and Pfannenstiel 2008) and flower mites use nectar odours to distinguish between host and non-host plants (Heyneman et al. 1991). The origin of FN scent has been linked to volatiles released by the petals that are absorbed and re-released by the nectar (Raguso 2004). However, a wide array of VOCs occurred in the nectar of wild tobacco (*Nicotiana attenuata*), and many of these compounds have not been detected in other flower parts, suggesting that in certain species, nectar emits its own scent (Kessler and Baldwin 2007).

3.2 *Protection from Nectar Thieves*

Nectar carbohydrates, amino acids and volatiles are apparently composed to provide an appetizing meal to legitimate nectar consumers and/or to signal the presence of nectar to these mutualists from a distance. Being a nutritionally valuable reward, however, nectar must also be protected from illegitimate consumers, which can be animals ("nectar thieves") but also microorganisms, such as bacteria, fungi and yeasts, which might use nectar as a suitable growing medium (Herrera et al. 2008, 2009; González-Teuber and Heil 2009a). The most well-known protection from non-legitimate nectar consumers in the case of FN is achieved by anatomical rather than chemical adaptations. Many flowers have evolved particular structures and shapes that allow access only to a few – or even one – coevolved species. This strategy is known in particular from bird- and butterfly-pollinated flowers, which produce their FN in nectar spurs: long, usually tube-shaped structures to which only their long-tongued specific pollinators have access. Most plant species, however, interact with a large number of different pollinators, and EFN in particular is usually openly presented. Moreover, long tubes cannot protect nectar from being infected by micro-organisms. Nectar, thus, requires also some kind of chemical protection. Mainly three different compounds classes are known to be involved in this protective function: secondary compounds, non-proteinogenic amino acids and proteins.

Surprisingly, nectar proteins appear to be one of the most important compound classes in the protective context. Nectar proteins were discovered more than

80 years ago (Buxbaum 1927; Lüttge 1961). One could suppose that these proteins supply nectar consumers with organic nitrogen, as described above for free amino acids. However, although a nutritive function of nectar proteins cannot be excluded at present, their main function appears to be protection. For example, the nectarins in the FN of ornamental tobacco (*Nicotiana langsdorffii* × *Nicotiana sanderae*) have been biochemically characterized and are likely to protect FN from microbial infestation through the Nectar Redox Cycle (Carter et al. 1999, 2006, 2007; Carter and Thornburg 2004; Park and Thornburg 2009). Floral nectar proteomes appear to be small: for example, five proteins have been found in the FN of ornamental tobacco, eight proteins have been detected in the FN of *Jacaranda mimosifolia* (Kram et al. 2008) and ten in the FN of *Rhododendron irroratum* (H.-G. Zha, personal communication). By contrast, more than 50 proteins are present in the EFN of *Acacia* myrmecophytes, which house ant colonies for their indirect defence (González-Teuber et al. 2009). The majority of these nectarins was identified as pathogenesis-related (PR) proteins, such as chitinases, glucanases and thaumatin-like proteins. The first two groups contributed more than 50% to the overall protein content, and similar numbers were then also found in nectars of other *Acacia* myrmecophytes (González-Teuber et al. 2010). Chitinases were also found in the FN of *Rhododendron irroratum* (H.-G. Zha, personal communication) and, thus, might be another common class of nectarins.

A role in antimicrobial defence has even been suggested for the GDSL lipase in the FN of blue jacaranda (*Jacaranda mimosifolia*) (Kram et al. 2008). Most nectarins, thus, appear to be involved in protecting nectar against microorganisms. The biological function of this effect is likely to be twofold. First, nectars are commonly infested by microorganisms, particularly yeasts, whose metabolic activities can dramatically change nectar chemistry (Herrera et al. 2008, 2009). The chemistry of nectar is closely linked to its function (see above). Therefore, although the presence of some nectar-infecting microorganisms or nectar robbers might have beneficial effects on the plant (Lara and Ornelas 2002), we can assume that most plants gain a benefit from keeping the nectar as sterile as possible, in order to maintain control over its chemical composition (Herrera et al. 2008). Nectaries might, however, also serve as an entrance for phytopathogens (Ivanoff and Keitt 1941; Keitt and Ivanoff 1941). Therefore, antimicrobial nectarins should serve to protect the nectary tissue from infection. More empirical evidence will be required to decide whether anti-microbial nectarins play an important role in reducing infections of tissues other than the nectary.

Protection of the nectar against non-mutualistic insects is often assumed to be the function of non-proteinogenic amino acids or of secondary metabolites (Baker 1977; Baker et al. 1978; Adler 2000). Several studies failed to demonstrate a negative effect of nectar robbers on plant fitness (Maloof and Inouye 2000; Lara and Ornelas 2002), but most authors assume that nectar consumption by non-mutualists represents a loss of a valuable resource. In fact, floral nectars of numerous plant families are "toxic". For example, FN of *Catalpa speciosa* contains iridoid glycosides that fend off nectar robbers but not legitimate pollinators (Stephenson 1982). However, toxic nectar components appear to function as a double-edged sword. For example, Adler and Irwin (2005) experimentally manipulated the concentration of gelsemine, the principal FN alkaloid of *Gelsemium sempervirens*. Gelsemine also deterred effective plant pollinators, thus decreasing the number of flowers probed and the time spent per flower by both pollinators and nectar robbers. Why should FN deter legitimate pollinators? Several scenarios are possible (Adler 2000; Adler and Irwin 2005). First, short visitation times by pollinators do not necessarily represent a fitness disadvantage for plants (Kessler et al. 2008). Second, preventing the loss of nectar to robbers might even be worth some reduction in pollinator visits. Third, gelsemine lowers the rate of selfing in *G. sempervirens* by reducing the proportion of pollen that is transferred from the same individual (Irwin and Adler 2008). Finally, secondary metabolites in nectar could inhibit microbial growth (Adler 2000) and, thus, complement the function of nectarins. However, following Adler and Irwin (2005), no clear fitness benefits for plants with toxic nectars have been demonstrated to date. An alternative explanation might be that secondary compounds, which are transported through the vascular system in a functional context of systemically induced defence, passively "leak" into the nectar (Adler 2000; Adler et al. 2006). Identifying the site of synthesis of toxic nectar components and elucidating their secretion mechanism would be the first steps towards an explanation of their function.

4 Nectary Structure and Evolution

Nectaries are structures, or simply areas on the plant surface, where nectar is being secreted. While floral nectaries are located within the floral organs (ovary, filaments or petals), extrafloral nectaries are found on all vegetative organs besides roots, commonly even within the inflorescences (outside sepals) (Elias 1983; Heil 2007, 2011). They may be located at the surface level, forming an outgrowth on the organ, or be sunken (Fahn and Rachmilevitz 1979). In some cases nectaries are engulfed very deeply, as in the case of gynopleural nectaries, which represent the most common type of floral nectaries in monocotyledons although they are almost absent from the dicotyledons (Smets et al. 2000).

4.1 Nectary Anatomy and Secretion Sites

Nectaries can be extremely diverse with respect to their localization, their structure and even the secretion mechanism (Elias 1983; Pate et al. 1985; Fahn 1988) (Sect. 5). For example, the anatomically most simple nectaries are "gestaltless" (Frey-Wyssling and Häusermann 1960) (i.e., without any externally visible structure), and in this case can only be identified as areas where nectar appears on the plant surface. Gestaltless nectaries appear to be more frequent among the extrafloral nectaries (Kirchoff and Kennedy 1985), whereas floral nectaries are normally well defined (Bernadello 2007). Because of their nature, however, such non-structural nectaries are difficult to identify when no nectar is observed, and gestaltless nectaries may indeed be strongly under-reported. Other nectaries form anatomically distinct and sometimes highly conspicuous structures with a complex ultrastructure (Schmid 1988), defined nectaries as a localized, multicellular glandular structure, although this definition does not cover nectaries that are formed by unicellular hairs.

Interestingly, many nectaries are characterized by a continuous cuticle present on the surface of the nectar epidermis (Gaffal et al. 1998), although the permeability for aqueous solutions of the cuticle is extremely low. Several different strategies have been evolved to reach high permeability during the active secretion. In some cases, the cuticle might simply break at the onset of the secretion process, and nectar is released through these ruptures (Durkee 1982). In other cases, specialized pores have been described, where the cuticle is forming "secretory pits" through which nectar can be exuded (Kronestedt-Robards et al. 1986; Arumugasamy et al. 1990; Stpiczyńska 2003b). Finally, the cuticle may have microchannels, narrow tubular interruptions of the cuticle (Stpiczyńska et al. 2005). In most cases, however, the nectar exits through modified stomata that remain permanently open, or through specialized trichomes (Figs. 1 b, d and 4) (Fahn 1988; Wist and Davis 2006; Gaffal et al. 2007; Vassilyev 2010).

4.2 Fine Structure and Vascularization

In general, the nectariferous tissue consists of two main components: an epidermis, with or without stomata or trichomes through which nectar is released to the exterior and a specialized parenchyma that produces and/or stores the pre-nectar (Fig. 4) (Fahn 1979). A third component can sometimes be distinguished by its more loosely packed cells as subnectary parenchyma (Stpiczyńska 2003a; Kaczorowski et al. 2008). The components of this general anatomical structure, however, are not necessarily recognizable in all nectaries. For example, gestaltless nectaries often do no exhibit a modified secretory tissue that can be clearly differentiated from the surrounding tissue (Zimmermann 1932; Elias 1983).

Nectaries can be connected to the phloem, the xylem, both or have no direct vascular connection (Fahn 1988; Wist and Davis 2006). Vascular bundles can be found in the nectariferous or the subnectariferous parenchyma. In most species studied, floral nectaries are vascularized by phloem only, or are not directly vascularized at all. For example, Wist and Davis (2006) reported that 50% of the Asteraceae lack direct vascular connections in their floral nectaries and also a taxonomically broader review found that nectaries of more than one-third of all plant species lack any direct vascularization (Fahn 1988). Only a minority of nectaries known receive direct vascular supply comprising xylem and phloem and, even in those ones, the last branches reaching the parenchyma or the epidermis are usually from the phloem (Elias et al. 1975; Davis et al. 1988). That nectar represents

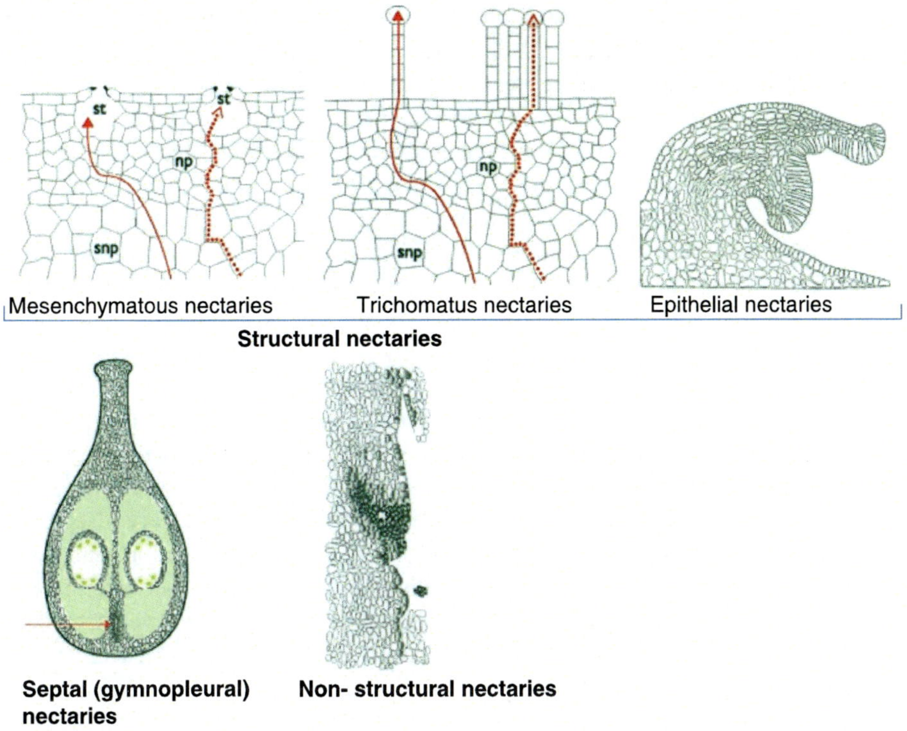

Fig. 4 Ultrastructure of nectaries. Nectaries can have different ultrastructural properties but most are characterized by a nectary parenchyma (np) and a sub-nectary parenchyma (snp). Nectar can be secreted to the outside via modified stomata (st), as illustrated in the case od mesenchymatous nectaries, via trichomes (tricomatous nectaries) or via the epidermis (epithelial nectaries). See text for further details

"secreted phloem sap" for most species (Agthe 1951; Frey-Wyssling et al. 1954; Zimmermann 1954; Lüttge 1961; Fahn 1988; de la Barrera and Nobel 2004) is further supported by an interesting characteristic of the companion cells of the nectary phloem. These companion cells commonly possess well developed wall ingrowths, which increase the surface area, thereby facilitating the uploading of pre-nectar component into the adjacent nectar parenchyma (Davis et al. 1988; Wist and Davis 2006).

4.3 Ultra Structure of Nectaries

Nectaries are structurally highly diverse and no "typical" or "representative" type of nectary can be defined. In the following, we shortly describe what appear to be the most widely distributed structural properties of nectariferous tissues (Fig. 4). Floral and extrafloral nectaries are not principally different, although several characteristics might be more common in one than in the other functional group.

Epidermal cells in the area of a nectary are in general small and polyhedric and have an anticlinal orientation. The storage of starch in the epidermis has been reported in some cases where nectaries are characterized by a particularly high energy requirement (Durkee et al. 1981; Nepi et al. 1996; Razem and Davis 1999), but does not appear to be a common phenomenon. As described above, the epidermis itself may represent the secretory complex, but more commonly the epidermis includes secretory structures, such as trichomes or modified stomata. Vogel (1997, 1998) defined four basic nectary types based on epidermal characteristics: mesenchymatous nectaries, epithelial nectaries, tricomatous nectaries and nectarioles. (1) Mesenchymatous nectaries form a very common type and are composed of glandular and storage tissues that usually secrete nectar via modified stomata (Davis 2003; Bernadello 2007; Ren et al. 2007b). The stomata can be localized on the surface of the nectary or in deep depressions and their regulation differs from that of leaf stomata: for instance, nectar stomata lack turgor- and ion-mediated movements (Davis and Gunning 1992, 1993). (2) Epithelial nectaries are basically formed by a glandular epidermis, whereas (3) tricomatus nectaries secrete via specialized trichomes. These trichomes show a variety of distributions and patterns (Fahn and Rachmilevitz 1970) and can be both of the uni- or the multicellular type. Most of the multicellular trichomes comprise a basal part at the level of epidermal cells, a stalk, and tip cells that possess all the characteristic traits of secretory cells such as an elaborate ER, dictyosomes, a high number of mitochondria and a high number of vesicles (Kronestedt-Robards et al. 1986; Sawidis et al. 1989). (4) Finally, an anatomically heterogenous group of small, discontinuous nectar-secreting structures, which are composed only of few cells, has been termed nectarioles (Vogel 1997, 1998).

Below the epidermis, a specialized parenchyma can be found in all cases of morphologically specialized nectaries. The nectary parenchyma is generally composed of several layers of small isodiametric cells with thin walls, dense granular cytoplasm, small vacuoles and relatively large nuclei (Heil 2011). These cells represent typical secretory cells and share their characteristics (D'Amato 1984) such as, for example, being rich in ribosomes and mitochondria and possessing an elaborate ER, many dictyosomes and – usually – a high number of vesicles. Another well-defined characteristic of this tissue is the high number of plasmodesmata, a trait that is supposed to ensure rapid trafficking of metabolites among the cells (Fahn 1979). Prior to and during secretion, the nectary parenchyma is subject to significant ultrastructural changes. For example, parenchyma cells frequently undergo continued cell division during secretion (Gaffal et al. 1998). While vacuoles in the pre-secretory phase are usually small, their volumes increase at the time of secretion. During this phase, a boost in the energy requirements leads to a more active endoplasmic reticulum and an accumulation of mitochondria and ribosomes in the cytoplasm of parenchymatous cells, while dictyosome numbers are reduced (Zhu and Hu 2002).

Below the nectary parenchyma, some nectaries present a subnectary parenchyma that can be differentiated by its large cells with bigger vacuoles, less dense cytoplasm and larger intracellular spaces (Durkee 1982; Cawoy et al. 2008). In contrast to the nectary parenchyma, minor or no changes take place in the subnectary parenchyma during secretion. Its main function appears the communication with the vascular system and perhaps some contribution to the synthesis of nectar sugars. When the nectary is vascularized, phloem and xylem bundles are always present in subnectary parenchyma. The xylem vessels in general terminate in this tissue, whereas phloem strands branch into the nectary parenchyma. The subnectary parenchyma is generally rich in chloroplasts and thereby might contribute to the production of nectar sugars, although it is commonly believed not to be directly involved in nectar production (Nepi et al. 2007).

4.4 Evolution of Nectaries

In summary, nectaries are structurally and functionally highly diverse, and they appear difficult, if not impossible, to define the "most typical" nectary. These great variations in nectary morphology and structure occur even at the intraspecific level: floral nectary structure might differ within individual flowers, among the flowers of the same plant, and among plants of a population. As an example for variability within individual flowers, Arabidopsis flowers contain two types of nectaries, lateral and meridian, which are characterized by significant morphological and structural differences. Lateral nectaries are more pronounced and more heavily innervated by sieve elements and realize the largest part of overall FN secretion (Davis et al. 1998). Variability has also been observed among extrafloral nectaries, where even central morphological traits might differ among nectaries of the same plant, only depending on their localization. For example, stipular extrafloral nectaries of cowpea (*Vigna unguiculata*) form an area of widely spaced secretory trichomes that lacks any direct connection to the vascular system, whereas extrafloral nectaries on the inflorescence stalk consist of a region with secretory, cone-shaped tissues that are connected to the phloem and release EFN through permanently open stomata (Kuo and Pate 1985). Extrafloral nectaries appear particularly variable, at least within the *Fabaceae*, where several species within the same plant posses both vascularized and non-vascuralized extrafloral nectaries of several different morphological types (Díaz-Castelazo et al. 2005).

Unfortunately, little is known on the evolution of nectaries, and their origin is still a matter of discussion (de la Barrera and Nobel 2004). Although floral nectaries are likely to have evolved from extrafloral nectaries (Matile 1965; Heil 2007), the high variability in contemporary nectary function and morphology might be related to a low evolutionary stability. It is likely that nectaries can be easily lost and gained during evolution (de la Barrera and Nobel 2004; Sugiura et al. 2006; Wooley et al. 2007). Moreover, extrafloral nectaries, and possibly floral nectaries, are phenotypically highly plastic (Mondor et al. 2006; Doak et al. 2007). For example, the number of extrafloral nectaries increased when plants of aspen (*Populus tremuloides*) or broad bean (*Vicia faba*) were damaged by herbivores or defoliated experimentally (Mondor and Addicott 2003; Wooley et al. 2007). The other way round, African *Acacia* myrmecophytes reduced the number of extrafloral nectaries when being released from herbivore pressure by large mammals over several years (Huntzinger et al. 2004; Palmer et al. 2008). *Malotus japonicus* plants cultivated under high light conditions produced larger extrafloral nectaries (Yamawo and Hada 2010), a trait that is likely related to higher EFN secretion rates (Petanidou et al. 2000; Yamawo and Hada 2010). Morphological traits also can vary with seasonal changes; Petanidou et al. (2000) compared various species of the Lamiaceae with different flowering times and observed lower floral nectary sizes, FN secretion rates and stomatal openings during Mediterranean summer. Although likely to minimize nectar flow when water supply is limited, it cannot be decided whether these differences were indicative of interspecific (adaptive) variability or of phenotypic plasticity.

It is likely that the ease with which nectaries are gained and lost, together with the different selective pressures on their functionality, provides explanations for their structural and mechanistic diversity. However, much more data will be required until we understand the evolution of nectaries. Few studies have so far focused on the degree of heritability in the morphological attributes of nectaries. Rudgers (2004) used sire-offspring regression analyses of *Gossypium thurberi* and reported that both the proportion of leaves with extrafloral nectaries and the size of these nectaries exhibit heritable variation. This observation is likely to represent a common case, because Mitchell (2004) reported high heritability values of nectary traits in a survey of various species. However, the high phenotypic plasticity of many – even morphological – traits of nectaries might cause that "even substantial amounts of genetic variation may be swamped out in the field (Mitchell 2004). Hints towards a possible selection by environmental parameters can also be seen in the observation that species located in hot and arid climates usually present stomatal secretion in their floral nectaries (Petanidou 2007), although it is again not possible to decide whether this pattern indicates differences among species or rather the effects of a high phenotypic plasticity. In summary, nectaries are phenotypically highly plastic in spite of the fact that central structural and morphological traits are clearly heritable, and many more studies will be required to understand their micro- and macro-evolution.

5 Secretion Mechanisms

The morphological diversity that we have described above is also resembled by a multitude of secretion mechanisms (Kram and Carter 2009). Scientists have discussed since more than a century whether or not the accumulation of starch represents a prerequisite of a functioning nectar secretion, and also the questions "holocrine or merocrine secretion?" and "apoplastic or symplastic transport?" date back to the thirties of the last century (Sect. 6). In fact, it is most likely that plants have evolved several solutions for the same problem and secrete nectar using various different mechanisms. In the following, we describe what appear to be the most common types of secretion mechanisms. In general, we can distinguish three different phases: (1) carbohydrate uploading, (2) processing of pre-nectar and synthesis of non-carbohydrate components and (3) nectar secretion (Heil 2011, see video on a putative mechanism of nectar secretion: http://www.youtube.com/watch?v=Nd8ryN_7BP8).

5.1 Uploading of Carbohydrates

Most authors agree that nectar basically represents "secreted phloem sap" for most species (Agthe 1951; Frey-Wyssling et al. 1954; Zimmermann 1954; Lüttge 1961; Fahn 1988; de la Barrera and Nobel 2004). However, nectar and phloem sap clearly differ in their chemical composition. The processing from phloem sap to pre-nectar and to the finally secreted, mature nectar certainly comprises more than a simple transport of phloem sap through the nectariferous tissue. In the most likely scenario, photosynthates are uploaded as sucrose from the phloem, either directly into the nectariferous tissue in the case of vascularized nectaries (about 60%), or into neighbouring parenchyma cells in non-vascularized nectaries, from where they might diffuse further into the secretory cells of the nectary (Vassilyev 2010). There are two alternative routes for the movement of the pre-nectar from the phloem into the nectary cells: a symplastic and an apoplastic route.

Although both pathways might exist and in fact could be active even within the same nectary, the symplastic component appears to be more common (Heil 2011). Fahn (1988) postulated that the pre-nectar is stored in vesicles and moves through the nectariferous tissue to the secretory cell via numerous plasmodesmata. During this process, pre-nectar is modified into mature nectar. This hypothesis is confirmed by the great number of plasmodesmata present between the phloem parenchyma and the nectar parenchyma cells (Sawidis et al. 1987) and by the high number of vesicles that can usually be observed in nectariferous tissues (Heil 2011). Furthermore, nectar secretion via trichomes excludes an apoplastic transport for the respective species, owing to apoplastic barriers in the external cell walls in the stalk and intermediate cells of the trichomes (Kuo and Pate 1985; Fahn 1988). For the symplastic pathway, the transport of sugars and amino acids is likely to be driven passively by the osmotically generated hydrostatic pressure differences between source and sink tissues (Lalonde et al. 2003).

The second possibility, an apoplasmic pathway, comprises a transport via intercellular spaces and cell walls to the secretory cells (Vassilyev 2010) where sucrose transporters are conceived to function as membrane complexes that respond to alterations in the source/sink balance. Effluxers that release sucrose and amino acids to the surrounding apoplasm during uploading from the phloem remain to be discovered. However, recent evidence indicates that cell wall invertases might play an important role in the uploading of sucrose from the phloem. An apoplastic invertase (AtCWINV4) has been discovered in *Arabidopsis thaliana* that is required to create the sink status for active nectar secretion (Ruhlmann et al. 2010). A mutant line, which lacked this activity, showed reduced levels of starch accumulation within the nectary (Kram and Carter 2009; Ruhlmann et al. 2010). However, at least nectar proteins are most likely transported in vesicles and their transport depends, thus, on the symplastic pathway.

A generally driving factor in the unloading of starch from the phloem is represented by the large transmembrane differences in sucrose concentration (Lalonde et al. 2003). Moreover, many floral nectaries (e.g., of tobacco and *Arabidopsis*) accumulate starch in the pre-flowering phase and thereby continuously remove sucrose from the cell plasma, thereby avoiding an equilibrium between sucrose concentrations in the phloem and the nectariferous tissue (Horner et al. 2007; Ren et al. 2007b). Thus, source–sink relations appear a major driving force during the uploading of carbohydrates to the nectariferous tissue.

5.2 Processing of Pre-nectar and Synthesis of Non-carbohydrate Components

As mentioned above, floral nectaries might rely on the accumulation of starch in order to reach peak nectar secretion rates. Besides Arabidopsis and tobacco, an accumulation and consecutive breakdown of starch have also been described for further, taxonomically unrelated species such as soya bean (*Glycine* max) and common foxglove (*Digitalis purpurea*) (Horner et al. 2003; Pacini et al. 2003; Gaffal et al. 2007). Many species possess amyloplasts in their nectary tissue (Pacini et al. 2003) that can become directly connected to the vacuole and consecutively emptied during the phase of most active FN secretion (Gaffal et al. 2007). All these processes appear linked to the mobilization of nectar carbohydrates. However, the accumulation of starch has not been reported for extrafloral nectaries so far and does therefore not represent an absolute requirement for a functioning secretion. In fact, the direct secretion of products of the current assimilation process has been shown repeatedly for FN, using girdling of flower shoots as well as darkening and defoliation experiments (von Czarnowski 1952; Gaffal et al. 2007). Experiments with ^{13}C-labelled CO_2 demonstrated that also EFN contains sugars that have been assimilated during the last hours before secretion (Radhika et al. 2008).

In summary, carbohydrates are uploaded as sucrose from the phloem to the secretory tissue where they are stored and/or processed (Wenzler et al. 2008; Kram and Carter 2009). During active secretion, sucrose is metabolized by cell wall invertases, which serve to produce hexose-rich nectars and to create the required source–sink relationships that prevent reloading of sucrose into the phloem (Agthe 1951; Zimmermann 1953; Frey-Wyssling et al. 1954; Peng et al. 2004). The above-mentioned apoplastic invertase, AtCWINV4, also plays a role in this context, as demonstrated by the absence of nectar in "lack of function" mutants (Ruhlmann et al. 2010). In fact, genes coding for the complete sucrose biosynthesis are upregulated in *A. thaliana* nectaries (Kram et al. 2009) and expression patterns of genes involved in starch metabolism allow a clear separation of an anabolic phase before anthesis and a catabolic phase during secretion in nectaries of ornamental tobacco (Ren et al. 2007a). A more primary role of extracellular invertases may be to double the osmotic contribution of uploaded sucrose that affects turgor and hence rates of different photoassimilate import (Roitsch 1999). However, secreted nectars are characterized by a wide range of concentrations and sucrose:hexose ratios (Baker and Baker 1975, 1982) and, therefore, cannot be produced only by a passive flow or transport mechanisms exclusively driven by sucrose-cleaving enzymes.

Unfortunately, research into the mechanisms of nectar production (de la Barrera and Nobel 2004; Gaffal et al. 2007; Wenzler et al. 2008; Kram and Carter 2009; Vassilyev 2010) has focused exclusively on FN carbohydrates. It remains completely unknown where non-carbohydrate nectar constituents are produced, where and how they are added to the pre-nectar and how they are secreted. Considering the abundance and chemical diversity of nectar proteins and the lack of reports of many of these nectarins from other tissues, synthesis in the nectary tissue appears likely. In fact, secretory cells of *Vigna unguiculata* extrafloral nectaries contain protein-rich inclusions (Kuo and Pate 1985) and all *NECTARIN* genes that correspond to nectar proteins in the FN of ornamental tobacco are expressed in the nectary tissue (Carter and Thornburg 2004; Thornburg 2007), where some of them are under the control of an MYB305 transcription factor (Liu et al. 2009). Furthermore, tobacco nectarins contain signal peptides for secretion and can therefore only be secreted by the fusion of vesicles with the plasma membrane. A similar evidence for other systems is lacking and is urgently needed to understand how plants produce nectar compounds and secrete them into the nectar. It appears likely, however, that pre-nectar is processed within vesicles and that nectar proteins are synthesized in the nectary parenchyma itself and then added to the pre-nectar during this process (Heil 2011).

5.3 Secretion

At the initiation of secretion, the vacuoles increase their volume, the number of dictyosomes is reduced, the endoplasmic reticulum becomes more active and the number of mitochondria increases (Zhu and Hu 2002). Together with this increase in the number of metabolically active organelles, the starch grains reduce their size or completely disappear (Zhu and Hu 2002; Horner et al. 2007; Ren et al. 2007b).

In principle, two different mechanisms of nectar secretion have been described so far: holocrine and merocrine. Holocrine secretion involves programmed cell death at the moment of secretion (Horner et al. 2003; Pacini et al. 2003; Gaffal et al. 2007). In this case, the nectar is produced within the cell and released by the rupture of the plasma membrane (Horner et al. 2003; Vesprini et al. 2008). The merocrine secretion allows the secreting cells to survive and to continue with their secretory activity (Nepi and Stpiczyńska 2007). Most extrafloral nectaries that have been investigated so far release EFN in response to herbivory or mechanical damage and can do so repeatedly (Heil et al. 2000, 2001; Heil 2009). By contrast, FN is usually secreted in an ontogenetically programmed manner. Therefore, holocrine secretion might be a common case in floral nectaries but we do not regard it as likely that this type of secretion has frequently evolved for EFN (Heil 2011).

The merocrine-type secretion of nectar usually occurs as a granulocrine secretion, which is characterized by the presence of vesicles that stem from the endoplasmic reticulum or the dictyosomes and that contain the pre-nectar. These vesicles move through plasmodesmata towards the epidermal cells and finally fuse with the plasma membrane to release their content into the extracellular space. Apparently, this secretion mechanism represents the most common and/or most accepted one (Jürgens and Geldner 2002; Heil 2011). Several aspects support this model. First, vesicles are common in nectariferous tissues (Benner and Schnepf 1975; Kuo and Pate 1985; Fahn 1988) and the number of dictyosomes in secreting and non-secreting nectary cells of *Billbergia nutans* (Bromeliaceae) differed in coincidence with secretory activity (Benner and Schnepf 1975). Second, non-carbohydrates such as lipids and proteins (Carter and Thornburg 2004; Kram et al. 2008) are likely to be synthesized in the nectariferous tissue and then must enter the pre-nectar before its secretion.

Although the secreted nectar is likely to be chemically mature in most cases, further changes to the biochemistry of nectar can occur in the post-secretory phase. For example, sucrose can also be eliminated from nectar by post-secretory hydrolysis, which is mediated by invertases that are secreted into the nectar itself (Zimmermann 1953; Heil et al. 2005). Second, nectars are commonly infested by microorganisms, particularly yeasts, whose metabolic activities can dramatically change nectar chemistry (Herrera et al. 2008, 2009). It has been assumed that the presence of some nectar-infecting microorganisms or nectar robbers might have beneficial effects on the plant (Lara and Ornelas 2002).

5.4 Regulation and Re-absorption

Although nectar sugars in some species represent less than 2% of the current net photosynthates (Pate et al. 1985), nectar production by *Asclepias syriaca* can consume up to 37% of daily assimilated carbon during blossoming (Southwick 1984). Nectar secretion undoubtedly can be costly, although many more studies on different plant groups will be required to obtain a realistic estimate of the "usual" costs of nectar production. In order not to waste energy, floral and extrafloral nectaries are commonly able to adjust the net secretion rates to the consumption rates and/or to re-absorb nectar that has not been consumed. An adjustment of FN net production to consumption rates has been demonstrated in various species (Corbet and Delfosse 1984; Gill 1988; Pyke 1991). *Macaranga tanarius* reduced EFN secretion in the absence of consumers and increased it immediately after consumers were allowed to feed for 1 day on the nectaries (Heil et al. 2000). However, quantitative adjustments of the apparent net production could result from either an inhibited de novo secretion or the re-absorption of accumulated nectar. Although it is not known whether plants really can adjust the de novo secretion, a re-absorption of FN has been shown unambiguously with different methods (Nepi and Stpiczyńska 2008), including labelling studies that demonstrated the uptake and allocation to other plant organs of externally applied ^{14}C-labelled sucrose or glutamine (Pederson et al. 1958; Ziegler and Lüttge 1959). Re-absorption of non-consumed FN appears common (see examples in Stpiczyńska 2003b), although this phenomenon has yet to be demonstrated for EFN (Nepi and Stpiczyńska 2008).

Nectar resorption has been hypothesized to be mediated by vesicles or by programmed cell death in the nectary tissue in combination with a phloem that remains active and the resulting changes in source–sink relationships (Kuo and Pate 1985; Gaffal et al. 2007). Under this scenario, extrafloral nectaries are likely to lack the capacity to re-absorb nectar because the regulation of EFN secretion is not dependent on ontogenetically programmed patterns. Numerous vesicles became associated with the plasma membrane at the moment of resorption in the nectar spurs of *Platanthera chlorantha*, and cell death was apparently not involved in this process (Stpiczyńska 2003b). Similarly, the observation of numerous mitochondria surrounding the plasmalemma may indicate active transport (Zhu and Hu 2002). It appears likely, thus, that the resorption of non-consumed nectar represents an active process in most species. Although most studies concern sugar resorption, other nectar components such as amino acids or proteins may also undergo resorption (Nepi et al. 1996). For example, water has been demonstrated to be reabsorbed as well (Nepi et al. 2001), a phenomenon that could be explained by cell turgor modifications and rapid changes in the osmolarity as a passive process (Castellanos et al. 2002). Many more studies will however be required until we truly understand the biochemical and physiological mechanisms that underlie the re-absorption of non-consumed nectar.

Another raising question is how plants sense the unconsumed nectar. The most accepted hypothesis is "sugar sensing"; sugars are able to act as central signalling molecules, interacting with hormones, light and stress signals (Rolland et al. 2006). Membrane-bound receptors are involved in sugar sensing and transduction (Loreti et al. 2001). It appears possible that the same mechanism is acting in nectaries. Alternatively, the re-absorption might be driven by changing source–sink relations; unconsumed nectar accumulates, becomes more concentrated due to evaporation and therefore the concentrations outside the nectary might become higher than inside. It will be a future goal to demonstrate whether similar or other mechanisms allow nectaries to control nectar homeostasis.

5.5 Regulation by Internal and External Factors

Obligate myrmecophytes among Central American *Acacia* species can adjust their EFN secretion to the identity of the inhabiting ants and provide high levels of the resource only to defending mutualists, but much lower amounts to non-defending parasites (Heil et al. 2009). Unfortunately, the cues that enable the defending partner to be identified are not yet known. In general, the regulation of nectar secretion seems to be complex and dependent on internal and external factors such as, for example, light availability, time of the day, nutrient availability and the presence of consumers or, in the case of EFN, herbivores.

Many nectaries secrete their nectar in a diurnal rhythm that appears to be adapted to the activity patterns of the respective consumers (Tilman 1978; Heil et al. 2000). On the other hand, EFN secretion with few exceptions represents an inducible defence mechanism that is activated by mechanical damage and/or herbivory. An induction of EFN secretion by herbivory has been demonstrated for 11 genera from six families (*Prunus* [Rosaceae], *Acacia, Phaseolus, Prosopis, Vicia* and *Leucaena* [Fabaceae], *Macaranga* and *Sapium* [Euphorbiaceae], *Populus* [Salicaceae], *Gossypium* [Malvaceae] and *Paulownia* [Scrophulariaceae]) (Wäckers and Wunderlin 1999; Heil et al. 2001, 2004; Mondor and Addicott 2003; Rogers et al. 2003; Wooley et al. 2007; Kobayashi et al. 2008; Pulice and Packer 2008) and, thus, appears to be very common.

Interestingly, FN and EFN secretion is, at least in part, under the control of the same hormone – jasmonic acid. Mechanical damage and herbivory increased endogenous jasmonic acid levels, and this change was correlated with an increase in EFN secretion, moreover, external application of jasmonic acid also increased EFN secretion rates by lima bean, *Macaranga tanarius* and several *Acacia* species (Heil et al. 2001, 2004; Heil 2004). Recently, it has been demonstrated that FN secretion by *Brassica napus* is influenced by the same hormone (Radhika et al. 2010). Interestingly, FN secretion by *Brassica* was not affected by folivory or JA application to the leaves, an observation that points to a functional separation of the hormonal controls of FN and EFN secretion (Radhika et al. 2010). In short, nectar secretion responds to multiple endogenous and exogenous triggers and appears generally adapted to ensure that plants obtain maximum mutualistic benefits from a minimum of investment in energy and other limiting resources. However, the physiological and genetic mechanisms by which this likely adaptive regulation is achieved remain to be investigated in most cases.

6 Nectar Research: Past and Future

Nectar serves in the interactions of plants with mutualistic animals and its presence and attractive function can be observed easily. It is likely for these reasons that research into nectar has a long history. However, the topic has turned into the "sleeping beauty" of plant sciences over the last 50 years and many recent observations represent in fact re-discoveries of formerly well-known phenomena (Heil 2011). In this last chapter, we present a short insight into the history of nectar research in order to acknowledge some of those brilliant scientists of the nineteenth and the first half of the twentieth century whose results still represent a considerable part of our current knowledge on nectar biology and physiology. We then discuss how contemporary "omics" techniques could be used to wake up the sleeping beauty and convert nectar research back into an active and lively component of the contemporary plant sciences.

6.1 A Short Historical Overview

The involvement of FN in the attraction of bees and other insects to flowers is likely known since thousands of years. Although being comparatively young, even the suggestion that EFN might be involved in plant defence has a history of more than a century, since it has been suggested by Delpino and Belt in 1874 (Belt 1874; Delpino 1874, 1886). Just like many other provocative ideas, this hypothesis has had a very hard life until being accepted: It was questioned by Darwin 2 years after its first publication (Darwin 1876) and was regarded as "finally rejected" by Nieuwenhuis von Uexküll-Güldenband in 1906 (Nieuwenhuis-von Uexküll-Güldenband 1906; Zimmermann 1953). By that time, most botanists agreed that EFN serves the excretion of "excess carboydrates" and it was not until the seventies of the last century that literally hundreds of ecological field studies were presented to support the defensive role of extrafloral nectar (Bentley 1977; Elias 1983; Koptur and Lawton 1988).

Do further recent reports also represent confirmatory work rather than truly novel discoveries? The answer is "yes". Matile (1965) suggested that floral nectaries are evolutionarily derived from extrafloral ones. Nectar proteins were discovered by Buxbaum (1927) and its invertase activity in the 1950s (Zimmermann 1953). Because the original publications were in German, they remained widely unknown, and the presence of invertase in the secreted nectar was published erroneously as a new discovery again more than 50 years later (Heil et al. 2005). Nectar re-absorption was suggested originally by Bonnier (1878) and already in experimentally demonstrated using radioactively marked amino acids and disaccharides (Pederson et al. 1958; Ziegler and Lüttge 1959). Another controversial discussion that dates back to the nineteenth century concerns the mechanisms of nectar secretion. Behrens related starch degradation to nectar secretion in 1879 (cited after Gaffal et al. 2007) whereas a direct involvement of the phloem was suggested by von Wettstein (1889). In summary, the current discoveries in nectar research represent a renaissance of a previously fashionable field, rather than a truly novel development.

6.2 Steps into the Future

Research into the biology and physiology of nectar was an important topic in plant research during the late nineteenth and the first half of the twentieth century.

However, little progress has been made since then, although central questions remained unresolved. Which factors have inhibited nectar research over the last 50 years? The field has probably from of the focus of the plant sciences on model plants that either do not produce nectar (i.e. wind-pollinated crops such as maize, rice and wheat), or do not rely on nectar because they are highly selfing (Arabidopsis). However, "omics" techniques have enabled non-model plants to be studied, and the first nectar-producing species (in particular, cotton, poplar, tobacco and *Petunia* species) have now reached the status of genetically tractable model systems.

Despite the number of publications related to nectaries and nectar, only few studies have treated the topic at a molecular level by now. Remarkably, the molecular events involved in the synthesis, regulation and secretion of nectar, as well as in the development of the nectary, remain poorly understood. Little information exists regarding genes that are involved in nectar synthesis and secretion and these mostly stem from *Arabidopsis thaliana* (Bowman and Smyth 1999; McKim et al. 2008; Kram et al. 2009; Ruhlmann et al. 2010). Three genes encoding for putative transcriptions factors have been established to be involved in nectary development: CRABS CLAW (CRC) and BLADE-ON-PETIOLE 1 and 2 (Bowman and Smyth 1999; McKim et al. 2008), whereas one gene coding for an invertase, CELL WALL INVERTASE 4, has been demonstrated to be essential for nectar secretion (Ruhlmann et al. 2010). As to our knowledge, these four genes are the only ones for which a function in the development of floral nectaries or in FN secretion has been demonstrated, and nothing appears to be known regarding genes involved EFN secretion. More research is needed to understand how plants produce nectar: the most important mediator of their interactions with mutualistic animals (Heil 2011).

Where is the future of nectar research, and which are the plant species and techniques to chose? Kram and Carter (2009) emphasized the genomic tools that are available for *Arabidopsis thaliana* and have therefore suggested Arabidopsis as a "model for functional nectary analysis". However, Arabidopsis is characterized by small flowers with tiny nectaries that release minute levels of nectar. Besides these technical problems, Arabidopsis is a highly selfing plant and does not rely on FN for its pollination, and it does not possess extrafloral nectaries. Thus, neither FN nor EFN play an important role in the life of *Arabidopsis* and many traits that characterize more important nectars might not be expressed in this species. In fact, nectar secretion schemes and nectar biochemistry can depend strongly on the biological importance of the nectar in the particular life history of a species and therefore can significantly vary even among closely related species (Heil et al. 2004, 2005). *Arabidopsis* is therefore not likely to possess all genes and mechanisms that are required for the full functioning of biologically important nectaries.

It is likely that specialized structural and functional properties of nectaries as well as nectar chemistry are based on the expression and regulation of a nectar-specific set of genes. Economically important plants such as orchard trees (apple, *Malus domestica*; pear, *Pyrus communis*; cherry, *Prunus avium*, etc.); crops (cotton, *Gossypium* sp.; soya, *Glycine max*; beans, *Vicia faba* and *Phaseolus* sp.; zucchini,

Cucurbita pepo, etc.) and timber trees (aspen, *Populus* sp.; cedar, *Cedrus* sp., ect.) depend on their nectaries for pollination or defence and should therefore express all genes that are required for a complete nectary development, a functioning nectar secretion and a complete and complex nectar biochemistry. New model plants such as poplar trees (which bear extrafloral nectaries: Wooley et al. 2007) or *Brassica rapa* (with its large floral nectaries: Hampton et al. 2010) combine benefits such as the availability of genomic and bioinformatic tools with relatively large nectaries and high nectar flow rates, which facilitates metabolomic studies. *B. rapa* can be a useful alternative to *Arabidopsis thaliana* not only for the size of its nectaries and the volumes of nectar secreted, but also because it is highly dependent on pollinators. First efforts to discover the genes involved in nectar development on this plant have already been made (Hampton et al. 2010). Regarding extrafloral nectaries, a member of the poplar genus, *Populus trichocarpa*, has been fully sequenced (Tuskan et al. 2006) and several tools are available (e.g., commercial microarrays, bioinformatic tools, mapman). Some characteristics making it useful for nectary/nectar research are its rapid growth, an easy vegetative propagation and tractability to *Agrobacterium*-mediated transformation.

Several research groups have initiated genomic projects over the recent years, projects that could help elucidate the molecular mechanisms involve in nectar biology (Table 1). The so-called next-generation sequencing methods (454 pyrosequencing, Ilumina Solexa, SOLiD) will provide a faster and more accurate option for genomic analyses. The possibility of affordable genomic studies and the advances in bioinformatics will promote solutions to address this gap in knowledge surrounding nectar specific genes. The "omics" approach has potential to provide answers to as yet unresolved questions: How is the phloem sap uploaded to the nectariferous tissue and how is the unidirectional transport ensured? How can this direction be reversed when nectaries start to re-absorb nectar? How is the pre-nectar modified to produce mature nectar, and where are the non-carbohydrate components of nectar synthesized? How is the "mature" nectar release on the apical side, and which transport processes are involved?

Future research should realize comparative biochemical analyses of nectar and phloem sap in order to identify nectar compounds that stem exclusively from the nectary itself. In general, the suggested use of large-scale untargeted attempts such as transcriptomics, proteomics and metabolomics will be most fruitful when

Table 1 Potential candidate species for genomic studies on nectaries

Name	Nectary	Scope	Taxon	Status	PI
Glycine max	Floral	Standard draft	Whole genome	Complete	Schutz J
Brassica rapa	Floral	Resequencing	Whole genome	Awaiting release	Rokhsar D
Salix purpurea	Floral	Standard draft	Whole genome	Assembly	Tuskan G
Eucalyptus sp.	Floral	Resequencing	Whole genome	Complete	Myburg A
Gossypium sp.	Extrafloral	Standard draft	Whole genome	Production	Paterson A
Phaseolus sp.	Extrafloral	Resequencing	Whole genome	Awaiting release	McClean P
Cedrus sp.	Extrafloral	EST	Transcriptome	Complete	Dean J

Source: JGI, DOE joint genome institute (http:\crwww.jgi.doe.gov)

following comparative strategies, using functionally similar nectaries of taxonomically unrelated species and functionally different nectaries within the same species. This strategy should enable an understanding of the general principles that underlie the synthesis of nectar components and the regulation of nectar secretion.

Acknowledgements We thank Frantisek Baluska for inviting this chapter, Harry T. Horner and Robert Thornburg and Rosanne Healy, Antonio Cisneros and Christian Kost for kindly providing photographs and Alejandro de León for designing Fig 3.

References

Adler LS (2000) The ecological significance of toxic nectar. Oikos 91:409–420

Adler LS, Irwin RE (2005) Ecological costs and benefits of defenses in nectar. Ecology 86:2968–2978

Adler LS, Wink M, Distl M, Lentz AJ (2006) Leaf herbivory and nutrients increase nectar alkaloids. Ecol Lett 9:960–967

Agthe C (1951) Über die physiologische Herkunft des Pflanzennektars. Ber schweiz Bot Ges 61:240–274

Arumugasamy K, Subramanian RB, Inamdar JA (1990) Structure, ontogeny and histochemistry of cyathial nectaries in *Euphorbia heterophylla* L. (Euphorbiaceae). Acta Soc Bot Polon 59:3–8

Baker HG (1977) Non-sugar chemical constituents of nectar. Apidologie 8:349–356

Baker HG, Baker I (1975) Studies of nectar-constitution and pollinator-plant coevolution. In: Gilbert F, Raven PH (eds) Coevolution of animals and plants. University of Texas Press, Austin, pp 100–140

Baker HG, Baker I (1982) Chemical constituents of nectar in relation to pollination mechanisms and phylogeny. In: Nitecki M (ed) Biochemical aspects of evolutionary biology. University of Chicago Press, Chicago, pp 131–171

Baker HG, Baker I (1983) A brief historical review of the chemistry of floral nectar. Columbia University Press, New York

Baker HG, Opler PA, Baker I (1978) A comparison of the amino acid complements of floral and extrafloral nectars. Bot Gaz 139:322–332

Belt T (1874) The naturalist in Nicaragua. J.M. Dent and Sons, London

Benner U, Schnepf E (1975) Die Morphologie der Nektarauscheidung bei Bromeliaceen: Beteiligung des Golgi-Apparates. Protoplasma 85:337–349

Bentley BL (1977) Extrafloral nectaries and protection by pugnacious bodyguards. Annu Rev Ecol Syst 8:407–427

Bernadello G (2007) A systematic survey of floral nectaries. In: Nicolson SW, Nepi M, Pacini E (eds) Nectaries and nectar. Springer, Dordrecht

Blüthgen N, Fiedler K (2004) Preferences for sugars and amino acids and their conditionality in a diverse nectar-feeding ant community. J Anim Ecol 73:155–166

Bonnier G (1878) Les nectaires Annales des Sciences Naturelles. Botanique 8:5–12

Bowman JL, Smyth DR (1999) CRABS CLAW, a gene that regulates carpel and nectary development in Arabidopsis, encodes a novel protein with zinc finger and helix-loop-helix domains. Development 126:2387–2396

Brandenburg A (2009) The effect of nectar reduction in *Petunia axillaris* on foraging and pollination behavior of nocturnal hawkmoths, observed in laboratory and field behavioral assays. PhD thesis. University of Neuchâtel, Neuchâtel, Switzerland, pp 153

Brandenburg A, Dell'Olivo A, Bshary R, Kuhlemeier C (2009) The sweetest thing: advances in nectar research. Curr Opin Plant Biol 12:486–490

Bronstein JL (2001) The exploitation of mutualisms. Ecol Lett 4:277–287
Buxbaum F (1927) Zur Frage des Eiweißgehaltes des Nektars. Planta (Berlin) 4:818–821
Carter C, Thornburg RW (2004) Is the nectar redox cycle a floral defense against microbial attack? Trends Plant Sci 9:320–324
Carter C, Graham R, Thornburg RW (1999) Nectarin I is a novel, soluble germin-like protein expressed in the nectar of *Nicotiana sp*. Plant Mol Biol 41:207–216
Carter C et al (2007) Tobacco nectaries express a novel NADPH oxidase implicated in the defense of floral reproductive tissues against microorganisms. Plant Physiol 143:389–399
Carter C et al (2006) A novel role for proline in plant floral nectars. Naturwissenschaften 93:72–79
Castellanos MC, Wilson P, Thomson JD (2002) Dynamic nectar replenishment in flowers of *Penstemon* (Scrophulariaceae). Am J Bot 89:111–118
Cawoy V, Kinet JM, Jacquemart AL (2008) Morphology of nectaries and biology of nectar production in the distylous species *Fagopyrum esculentum*. Ann Bot 102:675–684
Chamberlain SA, Holland JN (2009) Quantitative synthesis of context dependency in ant-plant protection mutualisms. Ecology 90:2384–2392
Corbet SA, Delfosse ES (1984) Honeybees and the nectar of *Echium plantagineum* L. in south-eastern Australia. Aust J Ecol 9:125–139
D'Amato F (1984) The role of polyploidy in reproductive organ tissue. In: Johri BM (ed) Embryology of Angiosperms. Springer, Berlin, Germany, pp 519–556
Darwin F (1876) On the glandular bodies on *Acacia sphaerocephala* and *Cecropia peltata* serving as food for ants. With an Appendix on the nectar-glands of the common brake fern, *Pteris Aquilina*. Bot J Linn Soc Lond 15:398–409
Davidson DW, McKey D (1993) The evolutionary ecology of symbiotic ant-plant relationships. J Hym Res 2:13–83
Davis AR (2003) Influence of elevated CO_2 and ultraviolet-B radiation levels on floral nectar production: a nectary-morphological perspective. Plant Syst Evol 238:169–181
Davis AR, Gunning BES (1992) The modified stomata of the floral nectary of *Vicia faba* L. 1. Development, anatomy and ultrastucture. Protoplasma 166:134–152
Davis AR, Gunning BES (1993) The modified stomata of the floral nectary of *Vicia faba* L. 3. Physiological aspects, including comparisons with foliar stomata. Bot Acta 106:241–253
Davis AR, Peterson RL, Shuel RW (1988) Vasculature and ultrastructure of the floral and stipular nectaries of *Vicia faba* (Leguminosae). Can J Bot 66:1435–1448
Davis AR, Pylatuik JD, Paradis JC, Low NH (1998) Nectar-carbohydrate production and composition vary in relation to nectary anatomy and location within individual flowers of several species of Brassicaceae. Planta 205:305–318
de la Barrera E, Nobel P (2004) Nectar: properties, floral aspects, and speculations on origin. Trends Plant Sci 9:65–69
Delpino F (1874) Rapporti tra insetti e nettari extranuziali nelle plante. Boll Soc Entomol ital 6:234–239
Delpino F (1886) Funzione mirmecofila nel regno vegetale. Mem R Accad Sci Bologna 4:215–323
Díaz-Castelazo C, Rico-Gray V, Ortega F, Angeles G (2005) Morphological and secretory characterization of extrafloral nectaries in plants of Coastal Veracruz, Mexico. Ann Bot 96:1175–1189
Doak P, Wagner D, Watson A (2007) Variable extrafloral nectary expression and its consequences in quaking aspen. Can J Bot 85:1–9
Durkee LT (1982) The floral and extra-floral nectaries of *Passiflora*. II. The extra-floral nectary. Am J Bot 69:1420–1428
Durkee LT, Gaal DJ, Reisner WH (1981) The floral and extra-floral nectaries of *Passiflora*.I. The floral nectary. Am J Bot 68:453–462
Elias TE, Rozich WR, Newcombe L (1975) The foliar and floral nectaries of *Turnera ulmifolia* L. Am J Bot 62:570–576
Elias TS (1983) Extrafloral nectaries: their structure and distribution. In: Bentley B, Elias TS (eds) The biology of nectaries. Columbia University Press, New York, NY, USA, pp 174–203

Fahn A (1979) Ultrastructure of nectaries in relation to nectar secretion. Am J Bot 66:977–985
Fahn A (1988) Secretory tissues in vascular plants. New Phytol 108:229–257
Fahn A, Rachmilevitz T (1970) Ultrastructure and nectar secretion in *Lonicera japonica*. In: Robson NKB, Cutler DF, Gregory M (eds) New research in plant anatomy. Academic, London, UK, pp 51–56
Fahn A, Rachmilevitz T (1979) Ultrastructure and nectar secretion in *Lonicera japonica*. In: Robson NKB, Cutler DF, Gregory M (eds) New research in plant anatomy. Academic, London, pp 51–56
Frey-Wyssling A, Häusermann E (1960) Deutung der gestaltlosen Nektarien. Ber schweiz Bot Ges 70:150–162
Frey-Wyssling A, Zimmermann M, Maurizio A (1954) Über den enzymatischen Zuckerumbau in Nektarien. Experientia 10:490–491
Gaffal KP, Friedrichs GJ, El-Gammal S (2007) Ultrastructural evidence for a dual function of the phloem and programmed cell death in the floral nectary of *Digitalis purpurea*. Ann Bot 99: 593–607
Gaffal KP, Heimler W, el-Gammal S (1998) The floral nectary of *Digitalis purpurea* L., structure and nectar secretion. Ann Bot 81:251–262
Gill FB (1988) Effects of nectar removal on nectar accumulation in flowers of *Heliconia imbricata* (Heliconiaceae). Biotropica 20:169–171
González-Teuber M, Heil M (2009a) Nectar chemistry is tailored for both attraction of mutualists and protection from exploiters. Plant Signal Behav 4:809–813
González-Teuber M, Heil M (2009b) The role of extrafloral nectar amino acids for the preferences of facultative and obligate ant mutualists. J Chem Ecol 35:459–468
González-Teuber M et al (2009) Pathogenesis-related proteins protect extrafloral nectar from microbial infestation. Plant J 58:464–473
González-Teuber M et al (2010) Glucanases and chitinases as causal agents in the protection of *Acacia* extrafloral nectar from infestation by phytopathogens. Plant Physiol 152:1705–1715
Hampton M et al (2010) Identification of differential gene expression in *Brassica rapa* nectaries through expressed sequence tag analysis. PLoS One 5:e8782
Heil M (2004) Induction of two indirect defences benefits Lima bean (*Phaseolus lunatus*, Fabaceae) in nature. J Ecol 92:527–536
Heil M (2007) Indirect defence – recent developments and open questions. In: Lüttge U, Beyschlag W, Murata J (eds) Progress in Botany, vol 69. Springer, Berlin, Heidelberg, New York, pp 360–395
Heil M (2008) Indirect defence via tritrophic interactions. New Phytol 178:41–61
Heil M (2009) Damaged-self recognition in plant herbivore defence. Trends Plant Sci 14:356–363
Heil M (2011) Nectar: generation, regulation and ecological functions. Trends Plant Sci 16: 191–200
Heil M, McKey D (2003) Protective ant-plant interactions as model systems in ecological and evolutionary research. Annu Rev Ecol Evol Syst 34:425–453
Heil M, Fiala B, Baumann B, Linsenmair KE (2000) Temporal, spatial and biotic variations in extrafloral nectar secretion by *Macaranga tanarius*. Funct Ecol 14:749–757
Heil M et al (2001) Extrafloral nectar production of the ant-associated plant, *Macaranga tanarius*, is an induced, indirect, defensive response elicited by jasmonic acid. Proc Natl Acad Sci USA 98:1083–1088
Heil M et al (2004) Evolutionary change from induced to constitutive expression of an indirect plant resistance. Nature 430:205–208
Heil M, Rattke J, Boland W (2005) Post-secretory hydrolysis of nectar sucrose and specialization in ant/plant mutualism. Science 308:560–563
Heil M et al (2009) Divergent investment strategies of *Acacia* myrmecophytes and the coexistence of mutualists and exploiters. Proc Natl Acad Sci USA 106:18091–18096
Herrera CM, Garcia IM, Perez R (2008) Invisible floral larcenies: microbial communities degrade floral nectar of bumble bee-pollinated plants. Ecology 89:2369–2376

Herrera CM, De Vega C, Canto A, Pozo MI (2009) Yeasts in floral nectar: a quantitative survey. Ann Bot 103:1415–1423

Heyneman A, Colwell R, Naeem S, Dobkin D (1991) Host plant discrimination: experiments with hummingbirds flower mites. In: Price PW, Lewinson T, Fernades G, Benson W (eds) Plant-animal interactions: evolutionary ecology in tropical and temperate regions. Wiley, New York, USA, pp 455–485

Hillwig MS et al (2010) Petunia nectar proteins have ribonuclease activity. J Exp Bot 61: 2951–2965

Horner H, Cervantes-Martinez T, Healy R, Palmer R (2003) Floral nectary development and structure in *Glycine max* (Leguminosae). Int J Plant Sci 164:675–690

Horner HT et al (2007) Amyloplast to chromoplast conversion in developing ornamental tobacco floral nectaries provides sugar for nectar and antioxidants for protection. Am J Bot 94:12–24

Huntzinger M, Karban R, Young TP, Palmer TM (2004) Relaxation of induced indirect defenses of acacias following exclusion of mammalian herbivores. Ecology 85:609–614

Irwin RE, Adler LS (2008) Nectar secondary compounds affect self-pollen transfer: Implications for female and male reproduction. Ecology 89:2207–2217

Ivanoff SS, Keitt GW (1941) Relations of nectar concentration to growth of *Erwinia amylovora* and fire blight infection of apple and pear blossoms. J Agric Res 62:0733–0743

Johnson SD, Nicolson SW (2008) Evolutionary associations between nectar properties and specificity in bird pollination systems. Biol Lett 4:49–52

Jürgens G, Geldner N (2002) Protein secretion in plants: from the trans-Golgi network to the outer space. Traffic 3:605–613

Kaczorowski RL, Juenger TE, Holtsford TR (2008) Heritability and correlation structure of nectar and floral morphology traits in *Nicotiana alata*. Evolution 62:1738–1750

Keitt GW, Ivanoff SS (1941) Transmission of fire blight by bees and its relation to nectar concentration of apple and pear blossoms. J Agric Res 62:0745–0753

Kessler D, Baldwin IT (2007) Making sense of nectar scents: the effects of nectar secondary metabolites on floral visitors of *Nicotiana attenuata*. Plant J 49:840–854

Kessler D, Gase K, Baldwin IT (2008) Field experiments with transformed plants reveal the sense of floral scents. Science 321:1200–1202

Kirchoff BK, Kennedy H (1985) Foliar, non-structural nectaries in the *Marantaceae*. Can J Bot 63:1785–1788

Kobayashi S, Asai T, Fujimoto Y, Kohshima S (2008) Anti-herbivore structures of *Paulownia tomentosa*: morphology, distribution, chemical constituents and changes during shoot and leaf development. Ann Bot 101:1035–1047

Kohler A, Leseigneur CDC, Verburgt L, Nicolson SW (2010) Dilute bird nectars: viscosity constrains food intake by licking in a sunbird. Am J Physiol-Regul Integr Comp Physiol 299: R1068–R1074

Koptur S, Lawton JH (1988) Interactions among vetches bearing extrafloral nectaries, their biotic protective agents, and herbivores. Ecology 69:278–283

Kost C, Heil M (2005) Increased availability of extrafloral nectar reduces herbivory in Lima bean plants (*Phaseolus lunatus*, Fabaceae). Basic Appl Ecol 6:237–248

Kram BW, Carter CJ (2009) Arabidopsis thaliana as a model for functional nectary analysis. Sexual Plant Reprod 22:235–246

Kram BW, Bainbridge EA, Perera M, Carter C (2008) Identification, cloning and characterization of a GDSL lipase secreted into the nectar of *Jacaranda mimosifolia*. Plant Mol Biol 68: 173–183

Kram BW, Xu WW, Carter CJ (2009) Uncovering the *Arabidopsis thaliana* nectary transcriptome: investigation of differential gene expression in floral nectariferous tissues. BMC Plant Biol 9:92

Kromer T, Kessler M, Lohaus G, Schmidt-Lebuhn AN (2008) Nectar sugar composition and concentration in relation to pollination syndromes in Bromeliaceae. Plant Biol 10:502–511

Kronestedt-Robards EC, Robards AW, Strak M, Olesen P (1986) Development of trichomes in the *Abutilon* nectary gland. Nordic J Bot 6:627–639
Kuo J, Pate JS (1985) The extrafloral nectaries of cowpea (*Vicia unguiculata* (L) Wapp). 1. Morphology, anatomy and fine-structure. Planta 166:15–27
Lach L, Hobbs RJ, Majer JD (2009) Herbivory-induced extrafloral nectar increases native and invasive ant worker survival. Popul Ecol 51:237–243
Lalonde S et al (2003) Phloem loading and unloading of sugars and amino acids. Plant Cell Environ 26:37–56
Lara C, Ornelas JF (2002) Effects of nectar theft by flower mites on hummingbird behavior and the reproductive success of their host plant, *Moussonia deppeana* (Gesneriaceae). Oikos 96:470–480
Limburg DD, Rosenheim JA (2001) Extrafloral nectar consumption and its influence on survival and development of an omnivorous predator, larval *Chrysoperla plorabunda* (Neuroptera: Chrysopidae). Environ Entomol 30:595–604
Liu GY, Ren G, Guirgis A, Thornburg RW (2009) The MYB305 transcription factor regulates expression of nectarin genes in the ornamental tobacco floral nectary. Plant Cell 21:2672–2687
Loreti E, De Bellis L, Alpi A, Perata P (2001) Why and how do plant cells sense sugars? Ann Bot 88:803–812
Lüttge U (1961) Über die Zusammensetzung des Nektars und den Mechanismus seiner Sekretion I. Planta 56:189–212
Maloof JE, Inouye DW (2000) Are nectar robbers cheaters or mutualists? Ecology 81:2651–2661
Martínez del Rio C (1990) Dietary, phylogenetic, and ecological correlates of intestinal sucrase and maltase activity in birds. Physiol Zool 63:987–1011
Matile P (1965) Über den Stoffwechsel und die Auxinabhängigkeit der Nektarsekretion. Ber schweiz Bot Ges 66
McKim SM et al (2008) The BLADE-ON-PETIOLE genes are essential for abscission zone formation in Arabidopsis. Development 135:1537–1546
Mitchell RJ (2004) Heritability of nectar traits: why do we know so little? Ecology 85:1527–1533
Mondor EB, Addicott JF (2003) Conspicuous extra-floral nectaries are inducible in *Vicia faba*. Ecol Lett 6:495–497
Mondor EB, Tremblay MN, Messing RH (2006) Extrafloral nectary phenotypic plasticity is damage and resource-dependent in *Vicia faba*. Biol Lett 2:583–585
Neiland MRM, Wilcock CC (1998) Fruit set, nectar reward, and rarity in the Orchidaceae. Am J Bot 85:1657–1671
Nepi M, Stpiczyńska M (2007) Nectar resorption and translocation in *Cucurbita pepo* L. and *Platanthera chlorantha* Custer (Rchb.). Plant Biol 9:93–100
Nepi M, Stpiczyńska M (2008) The complexity of nectar: secretion and resorption dynamically regulate nectar features. Naturwissenschaften 95:177–184
Nepi M, Ciampolini F, Pacini E (1996) Development and ultrastructure of *Cucurbita pepo* nectaries of male flowers. Ann Bot 78:95–104
Nepi M, Guarnieri M, Pacini E (2001) Nectar secretion, reabsorption, and sugar composition in male and female flowers of *Cucurbita pepo*. Int J Plant Sci 162:353–358
Nepi M, et al. (2007) Dynamics of nectar: new insights from *Cucurbita pepo* flowers. In: Ninth International Pollination Symposium on Plant-Pollinator Relationships. pp 34–35
Nepi M et al (2009) Nectar and pollination drops: how different are they? Ann Bot 104:205–219
Ness JH (2006) A mutualism's indirect costs: the most aggressive plant bodyguards also deter pollinators. Oikos 113:506–514
Nicolson SW (2007) Amino acid concentrations in the nectars of southern African bird-pollinated flowers, especially *Aloe* and *Erythrina*. J Chem Ecol 33:1707–1720
Nicolson SW, Nepi M, Pacini E (2007) Nectaries and nectar. Springer, Doordrecht, The Netherlands
Nieuwenhuis-von Uexküll-Güldenband M (1906) Extraflorale Zuckerausscheidungen und Ameisenschutz. Annales du Jardin botanique de Buitenzorg, Ser 2 6:195–328

Pacini E, Nepi M (2007) Nectar production and presentation. In: Nicolson SW, Nepi M, Pacini E (eds) Nectaries and nectar. Springer, Dordrecht, pp 167–205

Pacini E, Nepi M, Vesprini JL (2003) Nectar biodiversity: a short review. Plant Syst Evol 238: 7–21

Palmer TM et al (2008) Breakdown of an ant-plant mutualism follows the loss of large herbivores from an African savanna. Science 319:192–195

Park S, Thornburg RW (2009) Biochemistry of nectar proteins. J Plant Biol 52:27–34

Pate JS, Peoples MB, Storer PJ, Atkins CA (1985) The extrafloral nectaries of cowpea (*Vigna unguiculata* (L.) Walp.) II. Nectar composition, origin of nectar solutes, and nectary functioning. Planta 166:28–38

Patt JM, Pfannenstiel RS (2008) Odor-based recognition of nectar in cursorial spiders. Entomol Exp Applic 127:64–71

Pederson MW, Lefevre CW, Wiebe HH (1958) Absorption of C^{14} labelled sucrose by alfalfa nectaries. Science 127:758–759

Peng YB et al (2004) Nectar production and transportation in the nectaries of the female *Cucumis sativus* L. flower during anthesis. Protoplasma 224:71–78

Petanidou T (2007) Ecological and evolutionary aspects of floral nectars in Mediterranean habitats. In: Nicolson SW, Nepi M, Pacini E (eds) Nectaries and nectar. Springer, Dordrecht, pp 343–375

Petanidou T, Goethals V, Smets E (2000) Nectary structure of Labiatae in relation to their nectar secretion and characteristics in a Mediterranean shrub community-does flowering time matter? Plant Syst Evol 225:103–118

Petanidou T, Van Laere A, Ellis WN, Smets E (2006) What shapes amino acid and sugar composition in Mediterranean floral nectars? Oikos 115:155–169

Peumans WJ et al (1997) Lectin and alliinase are the predominant proteins in nectar from leek (*Allium porrum* L.) flowers. Planta 201:298–302

Potter CF, Bertin RI (1988) Amino acids in artificial nectar: feeding preferences of the flesh fly *Sarcophaga bullata*. Am Midl Nat 120:156–162

Pulice CE, Packer AA (2008) Simulated herbivory induces extrafloral nectary production in *Prunus avium*. Funct Ecol 22:801–807

Pyke GH (1991) What does it cost a plant to produce floral nectar? Nature 350:58–59

Radhika V et al (2008) Testing the optimal defence hypothesis for two indirect defences: extrafloral nectar and volatile organic compounds. Planta 228:449–457

Radhika V, Kost C, Boland W, Martin Heil M (2010) The role of jasmonate signalling in floral nectar secretion. PLoS One 5:e9265

Raguso RA (2004) Why are some floral nectars scented? Ecology 85:1486–1494

Razem FA, Davis AR (1999) Anatomical and ultrastructural changes of the floral nectar of *Pisum sativum* L. during flower development. Protoplasma 206:57–72

Ren G et al (2007a) Expression of starch metabolic genes in the developing nectaries of ornamental tobacco plants. Plant Sci 173:621–637

Ren G et al (2007b) Transient starch metabolism in ornamental tobacco floral nectaries regulates nectar composition and release. Plant Sci 173:277–290

Rogers WE, Siemann E, Lankau RA (2003) Damage induced production of extrafloral nectaries in native and invasive seedlings of Chinese tallow tree (*Sapium sebiferum*). Am Midl Nat 149: 413–417

Roitsch T (1999) Source-sink regulation by sugar and stress. Curr Opin Plant Biol 2:198–206

Rolland F, Baena-Gonzalez E, Sheen J (2006) Sugar sensing and signaling in plants: conserved and novel mechanisms. Annu Rev Plant Biol 57:675–709

Röse USR, Lewis J, Tumlinson JH (2006) Extrafloral nectar from cotton (*Gossypium hirsutum*) as a food source for parasitic wasps. Funct Ecol 20:67–74

Rudgers JA (2004) Enemies of herbivores can shape plant traits: selection in a facultative ant-plant mutualism. Ecology 85:192–205

Ruhlmann JM, Kram BW, Carter CJ (2010) Cell wall invertase 4 is required for nectar production in *Arabidopsis*. J Exp Bot 61:395–404

Sawidis T, Eleftheriou EP, Tsekos I (1987) The floral nectaries of *Hibiscus rosasinensis* I. Development of the secretory hairs. Ann Bot 59:643–652

Sawidis T, Eleftheriou EP, Tsekos I (1989) The floral nectaries of *Hibiscus rosasinensis* III. A morphometric and ultrastructural approach. Nord J Bot 9:63–71

Schmid R (1988) Reproductive versus extra-reproductive nectaries – historical perspective and terminological recommendations. Bot Rev 54:179–232

Smets EF, Ronse Decraene LP, Caris P, Rudall PJ (2000) Floral nectaries in monocotyledons: distribution and evolution. In: Wilson KL, Morrison DA (eds) Monocots: systematics and evolution. CSIRO Publishing, Collingwood, pp 230–240

Sobrinho TG, Schoereder JH, Rodrigues LL, Collevatti RG (2002) Ant visitation (Hymenoptera: Formicidae) to extrafloral nectaries increases seed set and seed viability in the tropical weed *Triumfetta semitriloba*. Sociobiology 39:353–368

Southwick EE (1984) Photosynthate allocation to floral nectar: a neglected energy investment. Ecology 65:1775–1779

Stephenson AG (1982) Iridoid gylcosides in the nectar of *Catalpa specisoa* are unpalatable to nectar thieves. J Chem Ecol 8:1025–1034

Stpiczyńska M (2003a) Floral longevity and nectar secretion of *Platanthera chlorantha* (Custer) Rchb. (Orchidaceae). Ann Bot 92:191–197

Stpiczyńska M (2003b) Nectar resorption in the spur of *Platanthera chlorantha* Custer (Rchb.) Orchidaceae – structural and microautoradiographic study. Plant Syst Evol 238:119–126

Stpiczyńska M, Milanesi C, Faleri C, Cresti M (2005) Ultrastructure of the nectar spur of *Platanthera chlorantha* (Custer) Rchb. (Orchidaceae) during successive stages of nectar secretion. Acta Biologica Cracoviensia Series Botanica 47:111–119

Sugiura S, Abe T, Makino S (2006) Loss of extrafloral nectary on an oceanic island plant and its consequences for herbivory. Am J Bot 93:491–495

Thornburg RW (2007) Molecular biology of the *Nicotiana* floral nectary. In: Nicolson SW, Nepi M, Pacini E (eds) Nectaries and nectar. Springer, Heidelberg, pp 265–287

Tilman D (1978) Cherries, ants and tent caterpillars: timing of nectar production in relation to susceptibility of caterpillars to ant predation. Ecology 59:686–692

Tuskan GA et al (2006) The genome of black cottonwood, *Populus trichocarpa* (Torr. & Gray). Science 313:1596–1604

Vassilyev AE (2010) On the mechanisms of nectar secretion: revisited. Ann Bot 105:349–354

Vesprini JL, Nepi M, Ciampolini F, Pacini E (2008) Holocrine secretion and cytoplasmic content of *Helleborus foetidus* L. (Ranunculaceae) nectar. Plant Biol 10:268–271

Vogel S (1997) Remarkable nectaries: structure, ecology, organophyletic perspectives.I. Substitutive nectaries. Flora 192:305–333

Vogel S (1998) Remarkable nectaries: structure, ecology, organophyletic perspectives - II. Nectarioles. Flora 193:1–29

von Czarnowski C (1952) Untersuchungen zur Frage der Nektarabsonderung. Arch Geflügelzucht Kleintierk 1:23–44

von Wettstein R (1889) Über die Compositen der österreichisch-ungarischen Flora mit zuckerabscheidenden Hüllschuppen. Sitzungsberichte der Kaiserlichen Akademie der Wissenschaften in Wien. Mathematisch-naturwissenschaftliche Classe 97:570–589

Wäckers FL, Wunderlin R (1999) Induction of cotton extrafloral nectar production in response to herbivory does not require a herbivore-specific elicitor. Entomol Exp Applic 91:149–154

Wagner RE et al (2007) Proteomic evaluation of gymnosperm pollination drop proteins indicates highly conserved and complex biological functions. Sex Plant Reprod 20:181–189

Weiss M (2001) Vision and learning in some neglected pollinators. In: Chittka L, Thomson JD (eds) Cognitive ecology of pollination, animal behavior and floral evolution. Cambridge University Press, Cambridge, pp 171–190

Wenzler M, Holscher D, Oerther T, Schneider B (2008) Nectar formation and floral nectary anatomy of *Anigozanthos flavidus*: a combined magnetic resonance imaging and spectroscopy study. J Exp Bot 59:3425–3434
Wist TJ, Davis AR (2006) Floral nectar production and nectary anatomy and ultrastructure of *Echinacea purpurea* (Asteraceae). Ann Bot 97:177–193
Wooley SC et al (2007) Extrafloral nectaries in aspen (*Populus tremuloides*): heritable genetic variation and herbivore-induced expression. Ann Bot 100:1337–1346
Yamawo A, Hada Y (2010) Effects of light on direct and indirect defences against herbivores of young plants of *Mallotus japonicus* demonstrate a trade-off between two indirect defence traits. Ann Bot 106:143–148
Zhu J, Hu ZH (2002) Cytological studies on the development of sieve element and floral nectary tissue in *Arabidopsis thaliana*. Acta Botanica Sinica 44:9–14
Ziegler H, Lüttge UE (1959) Über die resorption von C^{14} Glutaminsäure durch sezernierende Nektarien. Naturwissenschaften 46:176–177
Zimmermann J (1932) Über die extrafloralen Nektarien der Angiosperm. Beihefte Botanisches Centralblatt 49:99–196
Zimmermann M (1953) Paperchromatographische Untersuchungen über die pflanzliche Zuckersekretion. Ber schweiz Bot Ges 63:402–429
Zimmermann M (1954) Über die Sekretion saccharosespaltender Transglucosidasen im pflanzlichen Nektar. Experientia 15:145–146

Secretion in the Diatoms

Charlotte Aumeier and Diedrik Menzel

Abstract Diatoms, a class of heterokont algae, are among the most important primary producers. They may live attached to the ground as part of the benthos or buoyant in the oceans, in brackish water and in freshwater habitats. Their ability to stay in the water column as well as their propensity to adhere to structures is a result of secretion, even locomotary activity of pennate diatoms is a consequence of secretion. Diatoms make a cell wall of glass-like silicate. This feature, which sets diatoms apart from most other algae, also entails secretory activity. Secretion in diatoms has been studied by biochemical and immunological means, by histochemistry, light and electron microscopy and by physical methods. Recently, the genomes of two diatom species, one of the order Centrales and the other of the order Pennales, have been sequenced and transformation techniques have become available for the establishment of transgenic lines in some species. This has given diatom research an additional push forward, so that not surprisingly, tremendous progress has been made, particularly in the field of silica polycondensation and frustule morphogenesis. In trying to bring together classical biochemical, cell biological and structural work with modern molecular studies, the role of secretory processes in the diatoms is illuminated with particular emphasis on cell wall morphogenesis and locomotion.

1 Introduction

Diatoms (Bacillariophyceae) are unicellular, yellow brown algae, taxonomically grouped together with Xanthophyceae, Chrysophyceae and Phaeophyceae in the class Heterokontophyta. Diatoms either live in the water column as part of the

C. Aumeier • D. Menzel (✉)
Institute of Cellular and Molecular Botany, University of Bonn, Kirschallee 1, 53115 Bonn, Germany
e-mail: dmenzel@uni-bonn.de

J.M. Vivanco and F. Baluška (eds.), *Secretions and Exudates in Biological Systems*, Signaling and Communication in Plants 12, DOI 10.1007/978-3-642-23047-9_10,

phytoplankton or are attached to submerged surfaces as part of the benthos in marine, brackish and freshwater habitats around the world. Species diversity in this class of algae is very high coming up to 100,000 taxa.

Benthic diatoms may be perceived by the layman as a slimy, brownish layer on basically all surfaces exposed to water. Such biofilms constitute a special habitat of its own containing many other organisms such as bacteria, protists and lower metazoans (Cooksey and Wigglesworth-Cooksey 1995). Biofilms also cover the hulls of ships causing a constant problem worldwide to ship line owners, because more fuel is needed to compensate for the mounting drag that the growing layer of organisms on the ship hulls is creating. This phenomenon is also known as biofouling (Cooksey and Wigglesworth-Cooksey 1995).

Diatoms are pioneer species. Central to the diatom's ability to populate virgin surfaces is secretion of adhesive material also referred to as extracellular polymeric substances (EPS). It is therefore of great interest to know, what the composition of this material is, how it is secreted and how diatoms make use of secretory processes for the conditioning of their environment, for competition with their neighbours and defence against predators and for their adhesion to the substratum. Last but not least secretion is the diatom's unique ways to move along the surface of the substratum.

Diatoms engage in yet another type of activity, hardly recognized as a consequence of secretion. This is the intracellular "fabrication" and final delivery to the cell surface of their siliceous wall casing. This chapter will deal with these two major types of secretion, mucilage production including adhesive EPS and the formation of siliceous walls in the diatoms. For earlier reviews on this subject see Hoagland et al. (1993), Pickett-Heaps et al. (1990), Schmid (1994), Kröger and Poulsen (2008), Li and Volcani (1985), Zurzolo and Bowler (2001) to mention just a few; others will be cited throughout the text.

2 Biochemical and Structural Aspects of Diatom Secretion

2.1 *Extracellular Polymeric Substances and Soft Mucilage*

Even though it is not obvious by light microscopic examination, almost all diatoms secrete mucous material, which may spread over its silica cell surface thus forming a continuous organic coating. This surface coating is different from the adhesive polymer extruded through the raphe slit in pennate diatoms (see Sect. 4.2), or through polar pore fields in both, pennate and centric diatoms. In physical terms, the organic coating is easily extracted by hot water, it is essentially non-adhesive and so soft and movable, that it may gradually detach from the cell surface and get lost into the environment over time. In fact it can be readily removed from the cell surface by the scanning movement on the micrometer scale of a sharply pointed probe of an atomic force microscope (AFM). This is, in principle, an extremely fine and tapered mechanical instrument, which is made to move in successive lines over the surface of an object, while it is being held at a very narrow distance from this surface. A diatom cell wall scanned this way appears virtually wiped clean of mucilage, when monitored after some rounds of scanning at lower magnification. This was beautifully demonstrated by Crawford and coworkers (2001; see Fig. 1). The authors observed that such "cleaned" areas stay clean for as long as two hours, indicating a complete lack of elasticity and adhesiveness of the coating material.

The non-adhesive mucilage is secreted through pore fields in the silica shell. By AFM, Higgins et al. (2002) were able to image the actual act of secretion. After the cell surface in the girdle region of *Craspedostauros australis* had been "wiped clean" of mucilage, by scanning this area in so-called contact-mode (similar to what is shown in Fig. 1, Crawford et al. 2001), new material was seen to ooze out from the pores, when the same region was scanned again a little later. This material is high in sugar content including acidic and neutral hexoses, pentoses, 6-deoxyhexoses, *N*-acetylaminosugars and/or O-methylated sugars with a usually high sulphate ester content (Hoagland et al. 1993; Wustman et al. 1997, 1998; McConville et al. 1999; Chiovitti al. 2003). In addition, it almost always has a protein component associated with it, possibly constituting a proteoglycan or glycoprotein. The mucilage material secreted through the raphe and through the polar pore fields has an about 20 times higher adhesiveness (Lind et al. 1997). Therefore, the scanning probe of the AFM shows a completely different type of

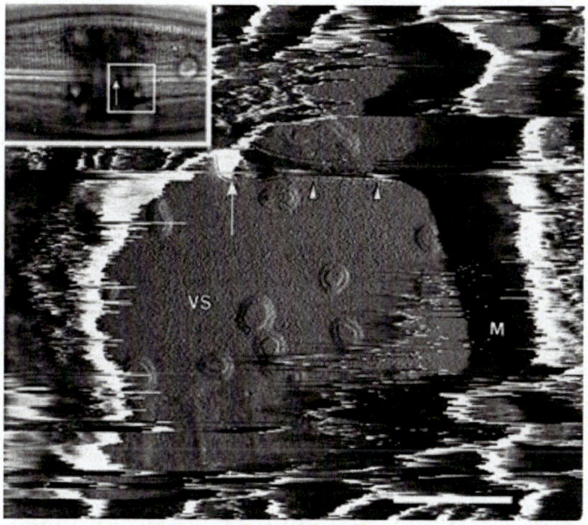

Fig. 1 *Cell surface structure of live Pinnularia viridis recorded by atomic force microscopy.* The scanned cell area is indicated in the *upper left* inset showing a cell from a related diatom species (Midian and Aumeier, unpublished data). In the centre of the main image, the finely granulated silica surface of the diatom frustule is visible, whereas the rest of the cell is covered under a thick layer of mucilage. This peculiar situation is caused by a high resolution scan, by which the central 20 square microns were swept clean, followed by a low resolution scan. The *arrow* points at the raphe terminal close to the central occlusion. Its position is more readily seen in the inset. The raphe slit is extending towards the *right* (*arrowheads*). The round, knob-like structures on the surface are ornamentations in the silica framework of the valve. Scale = 3 μm, modified from Crawford et al. (2001)

interaction (see also Chap. 4). Due to its sticky nature, the adhesive mucilage would not become swept away, but is rather smeared over the cell surface by the moving probe (Higgins et al. 2002).

Synthesis of EPS, of course, requires that diatoms contain the essential enzymes to link sugars to proteins, i.e., forming glycosidic bondages with N- or O-residues of the polypeptide backbone. Despite the overwhelming evidence for secretory activity in diatoms, little is known about the types of post-translational modifications and intracellular trafficking of proteins in diatoms. Only recently, essential parts of the N-glycosylation pathway have been reconstructed applying bioinformatic tools to the *Phaeodactylum* genome database (Baïet et al. 2011). According to this study the major enzymes initiating and modifying N-glycosylation in the ER and Golgi are present in *Phaedactylum* and also in *Fragilariopsis cylindrus* and *Thallasiosira pseudonana*, indicating that this metabolic pathway is indeed present in diatoms.

In the higher eukaryotes, the processing of N-linked glycans into complex oligosaccharide has been regarded as important factors for multicellular morphogenesis. Despite the fact that diatoms are capable of forming all sorts of differently shaped colonies, they essentially remain unicellular; therefore, this type of post-translational modification must have a different function. Baïet and coworkers suggest that it may be seen in relation to pleomorphism and adaptation to stress (Baïet et al. 2011).

Diatom EPS is the major cause of biofilm formation on solid surfaces; it is even an important factor modifying the surface of ice. For instance, the ice-dwelling centric diatom species, *Melosira arctica*, was shown to change the surface structure of ice by secreting copious amounts of EPS consisting of carbohydrate polymers as well as glycoproteins. In fact, by secreting this material, these algae produced a microporous surface on solid ice obviously transforming it into a habitable zone to other organisms (Krembs et al. 2010).

2.2 Secretion of Chrysolaminarin

Polysaccharide secretion is not only targeted to the cell surface but also to the vacuole. The dominant chemical species of storage carbohydrate is a type of 1,3-β-D-glucan, called chrysolaminarin, which differs from the phaeophytan laminarin by the lack of a terminal mannitol residue at the reducing end of the sugar chain (Beattie et al. 1961). Since this material escapes inadvertently into EPS fractions obtained from cell extracts for chemical analysis, i.e., after extracting live cells with warm water, and then also adheres to the cell surface as seen by histochemical staining, chrysolaminarin has originally been mistaken for a bona fide EPS component. However, Chiovitti and co-workers have shown that this is not so. Using cryofixation-freeze substitution methods followed by immunocytochemical staining with anti-β-1,3-glucan antibody, they unambiguously showed that chrysolaminarin is exclusively located in the vacuole (Chiovitti et al. 2004). Unfortunately, nothing is known yet as to how chrysolaminarin is targeted and delivered to the vacuole. It is likely that the Golgi is the major compartment responsible for the synthesis of matrix polysaccharides, however, it might just be the glycosyltransferase inserted into the membrane, which is targeted via the Golgi to the vacuolar compartment and that major synthesis takes place at the tonoplast of the mature vacuole.

2.3 Secretion of Chitin Fibres

Chitin is yet another type of polysaccharide secreted by some diatoms. Notably *Thalassiosira fluviatilis* (McLachlan et al. 1965) and *Cyclotella cryptica* (Herth 1978) have been studied in detail, which spin out tiny fibres made of chitin (*N*-acetylglucosamine in β-1,4-linkage) from pores around the edges of the cylindrical frustules (Fig. 2). *Thalassiosira* and some other species extrude thick composite chitin ropes from a pore field positioned in the middle of the valve, also called strutted processes or fultoportulae. The chitin ropes span between two recently divided daughter cells and keep them connected, creating long, flexible cell chains, which float in the water column (Blackwell et al. 1967).

Fine structural evidence indicates that the chitin fibres are made from a special membrane patch or grove underneath each of the pores in the frustule of pennates as well as centrics and underneath the central labiate or strutted process in some of the centric diatoms. The membrane in this grove is unusually thick in ultra structural cross-sections and has a high electron contrast indicating a high density of trans-membrane proteins. Just like it is well known for cellulose, chitin is not secreted by exocytosis. It is synthesized by a transmembrane enzyme complex, which spins out the glucan chains directly into the extracellular space, where they arrange into collinear, paracrystalline bundles, termed microfibrils. Herth (1978) has speculated that a significant part of the dense membrane region in the groves underneath the pores it made of the chitin synthase complexes. This speculation has received further credibility recently by a genomic study describing the occurrence of six genes encoding three different types of chitin synthases (Durkin et al. 2009). Obviously chitin-fibre production from pores is just a very special manifestation

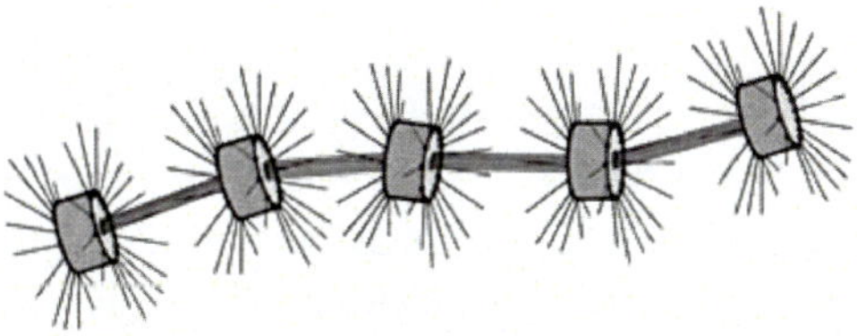

Fig. 2 *Secretion of chitin fibres in a floating chain of Thalassiosira cells, schematic drawing.* Chitin fibres emanate from the rims of these centric diatoms. Thicker ropes of chitin may be formed from a central pore field (termed strutted processes) on both valves, so that recently divided daughter cells remain connected with one another (see Pickett-Heaps et al. 2003 for a stunning video sequence showing cell division and rope formation)

of chitin synthesis in the diatoms. A genome comparison with the pennate diatom *Phaeodactylum* clearly identifies two candidates for chitin synthases, even though this diatom species and many others do not produce chitin fibres in the shape of spines and ropes as shown above for *Thalassiosira*. So there must be a more general role for chitin. All along it has been suspected that chitin might be part of the organic matrix of the silica frustule (see below). This has now been nicely shown by Brunner et al. (2009), who have been able to visualize a residual organic network mostly composed of cross-linked chitin fibres after the siliceous frustule of *Thalassiosira pseudonana* had been dissolved with NH_4F.

In addition, stress imposed on diatoms by nutrient starvation or by heavy metal exposure appears to induce chitin deposition in the girdle region (Durkin et al. 2009 and references cited there). This additional chitin deposition could serve as kind of reinforcement, when silica content is low. Alongside with chitin synthases, centric diatoms possess genes coding for chitin modifying and metabolizing enzymes such as chitinases and chitin-binding proteins (Durkin et al. 2009). One of these proteins is p150, which has three potential N-glycosylation sites and three chitin-binding domains (Davis et al. 2005). It appears in the girdle band region in copper-stressed cells, further strengthening the hypothesis, that it is part of a protection strategy under stress conditions.

One special issue awaits clarification in the future: There are two chitin synthase genes in the *Thalassiosira* genome coding for what appears to be a chimeric protein consisting of a myosin motor head at the N-terminus and chitin synthase domain for the rest of the molecule. This particular configuration of a chitin synthase is not unprecedented. The *Aspergillus nidulans* genome codes for two genes, CsmA and CsmB, which have exactly this type of chimeric construction, i.e., N-terminal myosin head and C-terminal chitin synthase (Horiuchi et al. 1999). Tsuizaki et al. (2009) showed that the products of both genes have important roles in polarized cell wall synthesis and maintenance of cell wall integrity. Moreover, Takeshita et al. (2005) demonstrated that the chitin synthase is guided to the proper location in the cell via the interaction of its myosin head with the underlying cellular actin cytoskeleton. Because of the striking similarity in the chimeric construction of these two chitin synthase isoforms between fungi and diatoms, a comparable function can be anticipated for the two diatom proteins as well.

2.4 Secretory Activities Associated with Frustule Formation

In abiotic nature, silicate (SiO_2) occurs as crystalline quartz and derivatives of it mixed with other minerals. In the biosphere, however, it occurs in amorphous form due to the fact that organic molecules such as small proteins, peptides and polymeric forms of unusual amino acids are associated with it. Spicules (fine needles in sponges), composite endoskeletons (silicoflagellates, sea urchin larvae), box-like exoskeletons (diatoms), are all made of this hybrid material. Amorphous silicate is also encountered as incrustations in chitinic insect mandibles and special types of plant cell walls (Coradin et al. 2004). Hence the formation of such silicate-containing material is more than "just" the consequence of silica sequestration (which would be remarkable enough), it is the consequence of the orchestrated secretion of all the necessary biomolecules required for guided polycondensation of silicic acid.

In diatoms we see the unique case, that the siliceous shell (frustule), which consists of the two valves plus the connecting girdle bands (Fig. 3), is intricately sculptured with pores, ridges and extensions. Each of these wall subcomponents

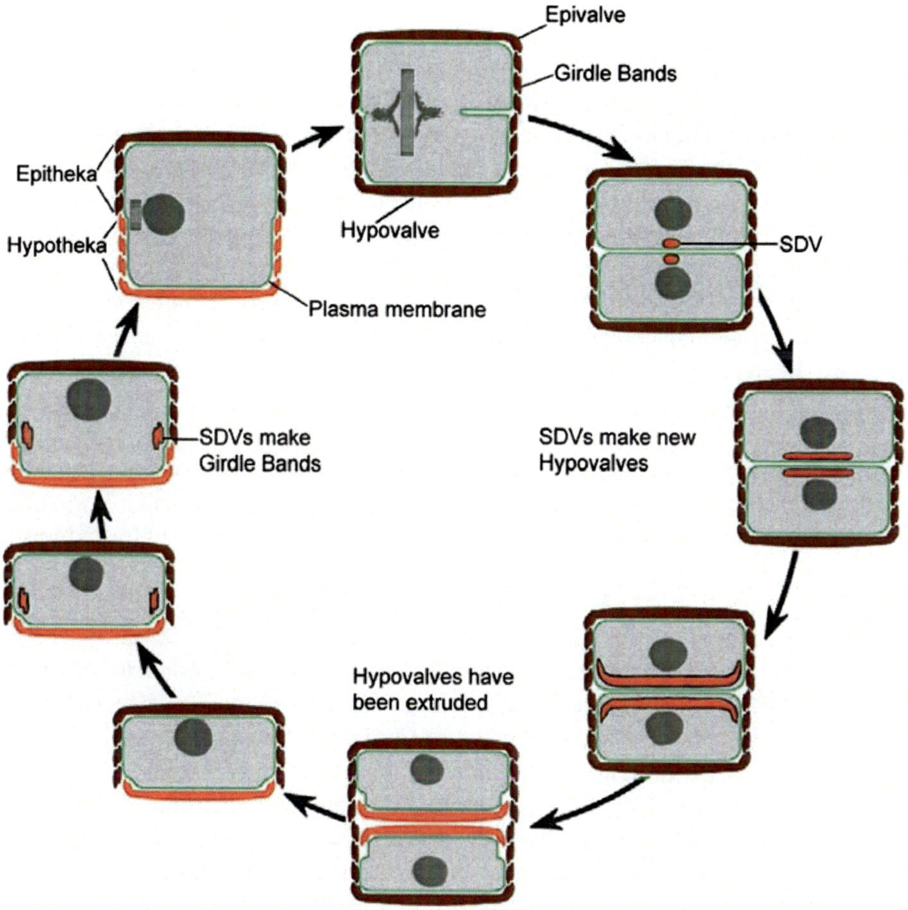

Fig. 3 *Making of the diatom frustule, shown here in cross-section through consecutive stages of the division cycle*. The frustule consists of epitheca and hypotheca fitting to each other like the *upper* and *lower* lid of a petri dish. Each daughter cell forms a new hypotheca (valve plus girdle bands) to be combined with the existing theca. The valves are made first. An SDV appears right after cytokinesis at centre position in each daughter cell, expands to span the entire width of the cell and becomes extruded to the cell surface by a giant, single step of exocytosis. Girdle bands are added one by one by the same mechanism to increase girth of the cells. Modified after Kröger (2007)

(i.e., upper and lower valve and connecting girdle bands) is sequentially fabricated, each in one piece by a shaping mechanism inside the so-called "silica depositing vesicle" (SDV) and then eventually deposited on the cell surface, each in one piece by a single act of exocytosis (see for instance Li and Volcani 1985; Pickett-Heaps et al. 1990 for comprehensive reviews).

2.4.1 Uptake and Utilization of Silicate

The key questions as to how silica enters the cell and eventually reaches the SDV are still not entirely resolved. Fresh water has a concentration of 100 μM monosilicic acid, sea water about 10–70 μM (DeMaster 2001). Above a threshold of 2 mM, silicic acid polymerizes and takes on the form of colloidal particles. So in order to manufacture a cell wall made of silica particles, a concentration mechanism should be operating in the cell, unless silica is taken up in the polymeric state as suggested by Gordon and Drum (1994). These authors proposed that silica is internalized into vesicles by receptor-coupled endocytosis, which would then be fed into the endocytic recycling pathway and redirected to their final target, the SDV, either directly or via the transgolgi system. Integral to this pathway would be an acidification of the vesicle lumen, which would help to resolubilize polymeric forms of silica and prevent any premature polymerization while in the transport phase. Interestingly, exactly that happens, when Germanium is present, and this is probably, why Germanium is a potent inhibitor of silification in diatoms (Lewin 1966; Azam 1974).

Alternatively, the uptake of silicic acid from the environment occurs through the plasma membrane either by carriers or by ionohores. By employing a low concentration of Silicium and Germanium isotopes, the uptake of the soluble acid form of both minerals in diatoms was proven a long time ago (Azam 1974). Sullivan and coworkers have observed silica transport in vitro into membrane vesicles isolated from *Nitzschia alba* (summarized in Sullivan and Volcani 1981).

In more recent times, a transcriptome study on *Phaeodactylum tricornutum* has shown that previously identified proteins, the silicic acid transporters (SIT-proteins), which have been implicated in silicic acid transport across membranes, constitute a larger protein family counting about 26 members (Sapriel et al. 2009; Thamatrakoln et al. 2006). So far nothing is known about their intracellular localization. The options are that they sit in the plasma membrane as well as in the silicalemma, which would have the consequence that silicic acid must pass through the cytoplasma, before it can be taken up by another set of SIT-proteins into the SDV. Or silicic acid gets concentrated in endocytic vesicles, which are eventually delivered to the SDV. In that case additional components would be required to keep silicic acid from polycondensating in the transport vesicles. Sumper and Brunner (2008) summarized the evidence for the hypothesis that in the intracellular stores silica exists in a polyamine-stabilized prepolymerized state in quantities large enough to allow the rapid formation of the complete hypovalve in a matter of not more than 15 min. The exact nature of these internal stores is not yet known, even though preliminary data point at a vesicular membrane fraction. It is also not yet understood, how the stored silica is mobilized and transported into the SDV. It would be a reasonable assumption that these processes are associated with the late endosomal compartment.

2.4.2 Biogenic Precipitation of Silicate in the SDV

The diatom frustule does not have the smooth surface appearance such as glass; rather it appears as a finely granulated material (Fig. 1) which, however, differs in detail from species to species. This structural feature is consistent with the assumption stated above that the frustule is made not by a crystallization process but by compacting silica nanoparticles (Crawford et al. 2001).

It has been speculated all along that silica is deposited as an organic-mineral hybrid material because the organic component has been revealed as discrete structural frameworks after hydrofluoric acid extraction (Reimann et al. 1966; Reimann and Volcani 1968; Robinson and Sullivan 1987). There has been a tremendous progress in the characterization of possible protein candidates involved in the process that is mostly confined to the lumen of the SDV (exceptions are the pleuralines, see below).

Major molecular players in the polycondensation process taking place in the SDV are the Silaffins, small peptides enriched in serine and lysine residues. Silaffins contain an N-terminal ER-import sequence, which is cleaved off during protein maturation, suggesting that these proteins arrive in the SDV through the secretory pathway (Poulsen and Kroger 2004). The amino groups of the lysines in these peptides are modified by long-chain polyamines and the hydroxy groups of all serines are phosphorylated (Kröger et al. 1999). As has been recently shown, phosphorylation takes place in the lumen of the ER and is catalysed by a novel kinase (Sheppard et al. 2010). As an additional component, free methylated polyamines are mixed in with these peptides derived from two alternative biosynthetic pathways either based on spermidine or 1,3-diaminopropane, respectively, and the latter may be modified with a terminally located quaternary ammonium group. The heterogeneity of the polyamine composition appears to be species-specific and one of the causes for the diversity of the frustule nanostructure. An essential component influencing the size of silica nanoparticles is a recently discovered strange peptide, coined silacidin, which is extremely rich in aspartic and glutamic acid and in phosphorylated serine. In vitro, silica precipitation is enhanced by the presence of phosphate alone; however, the poly-phosphorylated silacidin is remarkably more potent in that respect, by two to three orders of magnitude (Sumper et al. 2005).

A separate group of silaffin-like proteins, the cingulins, has very recently been discovered in the girdle bands of *Thalassiosira pseudonana* by Scheffel et al. (2011). GFP-fused silaffin 3 is found in the valves as expected, whereas the six known cingulines are restricted to the girdle bands, some of them to just one or two girdle bands, others to the entire cingulus. Cingulins, apparently, are capable of forming a prepatterned organic matrix in the girdle bands. This matrix cannot be removed by NH_4F-based extraction procedures as is the case for the regular silaffins, indicating a more intimate association of cingulins with the silica structure. It is also not associated with the chitin-containing scaffold, which has recently been imaged in extracted girdle bands (Brunner et al. 2009), because chitin can be removed by chitinase, whereas the cingulin matrix stays behind. Interestingly, polycondensation of silica on this cingulin-containing organic matrix can be achieved in vitro, when silicic acid plus a synthetic polyamine are added to it (Scheffel et al. 2011). So the hypothesis of an organic matrix as a nanostructured precursor for silica formation has been substantiated at least for the cingulins in the girdle bands.

Frustulins are yet another type of cell wall proteins contributing to the organic coating on the silica surface of the frustule. They are secreted into the extracellular matrix only after the frustule has been completed and released. They bind Ca^{2+}, most likely with their cystein-rich domains (ACR, van de Poll et al. 1999). Their function may be seen as kind of a protection against chemical etching of the frustule silica surface. Finally, Pleuralins, formally described as HEP (Kröger et al. 1997) are another protein component exclusively associated with the pleural bands, which form at the very last step, after all girdle bands are in place. The pleural bands may be perceived as sticky tapes, which hold epi- and hypotheca together. The pleuralins are secreted directly into the extracellular space, where they combine with the freshly extruded pleural bands (Kröger and Wetherbee 2000). It is conceivable that the pleuralins serve as the sticky coating on the surface of the pleural bands (Kröger and Poulsen 2008).

2.4.3 The SDV, a Giant Secretory Vesicle

The question as to how the siliceous wall of diatoms is laid down on the surface of the cell has intrigued diatom researchers early on. Transmission electron microscope image quality of diatom fine structure was still lacking some resolution at that time, which has made it difficult to precisely locate the membranes along the silica deposits. However, it was Reimann (1964), who proposed that the forming new valve was completely encased by membrane, even though there had been no concept yet, as to how the silica shell would be transported from the lumen of the membranous compartment, which has been termed SDV (Drum and Pankratz 1964), to the surface of the cell. In recent times, however, the remarkable feature is undisputed, that the newly formed siliceous valve remains wrapped in the silica lemma until completion and is then extruded to the cell surface in one giant step of exocytosis (Fig. 3). Considering this and the marvellously complex substructure and the shear size of the SDV, within which a complete valve is fabricated, the term "valve secreting machinery" as used by Pickett-Heaps (1983; see also Pickett-Heaps et al. 1990) seems to be most fitting to describe the nature of this organelle.

Exocytosis of each of the daughter cell hyovalves from the giant SDVs poses a serious problem, i.e., were does all the membrane go? At the moment, when the silica lemma and the plasma membrane fuse, so that the new hypovalve comes to lie outside of the cell, the membrane surface area would instantaneously double, unless the surplus of membrane is rapidly retrieved. If there were a burst of compensatory endocytosis, for example, the thousands of coated vesicles required to retrieve all of this membrane material would have been sighted in the extensive ultrastructural image material available so far. One interesting model assumes that surplus of membrane is being "reeled back" into the cell body at the proximal edge of the valve, however, the mechanism of internalization remains unclear (e.g., see Pickett-Heaps et al. 1990; Schmid 1994). Alternatively, the silicalemma fuses only at the margin of the SDV with the plasma membrane so that the entire silicalemma/plasma membrane sandwich covering the distal face of the new valve would get discarded so that the proximal area of the silicalemma becomes the new plasma membrane (see also Zurzolo and Bowler 2001). Obviously, nature has taken care in an as yet unknown way; otherwise we would see massive, convoluted membrane configurations piling up at this developmental stage in this region of the cell.

2.4.4 Frustule Sculpturing and Morphogenesis in the SDV

Looking at the intricate patterns of diatom frustules leaves the observer puzzled as to how this pattern is realized on the micrometer scale. Robinson and Sullivan (1987) have summarized some of the early hypotheses trying to explain the mechanism. Among them is the organic template model, which assumes that nucleation proteins located within or on the silica lemma lay down a template, which determines the shape of silicate deposition in the lumen of the SDV on the micrometer scale. This hypothesis has clearly been supported in the case of the pleural bands, where cingulins adopt a prepattern on to which silica is deposited (see above, Sect. 2.4.2). In the case of valve sculpturing this hypothesis is gradually losing grounds, whereas another type of shaping process is being more and more favoured. Fine structural observations have identified sub-membrane scaffold structures associated with the silicalemma. Even though these structures are of unknown composition, they are conspicuously appearing in regions, where a prominent surface structure is being sculptured in the frustule. A particular striking example is the hollow spine extending away from the middle of both valves in *Ditylum brightwellii*. Formation of this so-called Labiate Process (LP) occurs right after completion of cytokinesis and involves a scaffolding structure attached to the cytoplasmic face of the silicalemma, the Labiate Process Apparatus (LPA, Li and Volcani 1985). The particular location of this submembrane scaffold structure in the centre of the newly forming valve, where actually no silica deposition is visible in the adjacent lumen of the SDV, suggests that it may have a function as a placeholder obstructing silica deposition in this area of the SDV. Structures like the LPA should be suitable for anchoring conventional cytoskeletal elements on the cytoplasmic side of the silica lemma, i.e., microtubules and actin filaments, to exert force, which will result in shaping the contour of the silica lemma and in turn will influence the overall shape of the SDV. The central role of cytoskeletal elements in micrometer scale frustule morphogenesis (as opposed to nano-scale morphogenesis) has indeed been confirmed by Cohn et al. (1989, see also Pollock and Pickett-Heaps 2005) and recently by Tesson and Hildebrand (2010).

Another not less impressive example of a multipart scaffolding structure assembling along the cytoplasmic face of the silicalemma was described by Pickett-Heaps et al. (1994) at the extending ends of growing spines of the centric diatom *Chaetoceros peruvianus*. The molecular nature of this structure is yet completely unknown. It is a remarkable observation that post-cytokinetic valve formation involving secretion of component parts and manufacturing of the entire valve in the SDV and its extrusion/exocytosis onto the surface will be faithfully repeated, once the process has been interrupted. Pollock and Pickett-Heaps (2005) have demonstrated this for the centric diatom *Ditylum* by artificially separating the freshly completed valve from the cell body through plasmolysis. By an as yet unknown mechanism, both daughter cells recognize the loss of its hypovalve and immediately start replacing it with a new one.

The role of spacer vesicles or alveolar vacuoles in the complex moulding process of the silica lemma has been emphasized by Schmid (1994), who has compared wall morphogenesis between several diatoms of both the centrales and pennales. There is as yet little known of the actual release process of the valve after its manufacturing is completed. Schmid (1994) has pointed out that before release of the valve, organic coat material is exocytosed to mediate adhesion between the valve and the plasmalemma surface and subsequently the SDV expands. As the sculptured frustule separates from the cytoplasmic face, the negative imprint of its rims and knobs remains for some time as deep indentations in the cytoplasmic surface. This indicates that the surface is actively moulded. The enigmatic moment of fusion, which precedes delivery into the extracellular space has evaded observation so far (see Sect. 2.4.3).

3 Ecological Aspects of Diatom Secretion

It may not be immediately obvious, how the ability of diatoms to secrete a frustule composed of silica could possibly affect the world's climate. However, considering that diatoms together with other microalgae form large seasonal blooms in the world's oceans (Bowler et al. 2010), it is becoming clear that the shear scale of silicon and carbon fixed in biomass is overwhelming. Diatoms are estimated to contribute about 40% to the oceanic primary productivity (see Rabosky and Sorhannus 2009; and citations therein).

Survival of diatoms in the pelagic environment requires special strategies. One is the evolution of comparatively large cell bodies protected by a sturdy, siliceous cell wall. Second, these cells become decorated with sharp spikes and long spiny appendages made either of silicate setae or chitin fibrils (see Sect. 2.3). All this makes diatoms a bulky food item for predators, which have to be able to crack the glass frustules and somehow get rid of the waste; and then they have to be content with the relatively small amount of digestible cytoplasm per cell. When certain prey species such as *Thalassiosira* and its predator, the copepode *Calanus helgolandicus* are tested in vitro, it has been shown that the diatom adapts to heavy grazing by increasing the silica content in its frustule (Pondaven et al. 2007).

Many pelagic diatoms have the tendency to secrete adhesive materials at polar cell regions, which in the case of the centric diatoms such as *Biddulphia* species allows daughter cells to adhere to each other at the polar ends after cell division, thus forming long chains of many tens of individuals. As already shown above (Sect. 2.3), some other diatom species achieve the same result by chitin ropes between daughter cells. In the case of pennate diatoms, adherence between neighbouring cells is achieved by the extrusion of sticky material through the raphe slit, so that cells adhere to one another at their long sides. One particularly impressive example, where adherence and motility is combined, is *Bacillaria paradoxa*. This unique combination of features results in a chain of individuals, which constantly changes its size and shape like the unfolding and refolding of a Carpenter's rule (http://www.microscopy-uk.org.uk/mag//art98/bacill. html).

As a consequence of these adaptations and in contrast to other algae occurring in blooms, planktonic diatoms are not as much grazed by large predators such as copepods. Instead, their blooms are generally terminated by nutrient exhaustion, such as Nitrogen, Iron and Silicate. Due to the fact that many of the planktonic diatoms tend to aggregate even more as populations begin to die out, their bodies sink down fast in the water column and take with them silicate and carbon (Smetacek 1999). Fossil records of sea bed sediments reveal fluctuations worldwide in the scale of this process and the dynamics of species diversity evolution over geologic ages (Rabosky and Sorhannus 2009). Since Carbon and Silicon are removed from the environment for millions of years on a global scale, diatoms must indeed be considered major factors in the feedback mechanisms that govern world climate (deMaster 2001).

3.1 Secretion of Small Organic Compounds by Planktonic Diatoms

Another prevalent nature of pelagic diatoms is the secretion of water-soluble biomolecules into the environment, which turns out to be of importance for diatom survival. As in essentially all other cases of biogenic molecules released into the environment, this inevitably means that the producer of such kind of allelopathic material interacts in one way or the other with competing organisms in the same habitat or even controls entirely his environment (Vasconcelos and Leal 2008).

A good example are the short chain polyunsaturated aldehydes, collectively termed oxylipins, which are made by diatoms in order to deter copepode grazers and at the same time control the growth of bacterial communities during the time of diatom blooms (Leflaive and Ten-Hage 2009). How these substances are secreted, is essentially unknown. The chemical nature of these substances indicates that they are secreted directly through the plasma membrane by an eccrine mechanism. However, since many diatoms accumulate oily inclusion bodies, an alternative could be that they are secreted internally into these internal stores and become only liberated during grazing attack (Wichard et al. 2007).

3.2 Biofilm Formation by Benthic Diatoms

Virtually all surfaces of the aquatic habitats are covered with a layer/film of microorganisms. Diatoms are always a major component in these biofilms as long as light reaches down to these habitats. Man-made structures made of wood, concrete, metal or plastic surfaces submersed in water tend to age with time and deteriorate under the influence of microbial biofilms growing on them. This phenomenon has been termed biofouling (Molino and Wetherbee 2008).

The cause of biofilm formation is the production and secretion of polymeric substances (EPS) by the members in this microbial community. It has been shown that lose sediments become stabilized as a consequence of EPS secretion. Apparently, however, the amount of extracellular polymer was less important than the degree of cross-linking within the extracellular polymer network (de Brouwer et al. 2005). For the diatoms, these substances serve important functions. Some diatom taxa produce sticky patches (see Fig. 4), which enable them to attach to the substratum, thereby preventing them from being swept away by water currents (Wigglesworth-Cooksey et al. 2001). This should, in principle, be the same process as observed in the pelagic diatoms forming cell chains by excreting sticky pads between them (see above, Sect. 3).

Bacteria occurring in the same habitat contribute to surface film formation by secreting their own material and by interacting with the secretion of diatoms. For instance, bacteria may decrease diatom adhesion (Wigglesworth et al. 2001) or they may have a beneficial effect on diatom secretion. As reported by Bruckner et al.

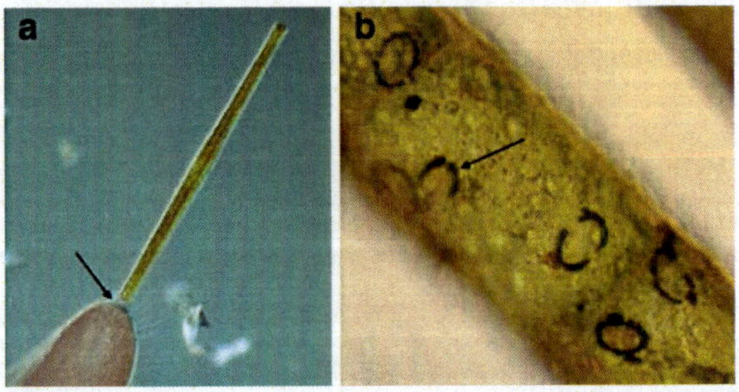

Fig. 4 *Epiphytic diatoms collected from the rocky reef of Helgoland, Germany.* (**a**) A single *Toxarium undulatum* cell has attached itself with an adhesive pad formed from one cell pole on the outermost tip of a red algal thallus. (**b**) Typical foot pads left on the surface of filamentous green alga by diatoms; staining with Hämatoxilin. Menzel and Mettbach unpublished

(2008), the growth of diatoms in biofilms of large freshwater lakes was enhanced in the presence of bacteria, and its quality was changed. In addition to polymeric substances, low molecular weight biomolecules such as amino acids also play a role in regulating the activities in the microbial consortium (Bruckner et al. 2011).

This general principle of forming adhesive structures has become diversified during evolution of benthic diatoms in several ways. The adhesive patch may become elongated by prolonged production of adhesive material from just one end of the diatom cell, so that a stalk-like holdfast is formed. If several individuals are involved in the production of the holdfast, it might become more elaborate by branching (Fig. 5).

Another important function for secretion of adhesive material is cell motility. Extrusion of adhesive material through the raphe slit enables pennate diatoms to glide along the surface of the substratum (see Sect. 4.2). Some pennate taxa such as *Achnathes* can sequentially engage in such kind of motility and later settle down and form an adhesive stalk. This has the advantage of roaming around for some time in order to find a favourable spot and than attach to this spot and proliferate by cell division. When daughter cells remain attached to one another, a colony may be formed, from which individuals can split off, in response to environmental cues. Several examples of stalk-forming diatoms are beautifully demonstrated in the movie made by Pickett-Heaps and co-workers (Pickett-Heaps et al. 2003).

A very interesting modification of the holdfast has evolved in the tube-dwelling diatoms, which shape the holdfast as a hollow tube, often quite elaborate as is shown in Fig. 6. The holdfast is then becoming a kind of hollow tree, which for one thing lifts the colony above the level of the substratum, but on the other hand serves as a protective coating, within which the diatoms travel forwards and backwards by the hundreds (Houpt 1994). Biofilms consist of several, sometimes many, different species of diatoms living as epiphytes on top of each other. As the seasons of the

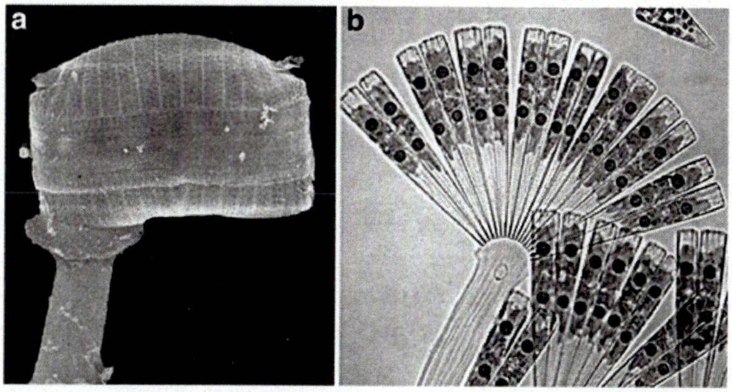

Fig. 5 (**a**) A scanning electon micrograph of an *Achnates longipes* cell sitting on its adhesive stalk that was produced by secretion of adhesive biopolymer from one end of the cell. From Roessler (2000), modified. (**b**) Fan forming pennate diatom, *Licmophora flabellata*. Many individuals have collectively formed a common stalk. This specimen was encountered as epiphyte on benthic marine algae collected from the rocky reef around Helgoland, Germany, Mettbach and Menzel (unpublished data)

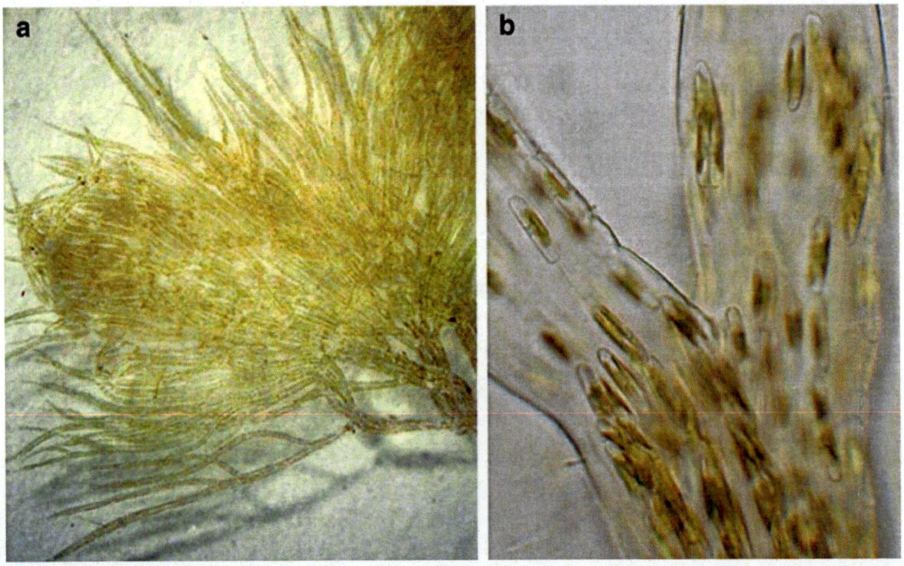

Fig. 6 *Tube-dwelling diatoms (Navicula) collected from the rocky reef of Helgoland, Germany.* (**a**) low magnification image showing the morphology of the tree-like assemblage of tubes. (**b**) Details of a tube branching point with diatoms, which were actively moving, when the picture was taken

year proceed, the marine coastal habitats, the river banks and the edges of a fresh or brackish water lake with all its plant and algal vegetation will become more and more covered with diatom populations competing for space, light and nutrition.

4 Secretion of Adhesive Biopolymers

Diatoms secrete different types of biopolymers, which by their chemical nature may be proteinaceous with a variable amount of carbohydrate side chains or they may be composed entirely of carbohydrates. In terms of their biophysical property they may be divided into two classes, i.e., adhesive or non-adhesive. Most the time diatoms produce both kinds at the same time. The non-adhesive type constitutes the organic covering (Sect. 2.1) that can be extracted by mild treatments and that is seen by AFM as a soft, several nm thick smear (Fig. 1). The adhesive type, which is secreted only from certain areas of the cell, either through pores at the cell poles or through the raphe slit, is also noticeable by AFM, however, as opposed to the first type, it adheres to the scanning probe requiring a considerably higher pulling force (3.5 versus 60 nN) to be retracted from it (Higgins et al. 2002). Gebeshuber et al. (2003) have pointed out that these adhesives have very interesting properties in the sense that they might lead the way for the design of a technical glue, which does not lose its adhesive properties under water.

4.1 Structural, Biochemical Aspects of Adhesive Stalks

The adhesive stalk of the monoraphid pennate diatom *Achnathes* (Figs. 5a and 7) is divided into three zones, the adhesive pad making contact with the substratum, the shaft of the stalk and the adhesive collar, which connects the stalk to the frustule at one polar end of the cell. The adhesive pad has a rather uniform, inconspicuous fine structure, whereas the stalk appears densely packed with fibrils oriented perpendicular to the long axis of the stalk. The fibrous material apparently converges in an electron-dense ribbon-like structure in the centre of the stalk (Wang et al. 1997).

Wustman et al. (1997) have shown that the stalk consists of a mixture of neutral and acidic polysaccharides, although there is some indication of glycoproteins. Determination of sugar residues in biochemical fractions isolated from the stalk material by various analytical methods (gas chromatography-mass spectrometry-analysis, ^{13}C-NMR Spectroscopy, IR-spectroscopy) and by immunofluorescence and lectin-binding revealed that alpha (1,2)-L-fucose units were present in the adhesive pads and outer layers of the shaft of *A. longipes* stalks, whereas the core region of the shaft contained sulfated galactosyl carbohydrates. The outer shaft

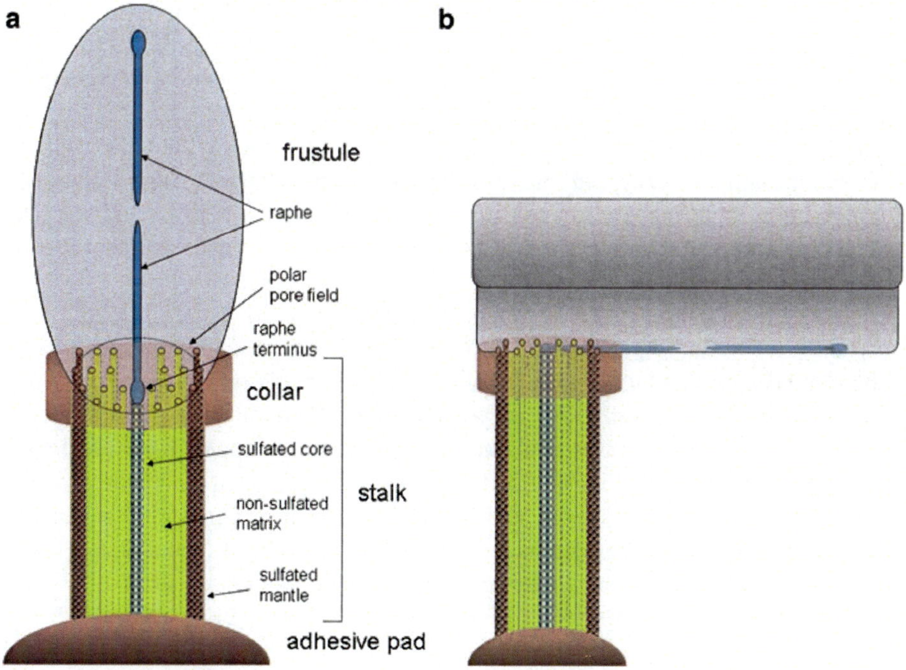

Fig. 7 *Structure and composition of adhesive stalks of Achnathes and related diatoms.* From Wustman et al. (1998), modified. Material constituting the sulphated core is very similar to that secreted through the raphe, when pennate diatoms engage in gliding motility. The frustule may sit head-on on the stalk or on an adhesive pad (**a**), or the stalk may be tilted by 90° from the long axis of the frustule (**b**). In this view the raphe slits and the pore field are obscured by the edge of the frustule. Their position is shown here for clarity

layers (mantle region) were also rich in terminal and 2-linked glucuronic acid, terminally linked fucose and nonsulfated D-galactose (Wustman et al. 1997, 1998). Neutral polysaccharides containing glucose, mannose and galactose residues were detected in all three compartments of the adhesive structure in *A. longipes*. Wustman et al. (1997) proposed that extracellular modification of the secreted material occurs, which could be due to cross-linking between polysaccharides or potentially also cross-linking between phenolic compound such as ferulic acid and dityrosine, which would require the activity of extracellular peroxidase in analogy to the higher plant cell wall. However, extracellular peroxidases have not been identified in diatoms. The presence of glucoronic acid rather suggests that Ca^{2+} could be a cross-linker (see below, Sect. 4.2).

In purified polymer fractions isolated from the stalk, Wustman et al. (1998) identified two mixed polysaccharides of the fucoglucuronogalactan-type with a molecular weight of 20,000,000 Mr and 100,000 Mr, respectively. The former had a high and the latter a low sulfate content. These polysaccharide moieties have a high affinity to a 10,000 Mr Mannose-containing protein and all three components may be O-glycosidically linked to one another forming a condensed proteoglycan assemblage. By immunocytochemical localization using monoclonal antibodies directed against fucosyl-residues, these authors showed that this high molecular weight proteoglycan material originated from the raphe terminus and contributed to both the central core region and the mantle region.

Stalk formation as well as gliding motility are inhibited by the cellulose synthase inhibitor 2,6-dichlorobenzonitrile (DCB), even though cellulose formation is not involved in either activity (Wang et al. 1997). The inhibitor does not disturb general metabolic and housekeeping function as demonstrated by the fact that cell divisions continue at a normal rate and daughter cells even adhere to one another after division leading to long chains of cells, only adhesive stalks do not form. The observation that an 18-kDa membrane protein of *Achnathes longipes* binds to a fluorescent analogue of DCB, supports the assumption that this inhibitor acts specifically by targeting just one protein. The function of this protein, however, has not been clarified until now. It has turned out recently that one molecular target of DCB in plants is a microtubule-associated protein, called MAP20 (Rajangam et al. 2008). This interesting finding points to microtubule-based mechanisms involved in the spatial control of carbohydrate biosynthesis in the diatoms, possibly beta 1,4 glycosyltransferases related to cellulose synthase. Obviously the control of localization by intracellular mechanisms to specific areas of the cell has a tremendous importance for polar secretion of adhesives in diatoms.

4.2 Physical and Molecular Properties of Adhesive Mucilage

Adhesiveness can be quantified using AFM by monitoring the force necessary to break connected adhesive threads as the probe is pulled back from the surface. Or vice versa, a diatom cell can be glued to the tip of a scanning probe and then this diatom cell, fixed in position, is touched down and allowed to form an adhesive connection to the surface of the substratum. The force that is required to break this connection can then be correlated with the adhesive strength of the material (Arce et al. 2004).

The unusual benthic, centric diatom *Toxarium undulatum* forms an adhesive pad at one cell pole by secreting adhesive mucilage through a polar pore field (Fig. 4). Strength of adhesion and recoiling behaviour of the adhesive nanofibres constituting this material has been studied by Dugdale et al. (2005) using AFM in the so-called fly-fishing mode, meaning that the scanning probe touches weakly on the surface of the adhesive pad, so that it picks up only the most adhesive component in it, often a single macromolecular fibre. The authors suggest that the behaviour of the adhesive fibre as it is breaking off from the AFM-cantilever in a series of uniformly sized small ruptures is consistent with a multimodal construction, i.e., a macromolecule consisting of regularly spaced units aligned with each other in register (Fig. 8).

This typical AFM fingerprint for a modular construction of an adhesive nanofibre most likely has its origin in a repetitive polypeptide pattern, where tightly folded domains are followed in a regular fashion by non-structured loops. Each time the AFM probe measures a peak, another folded domain has become stretched out (Figs. 8 and 9). When the cantilever of the scanning probe was lowered back, so as to relax the attached mucilage fibre, the fibre gets refolded. This had the effect that the same amount of force was necessary to stretch it again to the previous level. These data indicate that adhesive pads of diatoms are composed of multimodal nanofibres, which can stretch and recoil depending on the traction force that they are subjected to. Protease treatment of attached fibres weakens the recoiling activity and causes faster breakage, indicating that a significant portion of the modular fibre must be proteinaceous (Dugdale et al. 2005). The same force–distance relationship has been observed in nanofibres exuded from the raphe (see Sect. 4.3).

Ca^{2+} can be an important ionic regulator controlling the strength of adhesiveness in the diatom mucilage. This has been shown in a follow-up study on the same diatom species, again employing AFM-mediated stretching. Adding EGTA while adhesive nanofibres were pulled from the adhesive pads with the scanning probe

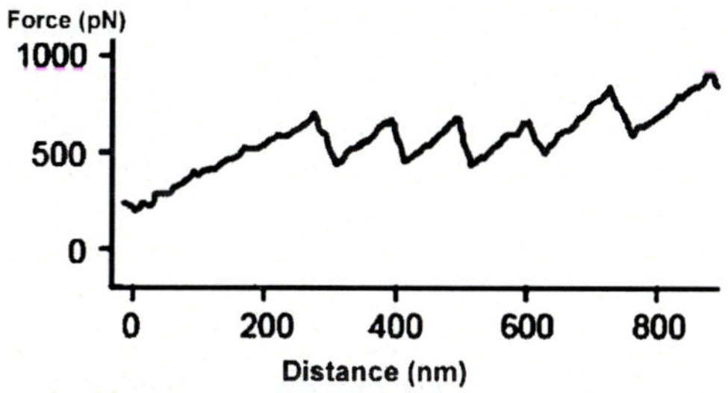

Fig. 8 *Typical saw tooth signature of a modular nanofibre*. Stretching of the adhesive fibre is more or less linear during the first 200–300 nm before maximal load is reached, then the first module unfolds leading to sudden relaxation, this process repeats with each module. In the case of *Toxarium undulatum*, about 27 modules comprise one adhesive fibre (Dugdale et al. 2005). This graph is redrawn with modification from Molino and Wetherbee (2008)

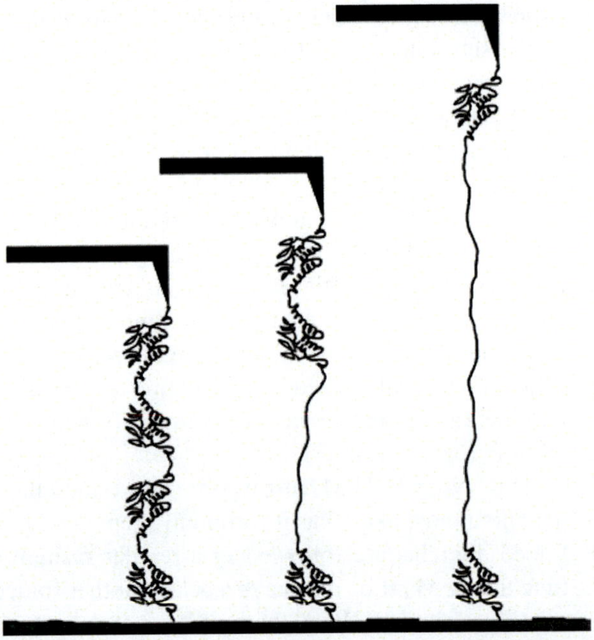

Fig. 9 *Idealized sketch of stepwise unfolding of a modular adhesive nanofibre under constant tensile force applied through the adhering tip of the AFM probe.* Unfolding of a peptide module results in a sudden increase in distance. Redrawn with modification from Molino and Wetherbee (2008)

caused rapid breakage (Chiovitti et al. 2008). Cohesiveness was restored when Ca^{++} was added back to the medium. The finding that treatment of attached cells with EDTA would eventually cause them to separate from their adhesive pads, opened the way to isolate and purify adhesive pads and biochemically analyse their composition. It was found that this material contained a single, high molecular weight (>220 kDa) sulphated glycoprotein. Amino acids prevalent in the protein backbone of this macromolecular species were glycine (22 mol%), aspartic acid/aspartamine (14 mol%) and histidine (11 mol%). Fourier transform infrared (FTIR) spectroscopic analysis indicated a high content of β-sheet conformation in the protein component, which is a conformation suitable for extensible proteins (Tatham and Shewry 2000). Small amount of cystein (5 mol%) and tyrosine (2 mol%) would potentially allow for either disulfide- or isodityrosine bridges between the protein backbones. However, the authors favour the model that connections between adjacent macromolecules in the adhesive nanofibre is due to ionic bridges of Ca^{2+} between the anionic sugar acid residues, and that during stretching these bridges break and reform as the glycoprotein modules slip past one another.

4.3 *Extrusion of Mucilage Through the Raphe*

Using an antibody directed against the major cell surface proteoglycan of the marine raphid diatom *Stauroneis decipiens*, Lind et al. (1997) have shown that this material is exuded through the raphe and deposited in the adhesive trails.

Furthermore, the antibody that was used for immunolabelling could actually inhibit the ability of the diatoms to adhere to the substratum and stopped the gliding motion, when added to live cells. Apparently the antibody is neutralizing the adhesive properties of the proteoglycan.

The monoraphid species, *Achnathes longipes*, is capable of roaming around by gliding motility for some time, before it settles down and exudes an adhesive stalk from one polar end of the cell, which will lift it up from the level of the substratum. For this scenario, Wustman et al. (1998) have shown that the same high molecular weight proteoglycan material present in the central core of the stalk is apparently also exuded from the raphe, when the diatom engages in gliding motility.

At a much more detailed molecular level, Dugdale et al. (2006) came to the same conclusion when comparing the biophysical properties of adhesive nanofibres obtained from adhesives pads of the marine, benthic diatom *Toxarium undulatum* with those obtained from the raphe adhesives of the motile stage of *Phaeodactylum tricornutum*. They proposed that the adhesive material in both cases is composed of modular proteinaceous macromolecules aligned with each other in register.

The adhesive material is highly glycosylated, therefore Alcian blue decoration and the PIA-TCH-Ag method (Post sectioning periodic acid-thiocarbohydrazide-silver proteinate staining) are both useful methods for their detection on the ultrastructural level. Using these methods, Edgar and Pickett-Heaps (1982) have shown that this material is present in small secretory vesicles originating from the Golgi compartment. These vesicles occur in copious numbers throughout the cell, but are particularly prominent alongside the raphe slit and in association with the actin bundles. The conclusion from this work is that the luminal content of these vesicles is deposited into the lumen of the raphe slit by secretion.

4.3.1 The Raphe Proteoglycan

The material that is exuded from the raphe was analysed biochemically as well as immunologically. After the protoplast of the marine raphid diatom *Stauroneis decipiens* had been removed from pelleted cells with detergent containing lysis buffer, four major frustule-associated proteoglycans (87 kDa, 112 kDa, $2\times >200$ kDa) have been extracted by urea and separated on SDS-PAGE. Apart from possible non-covalent interactions between the carbohydrate moieties of these four components, they are apparently capable of forming disulfide-bridges between them under natural condition, which can be broken by the addition of reducing agents (Lind et al. 1997). Monoclonal antibodies raised by these authors against the four components were used for immunofluorescence and immunogold labelling. The results showed that any of the four components was present in the trail, in the raphe and on the cell surface, which is consistent with the finding stated above that they combine by disulfide-bondage into one large macromolecular assemblage.

In a live cell assay, the epitopes of the proteoglycan assemblage exuded from the raphe could be saturated to a level that attachment to the glass surface, along which the diatoms were moving, was strongly reduced with the effect that motility in half of the cells stopped and was slowed down in the rest of the cells (Lind et al. 1997). The observation that although the epitopes recognized by the monoclonal antibodies were present in the raphe and trail as well as on the cell surface, only the raphe material was adhesive, whereas the cell surface was not, was explained by an as-yet undetermined modification in the raphe material. It is conceivable that another component co-secreted through the raphe specifically modifies the proteoglycan material at the moment of release).

4.3.2 Secretion-Based Mechanism of Gliding

Evidence for a causal link between gliding motility and secretion came from the early observation that pennate diatoms leave a trail of mucilage material as they glide on the surface of the substratum. This has been shown in many ways using fluorescence microscopy of lectin binding (Wigglesworth-Cooksey and Cooksey 2005), scanning electron microscopy (SEM) and light microscopic techniques (Edgar 1983; Edgar and Pickett-Heaps 1984). Trail formation can be demonstrated over large surface area using the dye "stains-all". Lind et al. (1997) have used silica microspheres added to a diatom sample. In a matter of a few seconds, these spheres adhered to the raphe and as the diatoms were beginning to move, the spheres collected into a band, which was deposited on the surface behind the moving diatom. Molino and Wetherbee (2008) reported that the marine *Amphora coffeaeformis* within 20 h deposit trails to the extent that the entire glass surface is completely covered with adhesive material.

The raphe adhering to the substratum is termed "driving raphe" (Fig. 10, Molino and Wetherbee 2008). The raphe on the opposite side, though not producing any driving force, is not entirely inactive. Microspheres attaching to it are seen to be moved forwards and backwards in an erratic fashion along both halves of the raphe separated by the central occlusion (see also Pickett-Heaps et al. 2003). This suggests that the motility mechanism may be capable of sensing mechanical load, i.e., as mucilage threads are becoming attached to the substratum, their movement is becoming synchronized (see Edgar and Pickett-Heaps 1984 for summary). Molino and Wetherbee (2008) have described a striking behaviour of pennate diatoms, illustrating the effect of load on the regulation of the motility apparatus. Usually the cells face the substratum with one of their valvar sides, so that the

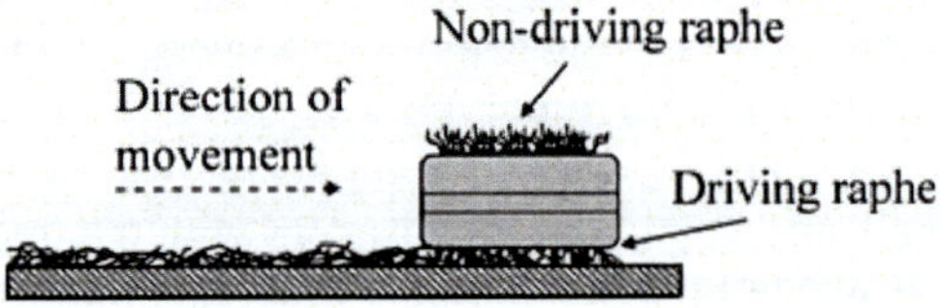

Fig. 10 *Deposition of an adhesive trail by a raphid pennate diatom as it is moving along the substratum*. The driving raphe is engaged in mucilage production, whereas the non-driving raphe is idle. Modified after Molino and Wetherbee (2008)

mucilage strands, secreted through the raphe slit, can make contact with the substratum. However, when cells happen to fall on the girdle side, none of the valvar sides has contact to the substratum. In this situation, adhesive material is secreted through the raphe, often from just one point along the slits, from both valvar sides, until one of the adhesive threads produced this way eventually makes contact with the substratum. At that moment motility sets in and the cell pulls itself towards the substratum until it tilts over. If both adhesive strands from both valvar sides happen to make contact to the substratum at the same time, a tug of war sets in, until one of the strands breaks, i.e., force increases as load increases. This is a remarkable example of regulation by mechanoperception, and there is currently not even a hypothesis as to how this is achieved in the diatoms.

Material exuding from the raphe can be observed by SEM as well as transmission electron microscopy (Edgar 1983). On the basis of these early observations, Edgar and Pickett-Heaps (1984) have put forward a mechanistic model as to how secretion could be an integral part of the gliding motility process in pennate diatoms. Instrumental in this model is the fact that two actin cables run along the raphe in parallel orientation. These are interpreted as intracellular rail ways, along which secretory vesicles move, driven by myosin molecules attached to their membrane surface. The model proposes further that once the secretory vesicle has fused with the plasma membrane in the vicinity of the raphe slit, the motor molecules remain attached to the patch of membrane, which is now a component part of the plasma membrane, and continue to move along the actin bundles, while the adhesive material released into the raphe slit remains attached to the outside of the membrane patch that originally constituted the inner vesicle membrane face. The model requires that the adhesive material swells upon contact with the water medium, and the increase in volume drives it through the raphe slit and brings it into contact with the substratum. So, in effect, a mechanical bridge is formed between the surface of the substratum outside of the cell and the actin cable inside the cell, and hence directional force generation by the myosins along the actin cable leads to translocation of the cell relative to the substratum (Fig. 11).

In an early study using a pharmacological approach, Webster et al. (1985) addressed the involvement of secretion and of the cytoskeleton in gliding movement of the marine diatom, *Amphora coffeaeformis*. The results obtained in this study, however, could not yet resolve the question, as to whether actin or microtubules or both were involved. At least, they clearly showed that secretion were involved, because monensin, a sodium proton exchanger, which affects Golgi function, was very effective in stopping gliding movements. Poulsen et al. (1999) were able to clarify some of the issues associated with the use of cytoskeletal drugs. They unambiguously demonstrated that cytochalasins were less effective than latrunculins in breaking down actin filaments in diatoms and they also used butandion monoxim (BDM), which clearly blocked movement indicating that myosin is involved. In their study, the actin filament bundles were visualized by means of fluorescently labelled phalloidin and hence the effects of inhibitors on the actin cytoskeleton could be monitored directly.

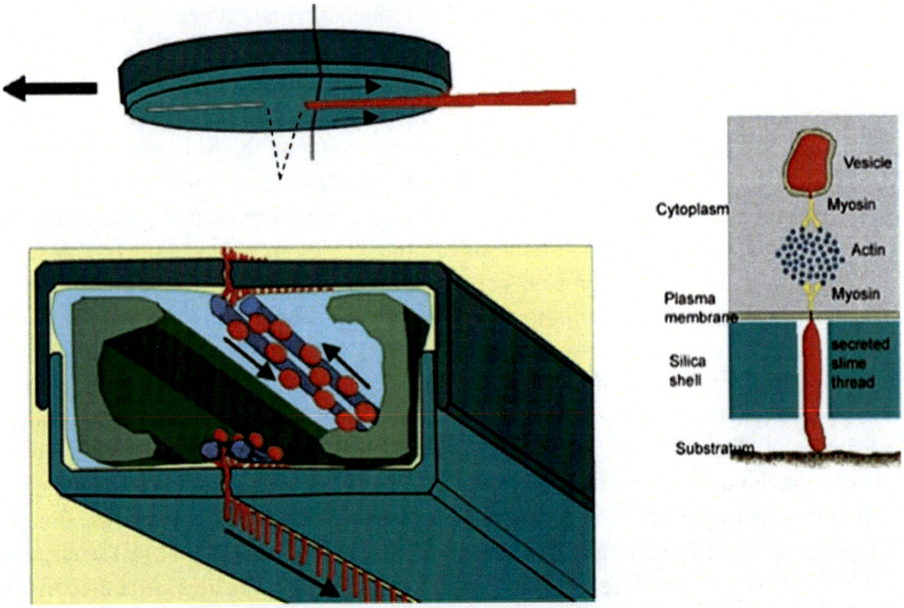

Fig. 11 *Model of the secretion-based motility machinery in pennate diatoms as described in Edgar and Pickett-Heaps* (1984). Secretory vesicles (*red*) move along the two raphe-associated actin filament bundles (*blue*). At some point, possibly at the leading edge of the driving raphe (see Fig. 10) they fuse with the plasma membrane. Mucilaginous content is released into the raphe slit, shown here as a continuous thread, but remains attached to the plasma membrane. Myosin (*yellow Y-shaped* structure) also remains attached and translocates along the actin filament bundle, pulling the mucilage thread with it. Because the thread adheres to the substratum with the other end, it remains stationary, whereas the frustule is moved relative to the ground into opposite direction. This mechanical bridge between substratum and actin cable has been termed "Adhesion Motility Complex" (AMC, Molino and Wetherbee 2008). From Menzel and Vugrek (1997), modified

Diatoms possess surprisingly many isoforms of myosin heavy chains (MHC). Recently, the domain structure of ten MHC isoforms has been characterized in some detail in the pennate diatom *Phaeodactylum tricornutum* (Heintzelman and Enriquez 2010). Which of these motor proteins is responsible for the gliding locomotion has not yet been determined, however, the high diversity, particularly in the tail portion of these myosins, leaves ample room for various types of cargo-binding and protein-to-protein crosstalk with elements of the intracellular signalling mechanisms. One of the MHC isoforms (PtMyo I) possesses a so-called FYVE-domain in the tail portion, which would enable it to bind to PtdIns(3)P-rich membrane domains, typical for the endosomal system. Therefore, PtMyo I is a promising candidate for motoring membrane recycling activities delivering the secretory vesicles to the raphe slit and even in the movement of the AMC (Fig. 11).

Since the pennate diatoms move bidirectionally, another most interesting question, namely, as to how reversal of direction is achieved and regulated, is awaiting clarification. Since there are two parallel actin cables running along the raphe, one possibility is (Model 1) that the polarity of the actin filaments in one cable is opposite to the polarity in the other cable. Alternatively (Model 2), the polarity in both actin cables is the same, and opposite directionality is mediated by different MHC isotypes, i.e., one moves in the regular direction towards the plus end of the actin filaments and the other moves in the opposite direction, as is the case for the animal myosin VI isoform (Wells et al. 1999).

Interestingly, the *Thallassiosira* genome sequencing project has identified one MHC isoform more than in *Phaeodactylum*, i.e., 11 MHC genes (cited in Heintzelman and Enriquez 2010), even though centric diatoms are not capable of performing gliding motility. So, the significance of MHC-isoform diversity in diatoms will remain an important question to be addressed in future research, because there is obviously more to it then cell locomotion. How secretion of mucilage is coordinated with the activity of the motor molecules is also yet a complete mystery. It is obvious that the secreted material must be kept from slipping sidewise between the inner face of the frustule and the protoplast. This is achieved by zones of tight connection between the plasma membrane and the frustule running parallel on both sides along the raphe. Cross-section through the raphe seen in many fine structural studies clearly demonstrate the presence of these "tight junctions" (for instance, see the images in Edgar and Pickett-Heaps 1984, Fig. 23; Schmid 1994, Fig. 12). In order to achieve straight and continuous movement, the process needs to be orchestrated in a way that at a given moment in time many such adhesive mucilage strands must be moved simultaneously into the same direction and get released from the cell, once the end of the raphe is reached, which is how the mucilage trails are being laid down on the surface of the substratum. At reversal of movement, either new mucilage strands must be inserted and hooked up to the opposite actin bundle via myosin (Model 1) or existing mucilage strands remain hooked to the same actin cable but switch motors (Model 2). These are extremely intriguing questions that should capture interest in a wider community of cell biologists for the coming years.

An additional level of complication is indicated by the observation, that the raphe-associated actin bundles support bidirectional movement of a uniform size class of small, intracellular organelles, most likely the secretory vesicles, and at the same time mucilage strands are moved along the raphe slit in a unidirectional fashion. This is brilliantly illustrated in the movie produced by Picket-Heaps and coworkers (Pickett-Heaps et al. 2003, Chap. 9)

Another type of secretion-based locomotion, apparently not involving the acto-myosin system, has been described in araphid diatoms (Pickett-Heaps et al. 1991). An example is *Ardissonia crystallina*, which possesses two slit-like pores at each end of the cell. Mucilage is continuously extruded from these slits resulting in a somewhat jerky forward and backward movement, as both ends may produce mucilage at the same time. Extrusion of two parallel trails of adhesive material has been beautifully pictured at high optical resolution in the movie made by Pickett-Heaps and coworkers (Pickett-Heaps et al. 2003, Chap. 9).

5 Prospects

Diatoms offer themselves as model systems to study several intriguing aspects of secretion. Integral to all of them is the question of temporal and spatial regulation. Particularly in the case of secretion-associated motility, the mechanisms of coordination between secretory processes and motility are as yet largely unexplored.

Diatoms are also excellent model systems to study the modular construction of adhesive biopolymers, which could give ultimate answers as to how technical glues could be designed to retain their property submerged under water. Complexity and reiterative nature of frustule morphology on the submicrometer scale has prompted theoretical biologists to develop and test mathematical models with some remarkable success in identifying algorithms, by which such structures may be generated (e.g., Parkinson et al. 1999; Bentley et al. 2005). The strong motive ever since behind these kinds of studies was the idea to technically reproduce material with such a substructure, because as has been summarized recently by Garcia and Buehler (2010), material with such a substructure has a remarkable toughness combined with an extreme light weight. Apparently, hardness of the silica frustule is not only a consequence of its composite properties, but also results from its architecture (Hamm et al. 2003).

Currently, nanotechnological research is looking for feasible approaches to manufacture such intricately fine-structured porous material in order to replace present-day linear lithographic techniques. Two major visions may be pursued in the future, either finding techniques to mimic from nature the mechanisms of nanofabrication in an artificial cell-free system or make diatoms do the work by genetically modifying their intracellular mechanisms (i.e., mineral sequestration, nanofabrication and secretion) with the effect that specific shapes and substructures with specific properties are formed, useful for nano-technological applications. Both approaches would, however, require that silica eventually is replaced by other meaningful materials such as ceramic/metallic alloys.

References

Arce FT, Avci R, Beech IB, Cooksey KE, Wigglesworth-Cooksey B (2004) A live bioprobe for studying diatom-surface interactions. Biophys J 87:4284–4297

Azam F (1974) Silicic-acid uptake in diatoms studied with [68Ge] Germanium acid as tracer. Planta 121:205–212

Baïet B, Burel C, Saint-Jean B, Louvet R, Menu-Bouaouiche L, Kiefer-Meyer M-C, Mathieu-Rivet E, Lefebvre T, Castel H, Carlier A, Cadoret J-P, Lerouge P, Bardor M (2011) N-glycans of *Phaeodactylum tricornutum* diatom and functional characterization of its *N*-acetylglucosaminyltransferase I enzyme. J Biol Chem 286:6152–6164

Beattie A, Percival E, Hirst EL (1961) Studies on metabolism of Chrysophyceae: comparative structural investigations on leucosin (Chrysolaminarin) separated from diatoms and laminarin from brown algae. Biochem J 79:531–537

Bentley K, Cox EJ, Bentley PJ (2005) Nature's Batik: a computer evolution model of diatom valve morphogenesis. J Nanosci Nanotechnol 5:25–34

Blackwell J, Parker KD, Rudall KM (1967) Chitin fibres of the diatoms *Thalassiosira fluviatilis* and *Cyclotella cryptica*. J Mol Biol 28:383–385

Bruckner CG, Bahulikar R, Rahalkar M, Schink B, Kroth PG (2008) Bacteria associated with benthic diatoms from lake Constance: phylogeny and influences on diatom growth and secretion of extracellular polymeric substances. Appl Environ Microbiol 74:7740–7749

Bruckner CG, Rehm C, Grossart HP, Kroth PG (2011) Growth and release of extracellular organic compounds by benthic diatoms depend on interactions with bacteria. Environ Microbiol 13:1052–1063. doi:10.1111/j.1462-2920.2010.02411.x

Brunner E, Richthammer P, Ehrlich H, Paasch S, Simon P, Ueberlein S, van Pée KH (2009) Chitin-based organic networks: an integral part of cell wall biosilica in the diatom *Thalassiosira pseudonana*. Angew Chem Int Ed 48:9724–9727

Bowler C, De Martino A, Falciatore A (2010) Diatom cell division in an environmental context. Curr Opin Plant Biol 13:623–630

Chiovitti A, Bacic A, Burke J, Wetherbee R (2003) Heterogeneous xylose-rich glycans are associated with extracellular glycoproteins from the biofouling diatom *Craspedostauros australis* (Bacillariophyceae). Europ J Phycol 38:351–360

Chiovitti A, Molinao P, Crawford SA, Teng R, Spurck T, Wetherbee R (2004) The glucans extracted with warm water from diatoms are mainly derived from intracellular chrysolaminaran and not extracellular polysaccharides. Eur J Phycol 39:117–128

Chiovitti A, Heraud P, Dugdale TM, Hodson OM, Curtain RCA, Dagastine RR, Wood BR, Wetherbee R (2008) Divalent cations stabilize the aggregation of sulfated glycoproteins in the adhesive nanofibers of the biofouling diatom *Toxarium undulatum*. Soft Matter 4:811–820

Cohn SA, Nash J, Pickett-Heaps JD (1989) The effect of drugs on diatom valve morphogenesis. Protoplasma 149:130–143

Cooksey KE, Wigglesworth-Cooksey B (1995) Adhesion of bacteria and diatoms to surfaces in the sea: a review. Aquat Microb Ecol 9:87–96

Coradin T, Lopez PJ, Gautier C, Livage J (2004) From biogenic to biomimetic silica. CR Palevol 3:443–452

Crawford SA, Higgins MJ, Mulvaney P, Whetherbee R (2001) Nanostructure of the diatom frustule as revealed by atomic force and scanning electron microscopy. J Phycol 37:543–554

Davis AK, Hildebrand M, Palenik B (2005) A stress-induced protein associated with the girdle band region of the diatom *Thalassiosira pseudonana* (Bacillariophyta). J Phycol 41:577–589

De Brouwer JF, Wolfstein K, Ruddy GK, Jones TE, Stal LJ (2005) Biogenic stabilization of intertidal sediments: the importance of extracellular polymeric substances produced by benthic diatoms. Microb Ecol 49:501–512

DeMaster DJ (2001) Encyclopedia of ocean sciences, vol 3, 1st edn. Elsevier Ltd, London, pp 1659–1667

Drum RW, Pankratz HS (1964) Post mitotic fine structure of *Gomphonema parvulum*. J Ultrastr Res 10:217–223

Dugdale TM, Dagastine R, Chiovitti A, Mulvaney P, Wetherbee R (2005) Single adhesive nanofibers from a live diatom have the signature fingerprint of modular proteins. Biophys J 89:4252–4260

Dugdale TM, Willis A, Wetherbee R (2006) Adhesive modular proteins occur in the extracellular mucilage of the motile, pennate diatom *Phaeodactylum tricornutum*. Biophys J 90:L58–60

Durkin CA, Mock T, Armbrust EV (2009) Chitin in diatoms and its association with the cell wall. Eukaryot Cell 8:1038–1050

Edgar LA (1983) Mucilage secretions of moving diatoms. Protoplasma 118:44–48

Edgar LA, Pickett-Heaps JD (1982) Ultrastructural localization of polysaccharides in the motile diatom *Navicula cuspidata*. Protoplasma 113:10–22

Edgar LA, Pickett-Heaps JD (1984) Diatom locomotion. Prog Phycol Res 3:49–88

Garcia AP, Buehler MJ (2010) Bioinspired nanoporous silicon provides great toughness at great deformability. Comput Mat Sci 48:303–309

Gebeshuber IC, Kindt JH, Thompson JB, Del Amo Y, Stachelberger H, Brzezinski MA, Stucky GD, Morse DE, Hansma PK (2003) Atomic force microscopy study of living diatoms in ambient conditions. J Microsc 212:292–299

Gordon R, Drum RW (1994) The chemical basis for diatom morphogenesis. Int Rev Cytol 150:243–372

Hamm CE, Merkel R, Springer O, Jurkojc P, Maler C, Pretchel K, Smetacek V (2003) Architecture and material properties of diatom shells provide effective mechanical protection. Nature 421:841–843

Heintzelman MB, Enriquez ME (2010) Myosin diversity in the diatom *Phaeodactylum tricornutum*. Cytoskeleton 67:142–151

Herth W (1978) A special chitin-fibril-synthesizing apparatus in the centric diatom Cyclotella. Naturwissenschaften 65:260–261

Higgins MJ, Crawford SA, Mulvaney P, Wetherbee R (2002) Characterization of the adhesive mucilages secreted by live diatom cells using atomic force microscopy. Protist 153:25–38

Hoagland KD, Rosowski JR, Gretz MR, Roemer SC (1993) Diatom extracellular polymeric substances: function, fine structure, chemistry, and physiology. J Phycol 29:537–566

Horiuchi H, Fujiwara M, Yamashita S, Ohta A, Takagi M (1999) Proliferation of intrahyphal hyphae caused by disruption of csmA, which encodes a class V chitin synthase with a myosin motor-like domain in *Aspergillus nidulans*. J Bacteriol 181:3721–3729

Houpt PM (1994) Marine tube-dwelling diatoms and their occurrence in the Netherlands. Aquatic Ecol 28:77–84

Krembs C, Eicken H, Deming JW (2010) Exopolymer alteration of physical properties of sea ice and implications for ice habitability and biogeochemistry in a warmer Arctic. Proc Natl Acad Sci USA 108:3653–3658

Kröger N (2007) Prescribing diatom morphology: toward genetic engineering of biological nanomaterials. Cur Opin Chem Biol 11:662–669

Kröger N, Lehmann G, Rachel R, Sumper M (1997) Characterization of a 200-kDa diatom protein that is specifically associated with a silica-based substructure of the cell wall. Eur J Biochem 250:99–105

Kröger N, Deutzmann R, Sumper M (1999) Polycationic peptides from diatom biosilica that direct silica nanosphere formation. Science 286:1129–1132

Kröger N, Wetherbee R (2000) Pleuralins are involved in theca differentiation in the diatom *Cylindrotheca fusiformis*. Protist 151:263–273

Kröger N, Poulsen N (2008) Diatoms - from cell wall biogenesis to nanotechnology. Annu Rev Genet 42:83–107

Leflaive J, Ten-Hage L (2009) Chemical interactions in diatoms: role of polyunsaturated aldehydes and precursors. New Phytol 184:794–805

Lewin J (1966) Silicon metabolism in diatoms. V. Germanium dioxide, a specific inhibitor of diatom growth. Phycologia 6:1–12

Li C-W, Volcani BE (1985) Studies on the biochemistry and fine structure of silica shell formation in diatoms VIII. Morphogenesis of the cell wall in a centric diatom, *Ditylum brightwelli*. Protoplasma 124:10–29

Lind JL, Heimann K, Miller EA, van Vliet C, Hoogenradd NJ, Wetherbee R (1997) Substratum adhesion and gliding in a diatom are mediated by extracellular proteoglycans. Planta 203:213–221

McConville MJ, Wetherbee R, Bacic A (1999) Subcellular location and composition of the wall and secreted extracellular sulphated polysaccharides/proteoglycans of the diatom Stauroneis amphioxys Gregory. Protoplasma 206:188–200

McLachlan J, McInnes AG, Falk M (1965) Studies on chitan (chitinpoly-n-acetylglucosamine) fibers of diatom *Thalassiosira fluviatilis* Hustedt. 1. Production and isolation of chitan fibers. Can J Bot 43:707–713

Menzel D, Vugrek O (1997) Muskelproteine in Pflanzen. Biol unserer Zeit 27:195–203
Molino PJ, Wetherbee R (2008) The biology of biofouling diatoms and their role in the development of microbial slimes. Biofouling 24:365–379
Parkinson J, Brechet Y, Gordon R (1999) Centric diatom morphogenesis: a model based on a DLA algorithm. Investigating the potential role of microtubules. Biochim Biophys Acta 1452:89–102
Pickett-Heaps JD (1983) Valve morphogenesis and the microtubule enter in three species of the Diatom Nitzschia. J Phycol 19:269–281
Pickett-Heaps JD, Schmid AM, Edgar L (1990) The cell biology of diatom valve formation. Prog Phycol Res 7:1–168
Pickett-Heaps JD, Hill DRA, Blaze KL (1991) Active gliding motility in the araphid marine diatom, Ardissonea (formerly Synedra) crystallina. J Phycol 27:718–725
Pickett-Heaps JD, Carpentr J, Koutoulis A (1994) Valve and seta (spine) morphogenesis in the centric ditom *Chaetoceros peruvianus* Brightwell. Protoplasma 181:269–282
Pickett-Heaps JD, Spurck T, Wetherbee R, Cohn S (2003) Diatoms: life in glass houses. Cytographics. http://www.cytographics.com
Pollock FM, Pickett-Heaps JD (2005) Spatial determinants in morphogenesis: recovery from plasmolysis in the diatom Ditylum. Cell Motil Cytoskeleton 60:71–82
Pondaven P, Gallinari M, Chollet S, Bucciarelli E, Sarthou G, Schultes S, Jean F (2007) Grazing-induced changes in cell wall silicification in a marine diatom. Protist 158:21–28
Poulsen N, Kroger N (2004) Silica morphogenesis by alternative processing of silaffins in the diatom *Thalassiosira pseudonana*. J Biol Chem 279:42993–42999
Poulsen NC, Spector I, Spurck TP, Schultz TF, Wetherbee R (1999) Diatom gliding is the result of an actin-myosin motility system. Cell Motil Cytoskeleton 44:23–33
Rabosky DL, Sorhannus U (2009) Diversity dynamics of marine planktonic diatoms across the Cenozoic. Nature 457:183–187
Reimann BEF (1964) Deposition of silica inside a diatom cell. Exp Cell Res 34:605–608
Reimann BEF, Lewin JC, Volcani BE (1966) Studies on the biochemistry and fine structure of the silica shell formation in diatoms. II. The structure of the cell wall of *Navicula pelliculosa* (Bréb.) Hilse. J Phycol 2:74–84
Reimann BEF, Volcani BE (1968) Studies on the biochemistry and fine structure of silica shell formation in diatoms III. The structure of the cell wall of *Phaecdactylum tricornutum* Bohlin. J Ultrastruct Res 21:182–193
Rajangam AS, Kumar M, Aspeborg H, Guerriero G, Arvestad L, Pansri P, Brown CJL, Hober S, Blomqvist K, Divne C et al (2008) MAP20, a microtubule-associated protein in the secondary cell walls of hybrid aspen, is a target of the cellulose synthesis inhibitor 2,6-dichlorobenzonitrile. Plant Physiol 148:1283–1294
Robinson DH, Sullivan CW (1987) How do diatoms make silicon biominerals? Trends Biochem Sci 12:151–154
Roessler PG (2000) More tools for diatom molecular biology research. J Phycol 36:259–260
Sapriel G, Quinet M, Heijde M, Jourdren L, Tanty V, Luo G, Le Crom S, Lopez PJ (2009) Genome-wide transcriptome analyses of silicon metabolism in Phaeodactylum tricornutum reveal the multilevel regulation of silicic acid transporters. PLoS One 4:e7458
Scheffel A, Poulsen N, Shian S, Kröger N (2011) Nanopatterned protein microrings from a diatom that direct silica morphogenesis. Proc Natl Acad Sci USA 108:3175–3180
Schmid A-MM (1994) Aspects of morphogenesis and function of diatome cell walls with implications for taxonomy. Protoplasma 181:43–60
Sheppard V, Poulsen N, Kröger N (2010) Characterization of an endoplasmic reticulum-associated silaffin kinase from the diatom *Thalassiosira pseudonana*. J Biol Chem 285:1166–1176
Smetacek V (1999) Diatoms and the ocean carbon cycle. Protist 150:25–32
Sumper M, Brunner E, Lehmann G (2005) Biomineralization in diatoms: characterization of novel polyamines associated with silica. FEBS Lett 579:3765–3769

Sumper M, Brunner E (2008) Silica biomineralisation in diatoms: the model organism *Thalassiosira pseudonana*. Chem Bio Chem 9:1187–1194

Sullivan CW, Volcani BE (1981) Silicon in the cellular metabolism of diatoms. In: Simpson TL, Volcani BE (eds) Silicon and siliceous structures in biological systems. Springer, New York, pp 15–42

Thamatrakoln K, Alverson AJ, Hildebrand M (2006) Comparative sequence analysis of diatom silicon transporters: toward a mechanistic model of silicon transport. J Phycol 42:822–834

Tatham AS, Shewry PR (2000) Elastomeric proteins: biological roles, structures and mechanisms. Trends Biochem Sci 25:567–571

Takeshita N, Ohta A, Horiuchi H (2005) CsmA, a class V chitin synthase with a myosin motor-like domain, is localized through direct interaction with the actin cytoskeleton in *Aspergillus nidulans*. Mol Biol Cell 16:1961–1970

Tesson B, Hildebrand M (2010) Extensive and intimate association of the cytoskeleton with forming silica in diatoms: control over patterning on the meso- and micro-scale. PLoS One 5:e14300

Tsuizaki M, Takeshita N, Ohta A, Horiuchi H (2009) Myosin motor-like domain of the class VI chitin synthase CsmB is essential to its functions in *Aspergillus nidulans*. Biosci Biotechnol Biochem 73:1163–1167

van de Poll WH, Vrieling EG, Gieskes WWC (1999) Location and expression of frustulins in the pennate diatoms *Cylindrotheca fusiformis*, Navicula pelliculosa, and *Navicula salinarum* (Bacillariophyceae). J Phycol 35:1044–1153

Vasconcelos MT, Leal MF (2008) Exudates of different marine algae promote growth and mediate trace metal binding in *Phaeodactylum tricornutum*. Mar Environ Res 66:499–507

Wang Y, Lu J, Mollet J-C, Cretz MR, Hoagland KD (1997) Extracellular matrix assembly in diatoms (Bacillariophyceae). II. 2,6-dichlorobenzonitrile inhibition of motility and stalk production in the marine diatom *Achnanfhes longipes*. Plant Physiol 113:071–1080

Webster DR, Cooksey KE, Rubin RW (1985) An investigation of the involvement of cytoskeletal structures and secretion in gliding motility of the marine diatom, *Amphora coffeaeformis*. Cell Motil 5:103–122

Wells AL, Lin AW, Chen L-Q, Safer D, Cain SM, Hasson T, Carragher BO, Milligan RA, Sweeney HL (1999) Myosin VI is an actin-basedmotor that moves backwards. Nature 401:505–508

Wichard T, Gerecht A, Boersma M, Poulet SA, Wiltshire K, Pohnert G (2007) Lipid and fatty acid composition of diatoms revisited: rapid wound-activated change of food quality parameters influences herbivorous copepod reproductive success. Chembiochem 8:1146–1153

Wigglesworth-Cooksey B, Cooksey KE (2005) Use of fluorophore-conjugated lectins to study cell-cell interactions in model marine biofilms. Appl Environ Microbiol 71:428–435

Wigglesworth-Cooksey B, Berglund D, Cooksey KE (2001) Cell–cell and cell–surface interactions in an illuminated biofilm: implications for marine sediment stabilization. Geochem Trans 10:75–82. doi:10.1039/b107814n

Wustman BA, Cretz MR, Hoagland KD (1997) Extracellular matrix assembly in diatoms (Bacillariophyceae) 1. A model of adhesives based on chemical characterization and localization of polysaccharides from the marine diatom *Achnanthes longipes* and other diatoms. Plant Physiol 113:1059–1069

Wustman BA, Lind J, Wetherbee R, Gretz MR (1998) Extracellular matrix assembly in diatoms (Bacillariophyceae). III. Organization of fucoglucurono-galactans within the adhesive stalks of Achnanthes longipes. Plant Physiol 116:1431–1441

Zurzolo C, Bowler C (2001) Exploring bioinorganic pattern formation in diatoms. A story of polarized trafficking. Plant Physiol 127:1339–1345

Bacterial Secretions

Brittany A. Barnett and Tiffany L. Weir

Abstract Prokaryotes are one of the oldest forms of life on earth and as such, have evolved the ability to survive in extremely varied environments that range from deep sea hydrothermal vents to Arctic permafrost and from the human gut to the roots of plants. One of the secrets of their success is the ability to secrete proteins and metabolites, which alter their environment, act as chemical communication signals, and allow them to ward of competitors. They can also exchange genetic material both within and between bacterial species, allowing them incredible adaptability and resulting in within species variability in characteristics such as pathogenicity, antibiotic resistance, and other factors that could influence their survival under different conditions. In this chapter, we will discuss the various types of bacterial secretions such as virulence factors like toxins and enzymes that break down host defenses and small molecules and peptides that bacteria use to communicate with one another. We will also cover many of the currently described secretion pathways that allow these components to move from their site of synthesis within the cells into the external milieu.

1 Types of Secretions

1.1 Virulence Factors

The ability of bacteria to cause disease in a host is attributed to multiple virulence factors acting individually or together during different stages of infection (Wu et al. 2008). Expression of virulence factors and behaviors associated with virulence must

B.A. Barnett • T.L. Weir (✉)
Department of Food Science and Human Nutrition, Colorado State University, Fort Collins, CO 80523-1571, USA
e-mail: Tiffany.Weir@colostate.edu

J.M. Vivanco and F. Baluška (eds.), *Secretions and Exudates in Biological Systems*, Signaling and Communication in Plants 12, DOI 10.1007/978-3-642-23047-9_11,

be coordinated for energy conservation, appropriate disease development, evasion of host defense, and eventual dispersal. Virulence factors include proteins that attach microbes to their hosts called adhesins, effectors that suppress host defenses, enzymes that break down host tissues known as invasins, secretion system proteins, polysaccharide capsules that protect pathogens from host defenses, and siderophores which increase a bacterium's ability to thrive in its host (Korves and Colosimo 2009).

Virulence factors apply to the elements that enable a microorganism to colonize a host niche where the organism proliferates and causes tissue damage or systemic inflammation, all in the name of proliferation (Chen et al. 2005) and can be grouped based on their mechanisms of action. Membrane proteins are involved in adhesion, colonization, and host invasions. They promote adherence to host cell surfaces, are imperative for resistance to antibiotics, and promote intercellular communication. Polysaccharide capsules surround the bacterial cell and have antiphagocytic properties, contain secretory proteins such as toxins, which can modify the host cell environment and are responsible for some host cell–bacteria interactions. Nearly all secreted proteins in Gram-negative bacteria are transported by distinct secretion systems I–VI which will be addressed later in the chapter. Cell wall and outer membrane components such as LPS or endotoxin, lipoteichoic acids can also act as virulence factors. Other components, such as biofilm forming proteins and siderophores, are primarily secreted as part of survival strategies in natural environments but also help augment bacterial virulence (Chen et al. 2005).

Virulence factor expression can be changed depending on a microbe's environment, for example, host defense systems such as polymorphonuclear leukocytes (PMNs) produce substantial amounts of superoxide anion and hydrogen peroxide as part of their oxygen-dependent bactericidal mechanisms. Thus, oxidative stress has a great effect on bacterial virulence. Comparing protein expression profiles of *Helicobacter pylori* under normal and oxidative stress conditions shows that the bacteria uses different virulence mechanisms under different environmental conditions. The production of a major virulence factor, called the urease accessory protein (UreE), as well as alkylhydroperoxide reductase (AhpC), which has antioxidant potential is greatly decreased under oxidative stress conditions (Wu et al. 2008). Identifying and targeting these proteins that are differentially expressed under changing conditions could provide novel drug targets against *H. pylori*.

Oxidative stress proteins and manganese transporters are beginning to be recognized as virulence factors. Metal ions like Fe^{2+} and Mn^{2+} are involved in oxidative stress. Unlike Fe^{2+}, Mn^{2+} and its transporters play important roles in protecting cells against reactive oxygen species. The full functionality of a microbial cell under certain conditions reveals what qualifies as virulence factors. Understanding that metal ion transporters are essential for microbial success exhibits a three-way relationship between Mn^{2+}/Mn^{2+} transporter, oxidative stress, and virulence (Wu et al. 2008). Ideally, researchers would combine genomic, proteomic, and structural assays to help understand complex roles of virulence factors.

Exotoxins are proteins secreted by microbes; they tend to be enzymes that kill host cells at impressively low concentrations. All gram-negative pathogens make endotoxin, although toxicity varies among species. Toxins alone can exhibit clinical features of cholera, tetanus, and diphtheria, but microbes rely on many other microbial mechanisms for complete pathogenic success (Finlay and Falkow 1997). Pore-forming toxins act by inserting themselves into the host cell plasma membrane, and this forms a pore that leads to lysis of the host cell. Other toxins, such as A–B toxins which have two components: the B subunit mediates binding to the host cell receptor and facilitates delivery of the toxin into the host cell while the A subunit contains the enzymatic and ultimately toxic activity that acts on the host cell. The B subunits can change for host specificity but the A subunit typically contains a conserved region that is important for enzymatic activity (Finlay and Falkow 1997).

Adherence for host invasion is an important part of microbial pathogenicity, a microbe can use pili which attach to host receptors, afimbrial adhesions, or it may adhere to the extracellular matrix. These attachments can be specific, for example, *H. pylori* adheres to GI epithelial cells but not CNS or urogenital tract epithelial cells (Finlay and Falkow 1997). After adhesion, many pathogenic microbes are capable of entering into and surviving within their eukaryotic hosts. This method ensures a protected cellular niche for the microbe to replicate or persist. Microbes use adhesion molecules, generally called invasins, which direct bacterial entry onto cells and can trigger eukaryotic cascades.

Genome sequencing has led to the development of other "high-throughput" approaches to defining virulence factors on a genomic level. Comparative genomics is a popular tool to identify virulence factors and genes involved in environmental persistence of pathogens. Researchers can correlate those differences to biological function and to gain insight into selective evolutionary pressures and patterns of gene transfer or loss, especially within the context of virulence in pathogenic species (Wu et al. 2008). Improving the genome annotation as well as functional and structural biology studies for characterizing these hypothetical proteins is needed to confirm the accuracy of predicting bacterial virulence factors within a genome.

While genomic strategies can help us to understand what bacteria are capable of, proteomic strategies are useful to identify proteins that are differentially expressed. Proteomics can define proteins that are differentially located or secreted to the outside of the cell, which can help to describe new virulence factors released by the cell. Only proteomics can show proteins that are posttranscriptionally modified (Wu et al. 2008). Understanding posttranscriptionally modified proteins is necessary as these proteins are those who are presented to a host cell for infection.

Nearly all bacterial virulence factors are tightly regulated with their expression linked to various environmental signals. Environmental signals that can affect virulence factor regulation include: temperature, ion concentrations, osmolatiry, iron levels, pH, carbon source availability, bacterial growth phase, and oxygen levels. Pathogens use one or more of these environmental factors to sense which microenvironment where they reside within a host or even within a specialized compartment inside a single host cell (Finlay and Falkow 1997). After obtaining the appropriate cues, virulence genes are activated and proteins are synthesized. These proteins are then transported to precise cellular locations, to be assembled into complexes. Upon formation, these protein complexes may activate upon contact with the host or in response to a unique cellular identifier. Bacteria may receive many different signals at once, so multiple virulence factors may be controlled in conjunction as several regulators that measure different parameters gather information. It is also possible for different environmental signals to regulate a single virulence factor. Thus, virulence expression is a sum of the signals that are sent by various cell regulators and sensing systems (Finlay and Falkow 1997). By gathering information to understand their surroundings, bacteria can regulate virulence factors according to what is necessary for a successful invasion.

1.2 Quorum-Sensing Signals

The notion that a bacterium survives alone and its success or failure is dependent upon good fortune has given way to a more complex view of microbial pathogenesis; successful invasion of a host is now understood to be a collective process, based upon microbial information sharing and active collaboration (Asad and Opal 2008). Infection strategies of phytopathogens, which often require swift global changes in gene expression and physiology in response to environmental cues, are quite reliant on cell-to-cell communication to coordinate crucial steps in pathogenesis (Mole et al. 2007). Bacteria not only use communication as a means of employing virulence factors, they also communicate extensively with each other to execute a communal approach of facilitating survival in their environments. A system of cell-to-cell signaling pathways controls bacterial growth, metabolism, biofilm formation, and many other essential functions in bacterial populations (Asad and Opal 2008).

Pectobacterium carotovorum subsp. *carotovorum* cell-to-cell communication systems are responsible for regulating the Type 3 secretion system (T3SS) plant cell-wall-degrading enzymes along with antibiotic production to keep competing microbes at bay. This pathogen employs quorum sensing systems that include up to three transcription activators that are responsive to two acyl homoserine lactone (AHL) molecules, which in turn are encoded by one synthase. A synthase is an enzyme which catalyzes a synthesis process, in this case a quorum sensing system. The other systems directly inhibit virulence in the absence of specific levels of AHL by up-regulating rsmA which is a member of the posttranscriptional Rsm system and destabilizes mRNA transcripts that encode plant cell wall-degrading enzymes, included cellulose, pectate lyase, and protease (Mole et al. 2007). Communication among bacteria is crucial for the activation or suppression of virulence factors, the bacteria do not act individually but as an entire group in their activation of virulence factors.

The ability to detect extracellular, small molecule signals and to alter gene expression in response to bacterial population densities is called quorum sensing (Asad and Opal 2008). A classic example of bacterial quorum sensing for use in a symbiotic relationship is *Vibrio fischeri* and Hawaiian bobtail squid.

The bioluminescent *V. fischeri* is taken up by carefully placed light organs on the outer surface of the squid. When the bacterial population reaches its threshold concentration, the bacterium activates its luciferase operon to generate visible light. The bacteria benefit from its association as the squid provides protection and a steady source of nutrients. The light source created by the bacterial enzymes provides the squid with clever system of camouflage. From below, the light organs look very much like the starry sky about the water surface, this tricks predatory fish and they do not seek to kill a squid they cannot sense.

Many bacteria use quorum sensing as a method to communicate with their coinhabitants. Surprisingly species and kingdom barriers may be crossed by quorum sensing molecules, communication signals may be processed by any number of receptors. Pathogens can sense human stress hormones as well as cytokines and accordingly act in a matter that is advantageous to bacterial survival (Asad and Opal 2008). Quorum sensing systems also control biofilm formation, growth potential, sporulation, antibiotic resistance expression, DNA transfer, virulence expression, autolysis, oxidative stress tolerance, metabolic activity, motility, antibiotic synthesis by antibiotic-producing bacteria, and sessile versus planktonic behavior.

Communication is an important part of bacterial everyday life. Quorum sensing is responsible for the autoinducer type 1 (AI-1) system which is widely used in multiple genera of gram negative bacteria and is very similar to the previously described *luxR*/*luxI* system of *V. fischeri*. Bacteria can use a series of *N*-AHL molecules for signaling. These AHL molecules are able to pass freely through cell membranes; when bacterial population densities are low, the small amounts of AHL are diluted away and the quorum sensing genes remain off. When population densities increase beyond a predetermined threshold level, enough AHL accumulates to interact with receptor within the cytosol and quorum sensing genes are stimulated (Asad and Opal 2008).

The precise role of autoinducer type 2 signaling in bacterial pathogenesis is not clear, since much of the transcriptional activity of autoinducer type 2 (AI-2) systems is directed at the regulation of metabolic pathways. An AI-2 inducer system expressed by the *luxS* gene encodes a dual-function enzyme that mediates the conversion of a toxic intermediate molecule, *S*-adenosyl-L-homocysteine, to homocysteine, a homologue of the amino acid cysteine. This enzymatic activity is central to an active methyl cycle in which methyl groups are attached to nucleic acid precursors, proteins, and other metabolites. Byproducts of LuxS enzyme activity are furanose structures that act as AI-2 signals. Many of the quorum sensing signaling events ascribed to AI-2 signaling may be the consequence of the LuxS as a detoxifying enzyme in the activated methyl cycle rather than a consequence of quorum sensing effects (Asad and Opal 2008).

Some Gram-positive bacterial pathogens possess another structurally different, but functionally similar, system of global regulation of genes based upon cell densities. This system is referred to as the accessory gene regulator system in *Staphylococci* spp. and as the *Enterococcus faecalis* regulator–QS system in *Enterococci* spp. Gram positive pathogens use short, cyclical peptides known as autoinducer peptides (AIP). Cell surface receptors sense these peptides then activate a histidine kinase that generates transcriptional activators for multiple gene loci. This system functions as a positive feedback loop to release more AIP and stimulate bacterial virulence (Asad and Opal 2008).

Bacteria use signal systems to direct movement of the cell; the high performance system of *E. coli* involves a limited number of components but is sophisticated (Hazelbauer et al. 2007). Transmembrane chemoreceptors, called methyl-accepting chemotaxis proteins (MCPs) direct cell locomotion by regulating the histidine kinase CheA; CheA phosphorylates a response regulator, which controls the rotational direction of the flagellar motor. Chemotactic sensitivity and the strength of the signal detection are regulated by modifications of chemoreceptors through reversible glutamyl methylation. The resulting interplay between motor control and sensory adaptation accounts for directed motile behavior. The mechanisms by which thousands of receptors localize in patches or the means by which those receptors communicate with one another are not fully understood. Receptor patches are likely built by interactions among trimers but how this occurs is not yet known. Signaling within a single chemoreceptor dimer involves specific, bidirectional conformational changes in its three modules. These intradimer changes might trigger changes in the trimer but, again, these effects are not fully defined (Hazelbauer et al. 2007).

Many different signals produced by bacteria have a common second messenger. Signal transduction pathways often use small molecule second messengers to integrate, amplify, and transmit information to intracellular sensors and effectors. Cyclic nucleotides such as cyclic adenosine monophosphate (cAMP) and cyclic guanosine monophosphate (cGMP) are among the most important second messengers. These molecules regulate a variety of functions ranging from sugar metabolism to ion channel conductance in prokaryotes and eukaryotes (Seshasayee et al. 2010).

The bacterial second messenger cyclic dimeric GMP (cyclic-di-GMP) was identified in 1987 as an allosteric activator of cellulose synthase in *Gluconacetobacter xylinus* and *Agrobacterium tumefaciens*. Cyclic-di-GMP has become recognized as an important regulator for convoluted cellular functions, including the switch between motile and sessile states (Seshasayee et al. 2010). High cellular levels of Cyclic-di-GMP promote exopolysaccharide production and surface adhesion, eventually leading to biofilm formation, whereas, low Cyclic-di-GMP result in flagellar gene expression and increased cellular motility. The relative absence of signal sensing partner domains in the EAL-only proteins in conjunction with the observation that EAL-only proteins are few in number in most genomes might be a sign that many EAL-only proteins are mainly regulated at the transcriptional level and may contribute towards maintaining cyclic-di-GMP homeostasis rather than respond directly to environmental or cellular cues (Seshasayee et al. 2010).

The EAL domain is a ubiquitous signal transduction protein domain in the bacteria. It is involved in hydrolysis of the second messenger cyclic-di-GMP. The domain name comes from an amino acid signature motif, EAL (Glu–Ala–Leu) which is highly conserved. EAL may have an important role in bacteria since the EAL domain is encoded by nearly all sequenced bacterial genomes. *Archaea* or

Eukarya genomes do not encode for EAL, with the exception of two reported proteins of *Anopheles gambiae*, which most likely came from bacterial contamination. EAL is often linked to sensory and/or output (signal transduction) domains although its biochemical activity has not been characterized yet (Schmidt et al. 2005).

1.3 Antibiotics

In 1928, Alexander Fleming noticed a spore from *Penicillium notatum* that landed on a plate that he had inoculated with *Staphylococci*. He left the plate to incubate while he was on vacation and returned to a plate covered with fungus and lysed bacteria. The beginning of microbially derived antibiotics had begun. Many antibiotics are derived from microbes, understandable, as microbes often have to compete with millions of other organisms. In order to survive within their environment, numerous bacteria secrete molecules that selectively kill or inhibit growth of their neighbors. *Streptomyces* spp., *Micromonospora* spp., and *Bacillus* spp. bacteria are all known to secrete effective antibiotics, as for fungi *Penicillium* spp. and *Cephalosporium* spp. secrete antibiotics as well. Antibiotics can work in a number of different ways: they inhibit cell wall synthesis, protein synthesis inhibition, nucleic acid synthesis inhibition, cell membrane disruption, and metabolic antagonism. In all cases, some small molecule enters the cell and inhibits an action necessary to maintain microbial life. Since microbial life must go on, bacterial resistance to antibiotics is necessary. Some may prevent entrance of the drug, while others pump the drug out of the cell, while others inactivate drugs by chemical modification (Prescott et al. 2005).

1.4 Chelators

Siderophores are low molecular weight compounds that are produced under iron-limiting conditions. These molecules chelate the ferric ion with a high specific activity and transport Fe (III) into a microbial cell. Nearly every plant pathogen and symbiont that has been studied produces siderophores (Loper and Buyer 1991). Iron is a necessary nutrient for the growth and survival of bacteria. In animals, essentially no iron exists in serum that is not bound to haem, iron-storage proteins, or as cofactors for various enzymes. Strict control of available iron in mammals serves as a barrier against infections. Siderophore biosynthesis alone is not a key identifier of virulence, but siderophores can play an important role in boosting the pathogenicity of a microbe. Patients with persistent bacterial infections may benefit from taking an iron chelator that could cross the cell membrane to bind Fe (III) which might otherwise be contributing to the growth of a bacterial pathogen (Hotta et al. 2010).

2 Bacterial Secretion Systems

Secreted proteins and other elements are crucial to a bacteria's ability to adapt and thrive in changing environments. Because secretion is central to their survival, bacteria have evolved varied and complex systems for transporting proteins into the external environment. Gram-negative bacteria contain an additional cell membrane not present in Gram-positive bacteria, and therefore have devised additional methods of protein transport. To date, there are six main secretion pathways (I–VI) that permit translocation of soluble proteins across the outer membrane of Gram-negative bacteria, and many bacteria have evolved/adopted several of these systems as a means of translocating different types of effector proteins. In fact, the opportunistic human pathogen *Pseudomonas aeruginosa* has been found to employ all of the secretion systems described in this chapter except Type IV, as well as several other systems , such as fimbrial usher porin (FUP) and two-partner secretion families (TPS) that are not covered here (Ma et al. 2003). The following sections provide a brief overview of the general secretory pathway (GSP) (Sec), which is mainly responsible for the secretion of peptides in Gram-positive bacteria as well as the export of proteins into the periplasm of Gram-negative bacteria. It also provides a description of six identified pathways of transporting peptides across the outer membrane in Gram-negative bacteria. Detailed descriptions of bacterial transport mechanisms are beyond the scope of this book, but can be found in numerous reviews.

2.1 General Secretory Pathway

Many bacterial peptides are translocated outside of the cytoplasm via a pathway that utilizes the secretory (Sec) translocon, which is present in all domains of life and is homologous to the eukaryotic system of protein translocation that occurs in the endoplasmic reticulum (Cao and Saier 2003). Briefly, peptides are synthesized in the cytoplasm with the appropriate N-terminal signals that direct them to the Sec pathway, where they are threaded through the transmembrane Sec channel in an unfolded state. On the *trans* side of the cytoplasmic membrane, the protein is further processed by cleaving the signal peptide and folded into its active form. Gram-positive bacteria, which lack the outer membrane present in Gram-negative species, rely primarily on the Sec translocon, sometimes referred to as GSP for protein secretion (Desvaux et al. 2004). This GSP is used to translocate a number of important virulence factors such as the exotoxins produced by *S. aureus* and lysteriolysin O of *Listeria monocytogenes* (Sibbald and van Dijl 2009). In Gram-negative bacteria, the Sec-dependent system is often referred to as the general export pathway (GEP), which exports peptides out of the cytoplasm and into the periplasm, where additional secretory systems described translocate them to the external milieu.

An additional pathway used for translocation of proteins out of the cytoplasm of both Gram-positive and Gram-negative bacteria is the Twin Arginine translocase (Tat). The signal peptide that directs proteins to Tat apparatus differs from those targeted to the Sec apparatus. Tat-directed signal peptides contain two arginines in the n-region and are typically longer, less hydrophobic signal peptides. Another distinction is that unlike the Sec apparatus, the Tat pathway transports fully folded proteins using an unknown chaperone mechanism (Teter and Klionsky 1999).

2.2 *Type I Secretion System*

The Type I secretion system (T1SS) is a Sec-independent system that exports proteins (usually proteases) across both the inner and outer membranes of Gram-negative bacteria. The T1SS is a simple system comprised of three protein complexes: an ATP-binding cassette (ABC) transporter that spans the inner membrane, a membrane fusion protein (MFP) in the periplasm, and an outer membrane protein (OMP). In addition to transporting proteins, it can transport a variety of molecules from β-glucans and polysaccharides to ions and xenobiotics (Wooldridge 2009). The *E. coli* hemolysin, HylA, transport system is the prototypical T1SS (Gentschev et al. 2002). The HylA export machinery is encoded by the highly regulated *hylCABD* operon, and consists of the ABC-transporter HylB and the MFP HylD, which are components specific to hemolysin transport; and the OMP TolC, which serves multiple functions in the cell (Wagner et al. 1983). Data suggests that HlyB has eight hydrophobic, α-helical transmembrane domains (TMDs) that inserted in the inner membrane (IM) with relatively large positively charged cytoplasmic loops and small periplasmic loops (Gentschev and Goebel 1992). HlyD is anchored in the cytoplasmic membrane by a single TMD and has a highly conserved periplasmic domain (Johnson and Church 1999). The OMP component TolC is part of at least four separate export systems (Zgurskaya and Nikaido 2000), and in its trimeric state forms a trans-periplasmic channel-tunnel comprised of an outer membrane (OM) barrel (channel) connected to an helical barrel (tunnel) projecting across the periplasmic space (Koronakis et al. 2000). Together these T1SS proteins form an unbroken tunnel from the cytoplasm across the inner and outer membranes to the exterior of the cell.

2.3 *Type II Secretion System*

The Type II secretion system (T2SS) is often referred to as the main terminal branch of the GSP, a misnomer since this pathway is more accurately referred to in Gram-negative bacteria as the GEP (Desvaux et al. 2004). T2SS includes at least 12 separate core proteins, including a complex that forms a pore through the outer membrane, a cytoplasmic ATPase, an inner TMD, and major and minor pseudopilins (Filloux 2004). Before proteins can enter the T2SS they must be transported from the cytoplasm via the Tat or Sec pathways to the periplasm where they are folded. Although the proteins translocated by this pathway are seemingly unrelated, a number of them are important virulence factors including cellulases, proteases, phospholipases, and toxins (Sandkvist 2001). Examples of virulence factors secreted by T2SS include heat labile toxins of enterotoxigenic *E. coli*, exotoxin A from *P. aeruginosa*, and *V. cholerae's* cholera toxin (Sandkvist 2001). This particular secretion system has only been found in the proteobacteria, including the *Aeromonadaceae*, *Alteromonadaceae*, *Legionellaceae*, *Enterobacteriaceae*, *Pseudomonaceae*, *Vibrionaceae*, and *Xanthomonadales* families and is required for virulence in many bacteria (Cianciotto 2005). There is also some evidence to suggest that T2SS assist bacteria to inhabit specific environmental niches. For example, the opportunistic human pathogen *P. aeruginosa* can live saprophytically in soil or on plants as well as infecting cross-kingdom hosts (Prithiviraj et al. 2006), and *V. cholerae* and *Legionella pneumophila* are both capable of surviving in aquatic habitats as well as in human hosts (Barker and Brown 1994). Finally, a number of nonpathogenic bacteria also contain T2SS responsible for secreting proteins that allow them to persist in various and sometimes extreme environments (Cianciotto 2005).

2.4 Type III Secretion Systems

The Type III secretion system (T3SS) in Gram-negative bacteria is structurally unique from T1SS and T2SS in that it includes an extracellular syringe-like appendage that is used to deliver effector proteins directly into host cells. As expected, this is a very important system for the delivery of virulence factors and is thus present in many plant and human pathogens (Hueck 1998). This system is also important for cell motility as there are corresponding T3SS that are used to export flagellar components such as flagellins (Aizawa 2001). The T3SS is one of the most complex secretion systems identified, is made up by approximately 30 different proteins, and the genetic machinery is usually arranged on operons referred to as pathogenicity islands or carried on plasmids and can be horizontally transferred between different bacterial species. The proteins involved in T3SS can be categorized as three different types: structural, chaperones, and effectors. The structural proteins make up the apparatus base, located in the bacterial cytoplasm, an inner rod that crosses through the two bacterial membranes, and the extracellular needle that crosses the eukaryotic cell membrane upon host cell contact to deliver the effector molecules. It is still unclear whether the needle complex actually punctures the host cell membrane or if secretes effectors that form a pore in the membrane that allows the needle to enter the host. In either case, injection of effectors is a rapid process to help the bacterium avoid phagocytosis (Cornelis 2006). Interestingly, this same rapid injection mechanism that helps the bacteria avoid phagocytes also facilitates their entry into host cells. In *Salmonella typhimurium*, injected effector proteins bind directly to host actin to interfere with the actin arrangement dynamics as well as activating host signal transduction cascades that indirectly promote actin rearrangement and allow bacterial entry into host cells (Zhou and Galán 2001). Chaperone proteins bind effectors in the bacterial cytoplasm, protecting and guiding them to the needle complex. Another unique aspect of the T3SS is that the signal peptide that directs secretion of effector molecules is never cleaved off the proteins (Cornelis 2006; Grynberg and Godzik 2009).

Although the majority of well-characterized T3SS effectors are from human pathogens, they are also an important aspect of the pathogenicity of most Gram-negative plant pathogens as well. The hypersensitive response and pathogenicity (Hrp and Hrp-like) T3SS are found in *Pseudomonas*, *Xanthomonas*, *Erwinia*, and *Ralstonia* spp. and are so important to bacterial pathogenesis that loss of one or more of the *hrp* genes can result in a complete loss of pathogenicity (Van Gijsegem et al. 1993). The needle structure in plant pathogens, *hrp* pili, is longer than the homologous structure in mammalian pathogens, presumably to allow breaching of the thick plant cell walls (Büttner and He 2009). It is thought that plant T3SS are responsible for the delivery of varied effector proteins, but the best known of these are the avirulence (Avr) proteins, named because of their failure to elicit disease in hosts with corresponding resistance genes (Alfano and Collmer 2004). Thus, T3SS effector molecules are thought to be important determinants in host specificity, and has been demonstrated in *Pantoea agglomerans* pv. *Gypsophilae*, whose avirulence on beet, which can be infected by the closely related *P. agglomerans* pv. *Betae*, appears to be solely limited by the presence of the effector protein PthG (Ezra et al. 2004).

2.5 Type IV Secretion Systems

Bacterial Type IV secretion systems (T4SS) are ancestrally related to bacterial plasmid conjugation systems and have played an important role in plant biology. The *Agrobacterium tumefaciens* VirB complex is a T4SS that facilitates the horizontal transfer of genes from the bacterium into a plant host. Under natural conditions, T-DNA carried on the tumor-inducing (Ti) plasmid is injected into a host which semirandomly inserts itself into the host genome, causing crown gall disease typified by tumor-like formations at the junction of the root and stem. Identification of the genetic mechanism behind this cross-kingdom genetic transfer has revolutionized plant molecular biology by allowing development of methods to intentionally transform plant genomes with foreign DNA (Schell and van Montagu 1977; Joos et al. 1983). Replacing the tumor-inducing genes carried on the Ti plasmid with DNA of interest allows the bacterium to insert the new DNA into the genome of compatible dicotyledonous plant hosts. This is done by simply soaking plant tissues in a solution containing *A. tumefaciens* carrying modified plasmids and growing new plants containing the DNA using tissue culturing techniques (Zupan and Zambryski 1995).

The T4SS can be classified into three subfamilies based on their function: conjugation systems, DNA uptake and release systems, and effector translocators (Cascales and Christie 2003). The conjugation subfamily is the largest and functions in the contact-dependent transfer of genetic material between the bacterium and recipient cells. The second subfamily has only been described in a handful of bacteria and also functions in the uptake or release of DNA, but in a contact-independent manner (Bacon et al. 2000; Dillard and Seifert 2001; Hofreuter et al. 2001). The T4SS in *A. tumefaciens* belongs to the third subfamily, which utilizes a syringe-like structure similar to that employed in T3SS to deliver virulence factors to a host. The *A. tumefaciens* T4SS is one of the best characterized with regards to structure and is made up of three main components: a coupling protein complex that provides energy to the transport, a transenvelope pore-forming complex for membrane translocation of DNA and effectors, and the conjugative pilus for delivery to host cells (Baron et al. 2002). This secretion system is important for bacterial environmental adaptation through the acquisition of genetic material and for pathogenesis in some species. In addition to the Ti plasmid necessary for the pathogenesis of *A. tumefaciens* on plant hosts, the pertussis toxin from *Bordetella pertussis* (Farizo et al. 2002) and the Dot/ICM substrates that facilitate intracellular replication of *Legionella pneumophila* (Conover et al. 2003) are also translocated using T4SS.

2.6 Type V or Autotransport Secretion System

The Type V secretion systems (T5SS) is a simple export mechanism that is similar to the T2SS in that both secrete proteins across the outer membrane utilizing a periplasmic intermediate step. However, unlike the T2SS and other systems we have discussed, the T5SS does not encode any inner membrane envelope-spanning complexes to harness energy from the inner membrane and drive transport. Instead, it is thought that the T5SS utilizes the Sec system to move proteins across the inner membrane and then draws energy from unfolded or partially folded higher energy intermediate states of the protein (Jacob-Dubuisson et al. 2004). These T5SS autotransporters (ATs) consist of large multidomain precursors that contain the N-terminal Sec-dependent signal sequence, a passenger domain that eventually becomes the secreted mature protein, and a C-terminal β-domain. Once transport across the inner membrane has occurred, the signal sequence is cleaved and it has been proposed that the β-domain inserts into the outer membrane forming a pore-like opening and driving transport of the passenger domain across the outer membrane. Outside the cell the passenger domain is then folded into its native conformation and may remain attached to the cell or cut and released into the external milieu. Many of the T5SS proteins that have been described are virulence factors from human pathogens, although the contribution to pathogen virulence of many of these has not been clearly demonstrated (Henderson and Nataro 2001). Examples include IgA proteases from *Neisseria* and *Haemophilus* (Pohlner et al. 1987; Poulsen et al. 1989), and serine protease autotransporters (SPATEs) that are the predominant secreted proteins of many enteric pathogens (Henderson et al. 1998).

Aside from the AT proteins, there is a two-partner secretion system (TPS), which is also classified as a T5SS. Like the AT secretion system, TPS is also dedicated to the transport of large protein virulence factors across the outer membrane. It differs from AT in that the secreted factor and the outer membrane component of the system are two separate proteins, usually encoded on a single operon (Thanassi et al. 2005). The outer membrane component of the TPS is one of the ubiquitously present transport proteins with homologues present in proteobacteria, cyanobacteria, and eukaryotes (Jacob-Dubuisson et al. 2001). The filamentous hemagglutinin adhesions (FHA) are important virulence factors from *B. pertussis* and *H. influenza* that utilize this mechanism of transport (Locht et al. 1993; St Geme 1994).

2.7 Type VI Secretion System

The Type VI secretion system (T6SS) is the most recent system to be discovered. Identification of a secreted protein from *V. cholerae* that lacked a signal peptide, unlike other secreted proteins from this bacteria, provided early evidence of the existence of an undescribed secretion system in 1996; however, it was not reported until a decade later (Bingle et al. 2008). Secretion of Hcp and three related VrgG proteins from *V. cholerae* were found to lack the N-terminal hydrophobic signal sequence required for transport by known systems and there was no nonflagellar T3SS or T4SS encoded in the genome that could account for transport of these proteins. The genes responsible for this mechanism were called "virulence-associated secretion" or VAS genes and were proposed to be a protypic T6SS (Pukatzki et al. 2006). Homologs of these genes are widespread among Gram-negative bacterial pathogens and in some cases attenuated virulence occurs when these genes are mutated. Currently, the components and mechanism of this secretion system are not well understood, but a central scaffolding component that interacts with itself, a DotU ortholog (associated with T4SS and virulence in *L. pneumophila*), and an outer membrane lipoprotein are all thought to play a role (Zheng and Leung 2007; Ma et al. 2009; Aschtgen et al. 2010). In *V. cholerae*, T6SS genes appear to be under regulation by quorum sensing and effector molecules, specifically translocation of VirG-1, directly results in fluid accumulation, altered host gene expression, and other host responses consistent with inflammatory diarrhea (Ma and Mekalanos 2010).

3 Conclusions

Bacteria secrete proteins as a means of interacting with other organisms, to respond to their environment, and to exchange genetic information that allows them to diversify and ensure continued survival of the species under varied conditions.

In addition to being potential targets for antimicrobials and vaccines, a better understanding of how and what bacteria are secreting can have important implications in a wide range of areas including biotechnology (as evidenced by plant transformations by *A. tumefaciens*), biofuel production, and the food industry. Despite the known diversity of secretion systems and their substrates, only a very small percentage of the total prokaryotic diversity has been cultured and studied. Culture-independent molecular methods of identifying microbial species in environmental samples have led to staggering new estimates of the total microbial diversity (Sogin et al. 2006; Roesch et al. 2007), suggesting that novel microbial secreted factors and their modes of transport also remain to be discovered. Metagenomics, an approach involving direct extraction and cloning of DNA from environmental samples containing assemblages of organisms, is one approach that could be used to identify novel secretion systems and effector molecules from uncultured microorganisms. A number of novel genes and gene products have already been identified by this approach including the first bacterial origin bacteriorhodopsin, new members of known protein families, and small molecule antibiotics (Handelsman 2004). Coupled with high throughput sequencing platforms, this is a promising approach to improve our understanding of the microbial world.

References

Aizawa S-I (2001) Bacterial flagella and type III secretion systems. FEMS Microbiol Lett 202:157–164

Alfano JR, Collmer A (2004) Type III secretion system effector proteins: double agents in bacterial disease and plant defense. Annu Rev Phytopathol 42:385–414

Asad S, Opal S (2008) Bench-to-bedside review: quorum sensing and the role of cell-to-cell communication during invasive bacterial infection. Crit Care 12:236

Aschtgen M-S, Gavioli M, Dessen A, Lloubès R, Cascales E (2010) The SciZ protein anchors the enteroaggregative *Escherichia coli* Type VI secretion system to the cell wall. Mol Microbiol 75:886–899

Bacon DJ, Alm RA, Burr DH, Hu L, Kopecko DJ, Ewing CP, Trust TJ, Guerry P (2000) Involvement of a plasmid in virulence of *Campylobacter jejuni* 81–176. Infect Immun 68:4384–4390

Barker J, Brown MRW (1994) Trojan Horses of the microbial world: protozoa and the survival of bacterial pathogens in the environment. Microbiology 140:1253–1259

Baron C, Callaghan O, Lanka DE (2002) Bacterial secrets of secretion: EuroConference on the biology of type IV secretion processes. Mol Microbiol 43:1359–1365

Bingle LEH, Bailey CM, Pallen MJ (2008) Type VI secretion: a beginner's guide. Curr Opin Microbiol 11:3–8

Büttner D, He SY (2009) Type III protein secretion in plant pathogenic bacteria. Plant Physiol 150:1656–1664

Cao TB, Saier MH (2003) The general protein secretory pathway: phylogenetic analyses leading to evolutionary conclusions. Biochim Biophys Acta 1609:115–125

Cascales E, Christie PJ (2003) The versatile bacterial type IV secretion systems. Nat Rev Microbiol 1:137–149

Chen L et al (2005) VFDB: a reference database for bacterial virulence factors. Nucleic Acid Research 33

Cianciotto NP (2005) Type II secretion: a protein secretion system for all seasons. Trends Microbiol 13:581–588

Conover GM, Derré I, Vogel JP, Isberg RR (2003) The Legionella pneumophila LidA protein: a translocated substrate of the Dot/Icm system associated with maintenance of bacterial integrity. Mol Microbiol 48:305–321

Cornelis GR (2006) The type III secretion injectisome. Nat Rev Microbiol 4:811–825

Desvaux M, Parham NJ, Scott-Tucker A, Henderson IR (2004) The general secretory pathway: a general misnomer? Trends Microbiol 12:306–309

Dillard JP, Seifert HS (2001) A variable genetic island specific for *Neisseria gonorrhoeae* is involved in providing DNA for natural transformation and is found more often in disseminated infection isolates. Mol Microbiol 41:263–277

Ezra D, Barash I, Weinthal DM, Gaba V, Manulis S (2004) pthG from *Pantoea agglomerans* pv. gypsophilae encodes an avirulence effector that determines incompatibility in multiple beet species. Mol Plant Pathol 5:105–113

Farizo KM, Fiddner S, Cheung AM, Burns DL (2002) Membrane localization of the S1 subunit of pertussis toxin in *Bordetella pertussis* and implications for pertussis toxin secretion. Infect Immun 70:1193–1201

Filloux A (2004) The underlying mechanisms of type II protein secretion. Biochim Biophys Acta 1694:163–179

Finlay BB, Falkow S (1997) Common themes in microbial pathogenicity revisited. Am Soc Microbiol 61:136–169

Gentschev I, Goebel W (1992) Topological and functional studies on HlyB of *Escherichia coli*. Mol Gen Genet 232:40–48

Gentschev I, Dietrich G, Goebel W (2002) The *E. coli* $\pm$ −hemolysin secretion system and its use in vaccine development. Trends Microbiol 10:39–45

Grynberg M, Godzik A (2009) The signal for signaling found. PLoS Pathog 5:e1000398

Handelsman J (2004) Metagenomics: application of genomics to uncultured microorganisms. Microbiol Mol Biol Rev 68:669–685

Hazelbauer GL, Falke JJ, Parkinson JS (2007) Bacterial chemoreceptors: high-performance signaling in networked arrays. Trends Biochem Sci 33:9–19

Henderson IR, Nataro JP (2001) Virulence functions of autotransporter proteins. Infect Immun 69:1231–1243

Henderson IR, Navarro-Garcia F, Nataro JP (1998) The great escape: structure and function of the autotransporter proteins. Trends Microbiol 6:370–378

Hofreuter D, Odenbreit S, Haas R (2001) Natural transformation competence in *Helicobacter pylori* is mediated by the basic components of a type IV secretion system. Mol Microbiol 41:379–391

Hotta K, Kim C-Y, Fox DT, Koppisch AT (2010) Siderophore-mediated iron acquisition in *Bacillus anthracis* and related strains. Microbiology 156:1918–1925

Hueck CJ (1998) Type III protein secretion systems in bacterial pathogens of animals and plants. Microbiol Mol Biol Rev 62:379–433

Jacob-Dubuisson F, Locht C, Antoine R (2001) Two-partner secretion in Gram-negative bacteria: a thrifty, specific pathway for large virulence proteins. Mol Microbiol 40:306–313

Jacob-Dubuisson F, Fernandez R, Coutte L (2004) Protein secretion through autotransporter and two-partner pathways. Biochem Biophys Acta 1694:235–257

Johnson JM, Church GM (1999) Alignment and structure prediction of divergent protein families: periplasmic and outer membrane proteins of bacterial efflux pumps. J Mol Biol 287:695–715

Joos H, Inze D, Caplan A, Sormann M, Van Montagu M, Schell J (1983) Genetic analysis of T-DNA transcripts in nopaline crown galls. Cell 32:1057–1067

Koronakis V, Sharff A, Koronakis E, Luisi B, Hughes C (2000) Crystal structure of the bacterial membrane protein TolC central to multidrug efflux and protein export. Nature 405:914–919

Korves T, Colosimo ME (2009) Controlled vocabularies for microbial virulence factors. Trends Microbiol 17:279–285

Locht C, Berlin P, Menozzi FD, Renauld G (1993) The filamentous haemagglutinin, a multifaceted adhesin produced by virulent Bordetella spp. Mol Microbiol 9:653–660

Loper JE, Buyer JS (1991) Siderophores in microbial interactions on plant surfaces. Mol Plant Microbe Interact 4:5–13

Ma AT, Mekalanos JJ (2010) In vivo actin cross-linking induced by *Vibrio cholerae* type VI secretion system is associated with intestinal inflammation. Proc Natl Acad Sci USA 107:4365–4370

Ma Q, Zhai Y, Schneider JC, Ramseier TM, Saier MH (2003) Protein secretion systems of *Pseudomonas aeruginosa* and *P. fluorescens*. Biochim Biophys Acta 1611:223–233

Ma L-S, Lin J-S, Lai E-M (2009) An IcmF family protein, ImpLM, is an integral inner membrane protein interacting with ImpKL, and its Walker A motif is required for Type VI secretion system-mediated Hcp secretion in *Agrobacterium tumefaciens*. J Bacteriol 191:4316–4329

Mole BM, Baltrus DA, Dangl JL, Grant SR (2007) Global virulence regulation networks in phytopathogenic bacteria. Trends Microbiol 15:363–371

Pohlner J, Halter R, Beyreuther K, Meyer TF (1987) Gene structure and extracellular secretion of Neisseria gonorrhoeae IgA protease. Nature 325:458–462

Poulsen K, Brandt J, Hjorth JP, Thøgersen HC, Kilian M (1989) Cloning and sequencing of the immunoglobulin A1 protease gene (iga) of *Haemophilus influenzae* serotype b. Infect Immun 57:3097–3105

Prescott LM, Harley JP, Klein DA (2005) Microbiology, 6th edn. McGraw-Hill, New York, NY

Prithiviraj B, Weir TL, Bais HP, Schweizer HP, Vivanco JM (2006) Plant models for animal pathogenesis. Cell Microbiol 7:315–324

Pukatzki S, Ma AT, Sturtevant D, Krastins B, Sarracino D, Nelson WC, Heidelberg JF, Mekalanos JJ (2006) Identification of a conserved bacterial protein secretion system in *Vibrio cholerae* using the Dictyostelium host model system. Proc Natl Acad Sci USA 103:1528–1533

Roesch LFW, Fulthorpe RR, Riva A, Casella G, Hadwin AKM, Kent AD, Daroub SH, Camargo FAO, Farmerie WG, Triplett EW (2007) Pyrosequencing enumerates and contrasts soil microbial diversity. ISME J 1:283–290

Sandkvist M (2001) Type II secretion and pathogenesis. Infect Immun 69:3523–3535

Schell J, van Montagu M (1977) The Ti-plasmid of Agrobacterium tumefaciens, a natural vector for the introduction of nif genes in plants. Basic Life Sci 9:159–179

Schmidt AJ, Ryjenkov DA, Gomelsky M (2005) The ubiquitous protein domain EAL is a cyclic diguanylate-specific phosphodiesterase: enzymatically active and inactive EAL domains. J Bacteriol 187:4774–4781

Seshasayee ASN, Fraser GM, Luscombe NM (2010) Comparative genomics of cyclic-di-GMP signaling in bacteria: post-translational regulation and catalytic activity. Nucleic Acids Res 38:1–12

Sibbald MJJB, van Dijl JM (2009) Secretome mapping in Gram-positive pathogens. In: Wooldridge K (ed) Bacterial secreted proteins: secretory mechanisms and role in pathogenesis. Horizon Scientific, Norwich, pp 193–224

Sogin ML, Morrison HG, Huber JA, Welch DM, Huse SM, Neal PR, Arrieta JM, Herndl GJ (2006) Microbial diversity in the deep sea and the underexplored "rare biosphere". Proc Natl Acad Sci USA 103:12115–12120

St Geme JW III (1994) The HMW1 adhesin of nontypeable Haemophilus influenzae recognizes sialylated glycoprotein receptors on cultured human epithelial cells. Infect Immun 62:3881–3889

Teter SA, Klionsky DJ (1999) How to get a folded protein across a membrane. Trends Cell Biol 9:428–431

Thanassi DG, Stathopoulos C, Karkal A, Li H (2005) Protein secretion in the absence of ATP: the autotransporter, two-partner secretion and chaperone/usher pathways of Gram-negative bacteria. Mol Membr Biol 22:63–72

Van Gijsegem F, Genin S, Boucher C (1993) Evolutionary conservation of pathogenicity determinants among plant and animal pathogenic bacteria. Trends Microbiol 1:175–180

Wagner W, Vogel M, Goebel W (1983) Transport of haemolysin across the outer membrane of Eschericia coli requires two functions. J Bacteriol 154:200–210

Wooldridge K (2009) Bacterial secreted proteins: secretory mechanisms and role in pathogenesis. Caister Academic, Norwich

Wu H-J, Want A, Jennings MP (2008) Discovery of virulence factors of pathogenic bacteria. Curr Opin Chem Biol 12:93–101

Zgurskaya HI, Nikaido H (2000) Multidrug resistance mechanisms: drug efflux across two membranes. Mol Microbiol 37:219–225

Zheng J, Leung KY (2007) Dissection of a type VI secretion system in *Edwardsiella tarda*. Mol Microbiol 66:1192–1206

Zhou D, Galán J (2001) Salmonella entry into host cells: the work in concert of type III secreted effector proteins. Microb Infect 3:1293–1298

Zupan JR, Zambryski P (1995) Transfer of T-DNA from Agrobacterium to the plant cell. Plant Physiol 107:1041–1047

Inter-Organ and -Tissue Communication via Secreted Proteins in Humans

Michael Pagliassotti

Abstract Adipose tissue is now recognized as an important source of bioactive molecules, so-called adipokines. These adipokines, in turn, modulate hormone action, nutrient metabolism, inflammation, and energy balance. Thus, adipokines are part of a complex biologic system that allows adipose tissue to communicate with other tissue/organ systems, including skeletal muscle, liver, and brain. Adiponectin is an abundant plasma adipokine that is secreted primarily from adipocytes. Unlike other adipokines, plasma concentrations of adiponectin are decreased in obesity and metabolic syndrome. Therefore, impaired regulation of adiponectin secretion and signaling appears to be important in the development and maintenance of obesity and obesity-related complications. In this chapter, adiponectin secretion and action will be examined, as an example of how organ-specific secretion is linked to biologic homeostasis in humans.

1 Introduction

Most proteins that are secreted from the cell are synthesized on membranes of the endoplasmic reticulum (ER) and transported through the membrane to the ER lumen. Proteins in the secretory pathway that are destined for compartments other than the ER or Golgi eventually interact with the trans-Golgi network. From this network, proteins can be loaded into one of at least three types of vesicles. The first type of vesicle that buds from the trans-Golgi network is directed to the lysosome, an organelle responsible for the intracellular degradation of macromolecules. Soluble proteins delivered by this pathway include lysosomal digestive enzymes, such as proteases, and membrane proteins, such as V-class protein pump proteins.

M. Pagliassotti (✉)
Department of Food Science and Human Nutrition, Colorado State University, Fort Collins, CO 80523, USA
e-mail: pagliasm@cahs.colostate.edu

J.M. Vivanco and F. Baluška (eds.), *Secretions and Exudates in Biological Systems*, Signaling and Communication in Plants 12, DOI 10.1007/978-3-642-23047-9_12,

The second type, moves to and fuses with the plasma membrane, releasing its contents by exocytosis in a continuous or constitutive manner. Examples of proteins released by this type of vesicle include collagen by fibroblasts and serum proteins by hepatocytes. The third type, secretory vesicle, is stored inside the cell until a signal for exocytosis causes release of their contents at the plasma membrane. Examples of proteins released in this regulated manner include insulin and glucagon from the pancreas, milk proteins from the mammary gland, and various proteins secreted by white adipose tissue.

2 White Adipose Tissue as a Secretory Organ

The traditional view of white adipose tissue was that this organ functioned as a fuel reservoir, storing energy in the form of lipid during periods of positive energy balance and mobilizing stored lipids to provide free fatty acids to the organism during periods of negative energy balance. This view also included a role for white adipose tissue in thermal insulation (for example, blubber of marine mammals). However, white adipose tissue also expresses a large number of secreted proteins and the identification of leptin in 1994 led to the general recognition that white adipose tissue could also function as an endocrine organ (Zhang et al. 1994; Trayhurn and Beattie 2001; Ailhaud 2006; Galic et al. 2010; Karastergiou and Mohamed-Ali 2010). The term adipokine is now used to identify proteins secreted by adipocytes from white adipose tissue.

3 Adiponectin

In 1995, a fat cell-specific protein was identified that contained homology to complement factor Clq and was termed adipocyte-complement related protein of 30 kDa (Acrp30) (Scherer et al. 1995; Berg et al. 2002). Additional studies reported the identification of this cDNA but named it AdipoQ (Hu et al. 1996), apM1 (Maeda et al. 1996), and GBP28 (Nakano et al. 1996). The most widely used name in the literature for this protein is adiponectin. Adiponectin can act on the liver to potentiate insulin suppression of hepatic gluconeogenesis (Berg et al. 2002) and on muscle where a proteolytic fragment of adiponectin can stimulate fatty acid oxidation (Fruebis et al. 2001; Iyengar and Scherer 2003). Thus, adiponectin serves to link the adipocyte with other peripheral organs and tissues (Iyengar and Scherer 2003; Wang et al. 2008a).

Adiponectin is synthesized as a 32-kDa monomeric protein and is then assembled into a low molecular weight trimer, a medium molecular weight hexamer, consisting of two trimers, and a high molecular weight (HMW) multimer, consisting of 4–6 trimers (Galic et al. 2010). HMW adiponectin is thought to be the active form in plasma (Hara et al. 2006). Posttranslational modifications, such

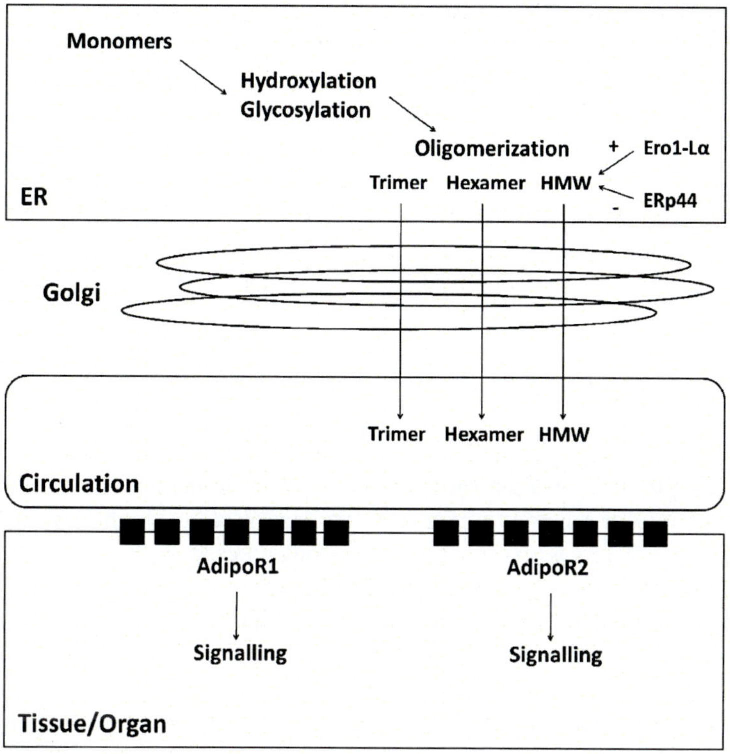

Fig. 1 Schematic of adiponectin processing in the endoplasmic reticulum (ER), forms present in the circulation, and receptor-mediated signaling in tissues and organs. See text for details

as lysine hydroxylation and glycosylation, appear to be critical to the formation of HMW complexes (Iyengar and Scherer 2003; Galic et al. 2010) (Fig. 1). Both mouse and bovine adiponectin contain four conserved proline residues that undergo hydroxylation, and five conserved lysine residues that undergo hydroxylation and glycosylation (Wang et al. 2008b). The distribution of circulating adiponectin oligomers appears to be regulated at the stage of secretion from adipocytes. Adiponectin secretion is controlled by the ER chaperone proteins, ER resident protein-44 (ERp44), and ER oxidoreductin-1 (Ero1-Lα), which cooperate to regulate adiponectin retention and release, respectively (Qiang et al. 2007; Wang et al. 2007) (Fig. 1). It appears that disulfide bond formation between ERp44 and Cys-39 facilitates the maturation and assembly of adiponectin into HMW complexes by extending the period of time that adiponectin is held in the ER lumen (Wang et al. 2007). ERo1-Lα appears to function by displacing adiponectin from ERp44 thereby promoting its release into the circulation (Qiang et al. 2007). However, it may also contribute to adiponectin maturation and folding within the ER through disulfide bond transfer (Wang et al. 2008b). Decreased expression of ERp44 and ERo1-Lα in adipose tissue of ob/ob mice is associated with a reduced ratio of HMW to total adiponectin in the circulation (Wang et al. 2007). Thiazolidinediones (TZDs), a class of insulin sensitizing agents that include rosiglitazone, may promote the selective secretion of the HMW form of adiponectin, in part, via upregulation of ERo1-Lα and downregulation of ERp44 (Maeda et al. 2001; Pajvani et al. 2004; Phillips et al. 2009).

4 Circulating Adiponectin as a Signal of Metabolic Status

In contrast to leptin, adiponectin levels are decreased in obese animals and human subjects (Ahima and Osei 2008). In human subjects with type 2 diabetes or cardiovascular disease, lower serum levels of total adiponectin appear to be almost entirely accounted for by selective reductions of the HMW oligomer (Kobayashi et al. 2004; Basu et al. 2007). Weight reduction, by calorie restriction or gastric bypass surgery, results in a selective elevation of the HMW oligomer (Bobbert et al. 2005; Salani et al. 2006; Swarbrick et al. 2006), whereas TZD treatment, which improves insulin sensitivity, increased the HMW oligomer in both diabetic mice and human subjects with type 2 diabetes (Pajvani et al. 2004).

In the first study to comprehensively examine the relationship between adiponectin and metabolic parameters, Lara-Castro and colleagues demonstrated that the amount of HMW adiponectin is highly correlated with multiple traits that characterize the metabolic syndrome (Lara-Castro et al. 2006). In particular, reduced levels of HMW adiponectin were associated with upper body fat distribution, insulin resistance, impaired lipid oxidation, and dyslipidemia in 68 subjects (33 women and 35 men) with and without type 2 diabetes (Lara-Castro et al. 2006). When combined with other studies, these results suggest that HMW adiponectin is an important biomarker for the metabolic syndrome (low levels linked to metabolic syndrome) and could play a pathogenic role in the development of impairments associated with obesity, cardiovascular disease, and type 2 diabetes (Araki et al. 2006; Hara et al. 2006; Lara-Castro et al. 2006; Liu et al. 2007; Torigoe et al. 2007).

5 Adiponectin Receptors Link Adiponectin to Biologic Activity

Adiponectin receptors 1 and 2 (AdipoR1, AdipoR2) were discovered by screening of cells that were transfected with a cDNA library from human skeletal muscle tissue for the binding of globular adiponectin (Yamauchi et al. 2003) (Fig. 1). The genes for human AdipoR1 and AdipoR2 are located on different chromosomes. The AdipoR1 protein consists of 375 amino acids with a predicted molecular mass of 42 kDa and the AdipoR2 protein consists of 386 amino acids with a predicted molecular mass of 43 kDa (Heiker et al. 2010). Together with the receptor for the steroid hormone progesterone, AdipoRs are currently classified into a progesterone and AdipoR receptor (PAQR) superfamily (Tang et al. 2005). PAQRs appear to be present in a wide variety of organisms including eubacteria and AdipoRs also appear to be highly conserved across species. PAQRs are expressed in a wide range of tissues and while both AdipoR1 and AdipoR2 are predominantly expressed in skeletal muscle and liver, they are also present in nearly every other human or mouse tissue and a variety of human and mouse cell lines (Yamauchi et al. 2003; Kadowaki et al. 2008; Heiker et al. 2010). In particular, studies have identified AdipoRs in human adipose tissue, pituitary gland, hypothalamus, and nucleus basalis of Meynert, thus it has been postulated that AdipoRs, in addition to regulating the endocrine actions of adiponectin, also may regulate the autocrine/paracrine actions of this protein (Nannipieri et al. 2009; Psilopanagioti et al. 2009; Savu et al. 2009). In addition, the presence of AdipoR1 in brain regions suggests that adiponectin may also integrate central neural signaling pathways (Psilopanagioti et al. 2009).

AdipoR1 and AdipoR2 have the ability to dimerize and internalize, and thus display characteristics of cell surface receptors that are linked to signal modulation (Heiker et al. 2010). In addition, the expression of AdipoR1 and AdipoR2 were significantly reduced in both skeletal muscle and adipose tissue of ob/ob mice, which exhibit obesity, insulin resistance, and hyperglycemia compared to control mice (Tsuchida et al. 2004). Studies have also found a direct correlation between adiponectin receptor gene expression and insulin sensitivity in nondiabetic Mexican Americans with or without a family history of type 2 diabetes (Civitarese et al. 2004). A separate study reported that adiponectin receptor expression in skeletal muscle was reduced in type 2 diabetic patients (Debard et al. 2004). Thus, there appears to be a coupling between white adipose tissue adiponectin secretion and adiponectin receptor expression and both appear to be reduced in diseases associated with disturbances in glucose and lipid homeostasis. This coupling implies that white adipose tissue communicates with other tissues and organs, in part, via secreted proteins (e.g., adiponectin) that interact with specific surface receptors (e.g., AdipoRs). The next section will summarize biologic outputs that result from the engagement of adiponectin with AdipoRs.

6 Biologic Actions of Adiponectin

Adiponectin has a number of potential biologic actions that influence nutrient metabolism in skeletal muscle and liver, endothelial and cardiac tissue function, and potentially food intake (Fig. 2).

6.1 Nutrient Metabolism

Adiponectin administration lowered plasma glucose, free fatty acid, and triglyceride levels in mice via enhanced insulin action and increased fatty acid oxidation (Fruebis et al. 2001; Nawrocki et al. 2006; Hopkins et al. 2007). The role of adiponectin in the regulation of circulating glucose and lipid levels is mediated,

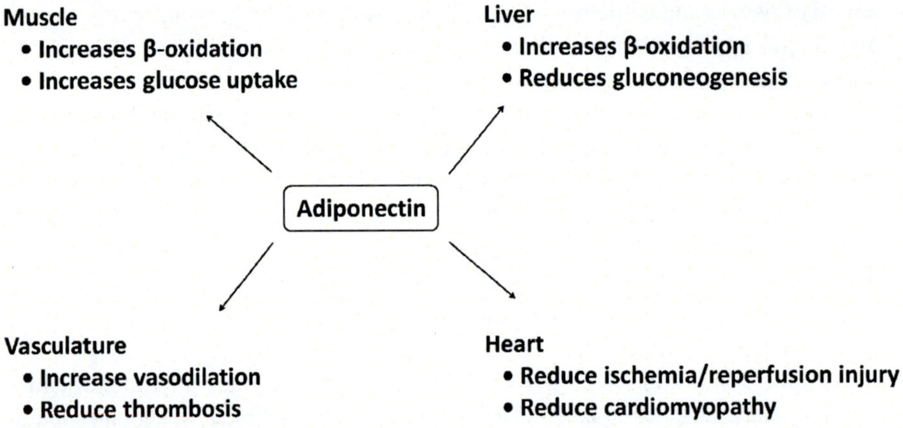

Fig. 2 Schematic of adiponectin action in skeletal muscle, liver, vasculature, and heart. See text for details

at least in part, by promoting the phosphorylation and activation of AMP-activated protein kinase (AMPK) in liver, skeletal muscle, and adipocytes (Tomas et al. 2002; Yamauchi et al. 2002; Wu et al. 2003). AMPK, which exists as a heterodimer comprising a catalytic α subunit and regulatory β and γ subunits, is a downstream component of a kinase cascade that acts as a sensor of cellular energy status (Kahn et al. 2005). Adiponectin-mediated activation of AMPK can lead to (1) GLUT4 translocation to the cell surface to promote glucose uptake in muscle and adipose tissue, (2) phosphorylation of phosphofructokinase-2 to increase glycolysis, (3) phosphorylation of acetyl-CoA carboxylase which can enhance fatty acid oxidation, and (4) suppression of hepatic glucose production, via inhibition of gluconeogenesis (Hopkins et al. 2007; Galic et al. 2010).

6.2 *Endothelial and Cardiac Tissue Function*

The activation of nuclear factor-κB (NF-κB), a protein involved in the inflammatory response, is a critical component of early stage atherosclerosis development (Mahadik et al. 2008). Adiponectin can reduce NF-κB activation and monocyte adhesion, expression of proinflammatory cytokines, such as IL-8 and TNF-α, and increase expression of anti-inflammatory cytokines, such as IL-10 in endothelial cells and/or macrophages (Ouchi et al. 2000, 2001; Yokota et al. 2000; Wulster-Radcliffe et al. 2004; Kobashi et al. 2005; Yamaguchi et al. 2005). Enforced overexpression of adiponectin also reduced the formation of atherosclerotic lesions in the aortic sinus of apolipoprotein E knockout mice (an animal model of atherosclerosis) (Okamoto et al. 2002). Adiponectin can also influence cardiac remodeling and may function to reduce or inhibit pathological cardiac growth, for example, in response to pressure overload (Hopkins et al. 2007). Obesity and related disorders are characterized by an increased incidence and severity of ischemic heart disease. Adiponectin may serve a cardioprotective role in this setting via effects on AMPK and apoptosis, and its ability to activate cyclooxygenase-2 in cardiac cells (Shibata et al. 2005).

6.3 Central Actions of Adiponectin

Recent studies have also suggested that adiponectin can regulate energy expenditure and food intake via activation of AMPK in the hypothalamus (Kubota et al. 2007). It is presently unclear what form of adiponectin influences hypothalamic signaling, since the HMW adiponectin may be too large to cross the blood–brain barrier (Galic et al. 2010).

7 Summary and Concluding Remarks

White adipose tissue is equipped with a well-organized vasculature and nerve supply, which enables it to interact with other tissue and organ systems throughout the body. Adipocytes vary in size, are embedded in a connective tissue matrix, and are uniquely adapted to store and release energy. However, white adipose tissue and adipocytes are also actively involved in the maintenance of cellular function of a diverse set of tissues/organs, including skeletal muscle, liver, and brain. These interactions are accomplished, in part, through endocrine-mediated mechanisms that involve secretion of biologically active adipokines. In this context, the secretion of adiponectin by white adipose tissue represents an important example of how adipose tissue secretory function is linked to whole body homeostasis. Adiponectin-mediated regulation involves a complex biologic system that includes protein synthesis, protein maturation and processing, vesicular transport and exocytosis, adiponectin–AdipoRs interaction, and engagement of tissue/organ signaling networks. Identification of the sites within this system that contribute to the decrease of circulating adiponectin and downregulation of AdipoRs in obesity and obesity-related disorders will lead to new insights into these diseases.

References

Ahima RS, Osei SY (2008) Adipokines in obesity. Front Horm Res 36:182–197

Ailhaud G (2006) Adipose tissue as a secretory organ: from adipogenesis to the metabolic syndrome. C R Biol 329:570–577

Araki S, Dobashi K, Kubo K, Asayama K, Shirahata A (2006) High molecular weight, rather than total, adiponectin levels better reflect metabolic abnormalities associated with childhood obesity. J Clin Endocrinol Metab 91:5113–5116

Basu R, Pajvani UB, Rizza RA, Scherer PE (2007) Selective downregulation of the high molecular weight form of adiponectin in hyperinsulinemia and in type 2 diabetes: differential regulation from nondiabetic subjects. Diabetes 56:2174–2177

Berg AH, Combs TP, Scherer PE (2002) ACRP30/adiponectin: an adipokine regulating glucose and lipid metabolism. Trends Endocrinol Metab 13:84–89

Bobbert T, Rochlitz H, Wegewitz U, Akpulat S, Mai K, Weickert MO, Mohlig M, Pfeiffer AF, Spranger J (2005) Changes of adiponectin oligomer composition by moderate weight reduction. Diabetes 54:2712–2719

Civitarese AE, Jenkinson CP, Richardson D, Bajaj M, Cusi K, Kashyap S, Berria R, Belfort R, DeFronzo RA, Mandarino LJ, Ravussin E (2004) Adiponectin receptors gene expression and insulin sensitivity in non-diabetic Mexican Americans with or without a family history of Type 2 diabetes. Diabetologia 47:816–820

Debard C, Laville M, Berbe V, Loizon E, Guillet C, Morio-Liondore B, Boirie Y, Vidal H (2004) Expression of key genes of fatty acid oxidation, including adiponectin receptors, in skeletal muscle of Type 2 diabetic patients. Diabetologia 47:917–925

Fruebis J, Tsao TS, Javorschi S, Ebbets-Reed D, Erickson MR, Yen FT, Bihain BE, Lodish HF (2001) Proteolytic cleavage product of 30-kDa adipocyte complement-related protein increases fatty acid oxidation in muscle and causes weight loss in mice. Proc Natl Acad Sci USA 98:2005–2010

Galic S, Oakhill JS, Steinberg GR (2010) Adipose tissue as an endocrine organ. Mol Cell Endocrinol 316:129–139

Hara K, Horikoshi M, Yamauchi T, Yago H, Miyazaki O, Ebinuma H, Imai Y, Nagai R, Kadowaki T (2006) Measurement of the high-molecular weight form of adiponectin in plasma is useful for the prediction of insulin resistance and metabolic syndrome. Diabetes Care 29:1357–1362

Heiker JT, Kosel D, Beck-Sickinger AG (2010) Molecular mechanisms of signal transduction via adiponectin and adiponectin receptors. Biol Chem 391:1005–1018

Hopkins TA, Ouchi N, Shibata R, Walsh K (2007) Adiponectin actions in the cardiovascular system. Cardiovasc Res 74:11–18

Hu E, Liang P, Spiegelman BM (1996) AdipoQ is a novel adipose-specific gene dysregulated in obesity. J Biol Chem 271:10697–10703

Iyengar P, Scherer PE (2003) Adiponectin/Acrp30, an adipocyte-specific secretory factor: physiological relevance during development. Pediatr Diabetes 4:32–37

Kadowaki T, Yamauchi T, Kubota N (2008) The physiological and pathophysiological role of adiponectin and adiponectin receptors in the peripheral tissues and CNS. FEBS Lett 582:74–80

Kahn BB, Alquier T, Carling D, Hardie DG (2005) AMP-activated protein kinase: ancient energy gauge provides clues to modern understanding of metabolism. Cell Metab 1:15–25

Karastergiou K, Mohamed-Ali V (2010) The autocrine and paracrine roles of adipokines. Mol Cell Endocrinol 318:69–78

Kobashi C, Urakaze M, Kishida M, Kibayashi E, Kobayashi H, Kihara S, Funahashi T, Takata M, Temaru R, Sato A, Yamazaki K, Nakamura N, Kobayashi M (2005) Adiponectin inhibits endothelial synthesis of interleukin-8. Circ Res 97:1245–1252

Kobayashi H, Ouchi N, Kihara S, Walsh K, Kumada M, Abe Y, Funahashi T, Matsuzawa Y (2004) Selective suppression of endothelial cell apoptosis by the high molecular weight form of adiponectin. Circ Res 94:e27–e31

Kubota N, Yano W, Kubota T, Yamauchi T, Itoh S, Kumagai H, Kozono H, Takamoto I, Okamoto S, Shiuchi T, Suzuki R, Satoh H, Tsuchida A, Moroi M, Sugi K, Noda T, Ebinuma H, Ueta Y, Kondo T, Araki E, Ezaki O, Nagai R, Tobe K, Terauchi Y, Ueki K, Minokoshi Y, Kadowaki T (2007) Adiponectin stimulates AMP-activated protein kinase in the hypothalamus and increases food intake. Cell Metab 6:55–68

Lara-Castro C, Luo N, Wallace P, Klein RL, Garvey WT (2006) Adiponectin multimeric complexes and the metabolic syndrome trait cluster. Diabetes 55:249–259

Liu Y, Retnakaran R, Hanley A, Tungtrongchitr R, Shaw C, Sweeney G (2007) Total and high molecular weight but not trimeric or hexameric forms of adiponectin correlate with markers of the metabolic syndrome and liver injury in Thai subjects. J Clin Endocrinol Metab 92:4313–4318

Maeda K, Okubo K, Shimomura I, Funahashi T, Matsuzawa Y, Matsubara K (1996) cDNA cloning and expression of a novel adipose specific collagen-like factor, apM1 (AdiPose Most abundant Gene transcript 1). Biochem Biophys Res Commun 221:286–289

Maeda N, Takahashi M, Funahashi T, Kihara S, Nishizawa H, Kishida K, Nagaretani H, Matsuda M, Komuro R, Ouchi N, Kuriyama H, Hotta K, Nakamura T, Shimomura I, Matsuzawa Y (2001) PPARgamma ligands increase expression and plasma concentrations of adiponectin, an adipose-derived protein. Diabetes 50:2094–2099

Mahadik SR, Deo SS, Mehtalia SD (2008) Association of adiposity, inflammation and atherosclerosis: the role of adipocytokines and CRP in Asian Indian subjects. Metab Syndr Relat Disord 6:121–128

Nakano Y, Tobe T, Choi-Miura NH, Mazda T, Tomita M (1996) Isolation and characterization of GBP28, a novel gelatin-binding protein purified from human plasma. J Biochem 120:803–812

Nannipieri M, Cecchetti F, Anselmino M, Mancini E, Marchetti G, Bonotti A, Baldi S, Solito B, Giannetti M, Pinchera A, Santini F, Ferrannini E (2009) Pattern of expression of adiponectin receptors in human liver and its relation to nonalcoholic steatohepatitis. Obes Surg 19:467–474

Nawrocki AR, Rajala MW, Tomas E, Pajvani UB, Saha AK, Trumbauer ME, Pang Z, Chen AS, Ruderman NB, Chen H, Rossetti L, Scherer PE (2006) Mice lacking adiponectin show decreased hepatic insulin sensitivity and reduced responsiveness to peroxisome proliferator-activated receptor gamma agonists. J Biol Chem 281:2654–2660

Okamoto Y, Kihara S, Ouchi N, Nishida M, Arita Y, Kumada M, Ohashi K, Sakai N, Shimomura I, Kobayashi H, Terasaka N, Inaba T, Funahashi T, Matsuzawa Y (2002) Adiponectin reduces atherosclerosis in apolipoprotein E-deficient mice. Circulation 106:2767–2770

Ouchi N, Kihara S, Arita Y, Okamoto Y, Maeda K, Kuriyama H, Hotta K, Nishida M, Takahashi M, Muraguchi M, Ohmoto Y, Nakamura T, Yamashita S, Funahashi T, Matsuzawa Y (2000) Adiponectin, an adipocyte-derived plasma protein, inhibits endothelial NF-kappaB signaling through a cAMP-dependent pathway. Circulation 102:1296–1301

Ouchi N, Kihara S, Arita Y, Nishida M, Matsuyama A, Okamoto Y, Ishigami M, Kuriyama H, Kishida K, Nishizawa H, Hotta K, Muraguchi M, Ohmoto Y, Yamashita S, Funahashi T, Matsuzawa Y (2001) Adipocyte-derived plasma protein, adiponectin, suppresses lipid accumulation and class A scavenger receptor expression in human monocyte-derived macrophages. Circulation 103:1057–1063

Pajvani UB, Hawkins M, Combs TP, Rajala MW, Doebber T, Berger JP, Wagner JA, Wu M, Knopps A, Xiang AH, Utzschneider KM, Kahn SE, Olefsky JM, Buchanan TA, Scherer PE (2004) Complex distribution, not absolute amount of adiponectin, correlates with thiazolidinedione-mediated improvement in insulin sensitivity. J Biol Chem 279:12152–12162

Phillips SA, Kung J, Ciaraldi TP, Choe C, Christiansen L, Mudaliar S, Henry RR (2009) Selective regulation of cellular and secreted multimeric adiponectin by antidiabetic therapies in humans. Am J Physiol Endocrinol Metab 297:E767–E773

Psilopanagioti A, Papadaki H, Kranioti TK, Alexandrides EF, Varakis JN (2009) Expression of adiponectin and adiponectin receptors in human pituitary gland and brain. Neuroendocrinology 89:38–47

Qiang L, Wang H, Farmer SR (2007) Adiponectin secretion is regulated by SIRT1 and the endoplasmic reticulum oxidoreductase Ero1-L alpha. Mol Cell Biol 27:4698–4707

Salani B, Briatore L, Andraghetti G, Adami GF, Maggi D, Cordera R (2006) High-molecular weight adiponectin isoforms increase after biliopancreatic diversion in obese subjects. Obesity (Silver Spring) 14:1511–1514

Savu MK, Phillips SA, Oh DK, Park K, Gerlan C, Ciaraldi TP, Henry RR (2009) Response of adiponectin and its receptors to changes in metabolic state after gastric bypass surgery: dissociation between adipose tissue expression and circulating levels. Surg Obes Relat Dis 5:172–180

Scherer PE, Williams S, Fogliano M, Baldini G, Lodish HF (1995) A novel serum protein similar to C1q, produced exclusively in adipocytes. J Biol Chem 270:26746–26749

Shibata R, Sato K, Pimentel DR, Takemura Y, Kihara S, Ohashi K, Funahashi T, OuchiN WK (2005) Adiponectin protects against myocardial ischemia-reperfusion injury through AMPK- and COX-2-dependent mechanisms. Nat Med 11:1096–1103

Swarbrick MM, Austrheim-Smith IT, Stanhope KL, Van Loan MD, Ali MR, Wolfe BM, Havel PJ (2006) Circulating concentrations of high-molecular-weight adiponectin are increased following Roux-en-Y gastric bypass surgery. Diabetologia 49:2552–2558

Tang YT, Hu T, Arterburn M, Boyle B, Bright JM, Emtage PC, Funk WD (2005) PAQR proteins: a novel membrane receptor family defined by an ancient 7-transmembrane pass motif. J Mol Evol 61:372–380

Tomas E, Tsao TS, Saha AK, Murrey HE, Zhang CCC, Itani SI, Lodish HF, Ruderman NB (2002) Enhanced muscle fat oxidation and glucose transport by ACRP30 globular domain: acetyl-CoA carboxylase inhibition and AMP-activated protein kinase activation. Proc Natl Acad Sci USA 99:16309–16313

Torigoe M, Matsui H, Ogawa Y, Murakami H, Murakami R, Cheng XW, Numaguchi Y, Murohara T, Okumura K (2007) Impact of the high-molecular-weight form of adiponectin on endothelial function in healthy young men. Clin Endocrinol 67:276–281

Trayhurn P, Beattie JH (2001) Physiological role of adipose tissue: white adipose tissue as an endocrine and secretory organ. Proc Nutr Soc 60:329–339

Tsuchida A, Yamauchi T, Ito Y, Hada Y, Maki T, Takekawa S, Kamon J, Kobayashi M, Suzuki R, Hara K, Kubota N, Terauchi Y, Froguel P, Nakae J, Kasuga M, Accili D, Tobe K, Ueki K, Nagai R, Kadowaki T (2004) Insulin/Foxo1 pathway regulates expression levels of adiponectin receptors and adiponectin sensitivity. J Biol Chem 279:30817–30822

Wang ZV, Schraw TD, Kim JY, Khan T, Rajala MW, Follenzi A, Scherer PE (2007) Secretion of the adipocyte-specific secretory protein adiponectin critically depends on thiol-mediated protein retention. Mol Cell Biol 27:3716–3731

Wang P, Mariman E, Renes J, Keijer J (2008a) The secretory function of adipocytes in the physiology of white adipose tissue. J Cell Physiol 216:3–13

Wang Y, Lam KS, Yau MH, Xu A (2008b) Post-translational modifications of adiponectin: mechanisms and functional implications. Biochem J 409:623–633

Wu X, Motoshima H, Mahadev K, Stalker TJ, Scalia R, Goldstein BJ (2003) Involvement of AMP-activated protein kinase in glucose uptake stimulated by the globular domain of adiponectin in primary rat adipocytes. Diabetes 52:1355–1363

Wulster-Radcliffe MC, Ajuwon KM, Wang J, Christian JA, Spurlock ME (2004) Adiponectin differentially regulates cytokines in porcine macrophages. Biochem Biophys Res Commun 316:924–929

Yamaguchi N, Argueta JG, Masuhiro Y, Kagishita M, Nonak K, Saito T, Hanazawa S, Yamashita Y (2005) Adiponectin inhibits Toll-like receptor family-induced signaling. FEBS Lett 579:6821–6826

Yamauchi T, Kamon J, Minokoshi Y, Ito Y, Waki H, Uchida S, Yamashita S, Noda M, Kita S, Ueki K, Eto K, Akanuma Y, Froguel P, Foufelle F, Ferre P, Carling D, Kimura S, Nagai R, Kahn BB, Kadowaki T (2002) Adiponectin stimulates glucose utilization and fatty-acid oxidation by activating AMP-activated protein kinase. Nat Med 8:1288–1295

Yamauchi T, Kamon J, Ito Y, Tsuchida A, Yokomizo T, Kita S, Sugiyama T, Miyagishi M, Hara K, Tsunoda M, Murakami K, Ohteki T, Uchida S, Takekawa S, Waki H, Tsuno NH, Shibata Y, Terauchi Y, Froguel P, Tobe K, Koyasu S, Taira K, Kitamura T, Shimizu T, Nagai R, Kadowaki T (2003) Cloning of adiponectin receptors that mediate antidiabetic metabolic effects. Nature 423:762–769

Yokota T, Oritani K, Takahashi I, Ishikawa J, Matsuyama A, Ouchi N, Kihara S, Funahashi T, Tenner AJ, Tomiyama Y, Matsuzawa Y (2000) Adiponectin, a new member of the family of soluble defense collagens, negatively regulates the growth of myelomonocytic progenitors and the functions of macrophages. Blood 96:1723–1732

Zhang Y, Proenca R, Maffei M, Barone M, Leopold L, Friedman JM (1994) Positional cloning of the mouse obese gene and its human homologue. Nature 372:425–432

Index

J.M. Vivanco and F. Baluška (eds.), *Secretions and Exudates in Biological Systems*, Signaling and Communication in Plants 12, DOI 10.1007/978-3-642-23047-9,